CONTENTS

1 WHOLE NUMBERS

2 ADDITION

3 SUBTRACTION

CONTENTS

4 MULTIPLICATION

5 DIVISION

6 FRACTIONS

7 DECIMALS

8 PATTERNS AND ALGEBRA

9 CHANCE

10 DATA

ANSWERS

ABOUT CATCH-UP MATHS

The Catch-Up Maths series enables students to start from scratch when they are struggling with their year level maths. Each book takes maths back to the foundation and ensures that all basic concepts are consolidated before moving forward. Lots of revision and opportunities to practise and build confidence are provided before moving on to new topics.

Each new strand and sub-strand of the primary maths curriculum is introduced clearly with simple explanations, examples and trial questions (with answers), before children move to the Practice section. To ensure concepts are understood fully, videos of the author working through and explaining new concepts are included in every chapter.

A QR code on the topic page provides access to the subtitled video.

This book has 10 chapters divided between the Year 6 Number & Algebra and Statistics & Probability strands of the Australian Maths Curriculum. The chapters are:

1 Whole Numbers
2 Addition
3 Subtraction
4 Multiplication
5 Division
6 Fractions
7 Decimals
8 Patterns & Algebra
9 Chance
10 Data

★ Review section that can be used as assessment and to check students' progress is included at the end of each chapter.

★ Answers are at the back of the book.

How to use this book

Children can work through the pages from front to back, or choose individual topics to reinforce areas where they are struggling.

The topics are introduced with:

- clear instructions, using simple language
- completed examples and incomplete examples for students to tackle before moving on to the **Your Turn** section
- a video linked by QR code that shows the incomplete example and has the author giving extra instruction about the page.

HOW TO USE THE QR CODES IN CATCH-UP MATHS

A unique aspect of the **Catch-Up Maths** series is the **instructional video** created by the author for every new topic.

Access the video using a QR code reader app

SCAN to watch video

The videos give a step-by-step explanation of the examples and work through them to provide a helpful lesson for the student. The videos are simply accessed via the QR code on the same page and can be watched on a phone, tablet, computer or interactive whiteboard.

Each video shows the page from the book. The author talks through the concepts and examples, and demonstrates what students need to do. Her words are shown as easy-to-read captions so that the audio can be turned down in noisy classrooms, and for students with hearing difficulties. The solutions to the examples are presented before students are expected to tackle the 'Your Turn' section. This careful instruction ensures that students can confidently move on to the following Practice questions. Those assisting students should encourage them to check their 'Your Turn' answers before moving on.

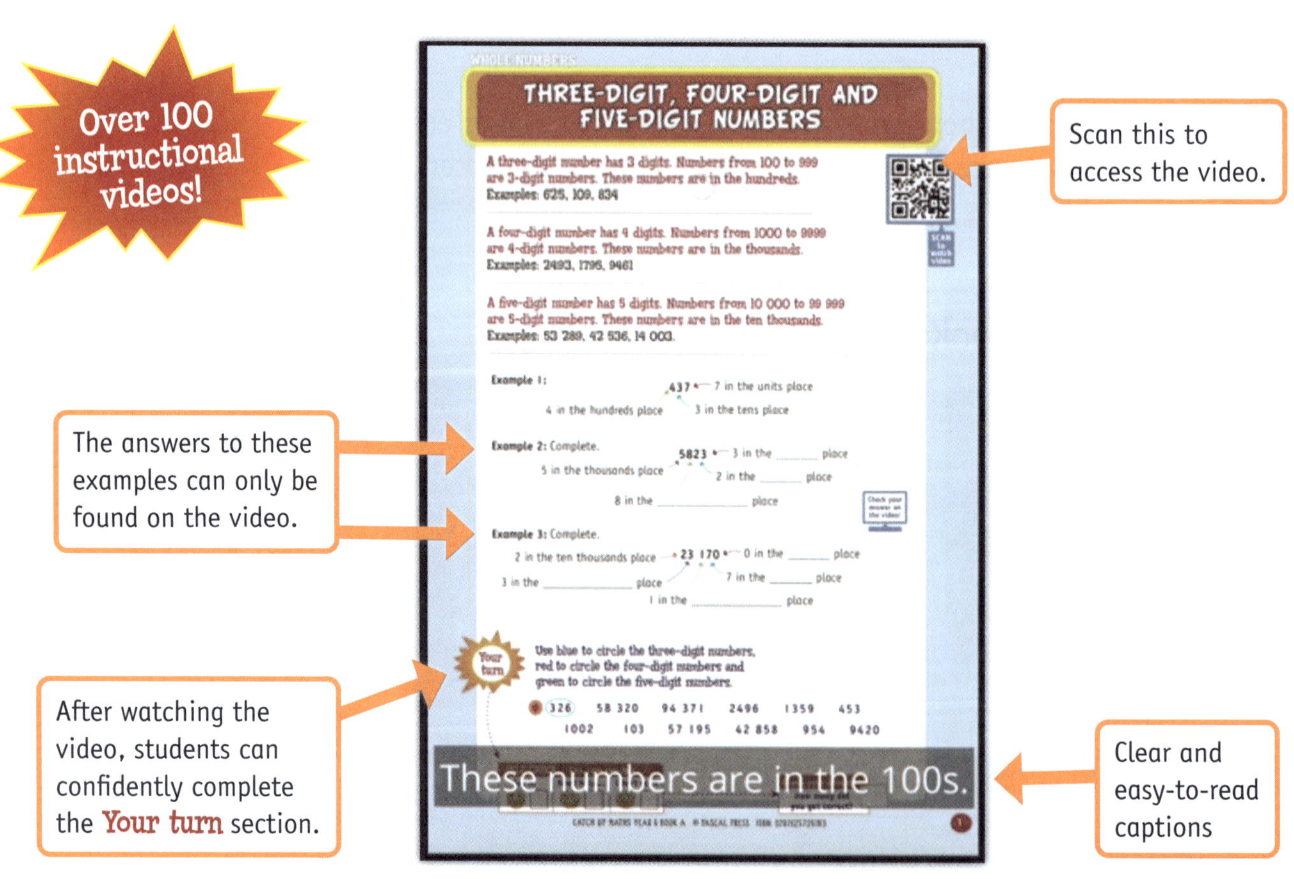

CATCH UP MATHS YEAR 6 BOOK A © PASCAL PRESS ISBN: 9781925726183

AUSTRALIAN CURRICULUM CORRELATIONS

ACARA CODE	Content Description	Pages
1. Whole Numbers		**PAGES 1–59**
ACMNA052 Year 3	Recognise, model, represent and order numbers to at least 10 000	**1, 2, 11, 12**
ACMNA053 Year 3	Apply place value to partition, rearrange and regroup numbers to at least 10 000 to assist calculations and solve problems	**3, 4, 5, 6, 7, 8, 9, 10**
ACMNA071 Year 4	Investigate and use the properties of odd and even numbers	**21, 22**
ACMNA072 Year 4	Recognise, represent and order numbers to at least tens of thousands	**1, 2, 13, 14, 23, 24, 25, 26, 29, 30, 31, 32**
ACMNA073 Year 4	Apply place value to partition, rearrange and regroup numbers to at least tens of thousands to assist calculations and solve problems	**3, 4, 5, 6, 7, 8, 9, 10, 13, 14, 15, 16, 17, 18, 19, 20**
ACMNA098 Year 5	Identify and describe factors and multiples of whole numbers and use them to solve problems	**33, 34, 35, 36, 37, 38, 39, 40**
ACMNA099 Year 5	Use estimation and rounding to check the reasonableness of answers to calculations	**11, 12, 27, 28**
ACMNA122 Year 6	Identify and describe properties of prime, composite, square and triangular numbers	**43, 44**
ACMNA124 Year 6	Investigate everyday situations that use integers. Locate and represent these numbers on a number line	**41, 42**
2. Addition		**PAGES 60–101**
ACMNA055 Year 3	Recall addition facts for single-digit numbers and related subtraction facts to develop increasingly efficient mental strategies for computation	**68, 69, 74, 75, 76, 77**
ACMNA059 Year 3	Represent money values in multiple ways and count the change required for simple transactions to the nearest five cents	**78, 79**
ACMNA073 Year 4	Apply place value to partition, rearrange and regroup numbers to at least tens of thousands to assist calculations and solve problems	**60, 61, 62, 63, 64, 65, 66, 67, 70, 71, 72, 73, 74, 75, 76, 77**
ACMNA080 Year 4	Solve problems involving purchases and the calculation of change to the nearest five cents with and without digital technologies	**78, 79, 80, 81**
ACMNA098 Year 5	Identify and describe factors and multiples of whole numbers and use them to solve problems	**78, 79**
ACMNA099 Year 5	Use estimation and rounding to check the reasonableness of answers to calculations	**68, 69, 94**
ACMNA106 Year 5	Create simple financial plans	**82, 83, 88, 89, 90**

ACARA CODE	Content Description	Pages
3. Subtraction		**PAGES 102–139**
ACMNA052 Year 3	Recognise, model, represent and order numbers to at least 10 000	**118, 119**
ACMNA053 Year 3	Apply place value to partition, rearrange and regroup numbers to at least 10 000 to assist calculations and solve problems	**110, 111, 112, 113, 114, 115**
ACMNA055 Year 3	Recall addition facts for single-digit numbers and related subtraction facts to develop increasingly efficient mental strategies for computation	**102, 103, 104, 105, 106, 107, 108, 109, 110, 111, 112, 113**
ACMNA073 Year 4	Apply place value to partition, rearrange and regroup numbers to at least tens of thousands to assist calculations and solve problems	**102, 103, 104, 105, 106, 107, 108, 109, 110, 111, 112, 113, 120, 121, 122, 123**
ACMNA099 Year 5	Use estimation and rounding to check the reasonableness of answers to calculations	**126, 127**
4. Multiplication		**PAGES 140–165**
ACMNA074 Year 4	Investigate number sequences involving multiples of 3, 4, 6, 7, 8, and 9	**140, 141**
ACMNA075 Year 4	Recall multiplication facts up to 10 × 10 and related division facts	**140, 141**
ACMNA076 Year 4	Develop efficient mental and written strategies, and use appropriate digital technologies for multiplication and for division where there is no remainder	**142, 146, 146, 147, 148, 149**
ACMNA098 Year 5	Identify and describe factors and multiples of whole numbers and use them to solve problems	**140, 141**
ACMNA100 Year 5	Solve problems involving multiplication of large numbers by one- or two-digit numbers using efficient mental, written strategies and appropriate digital technologies	**144, 145, 146, 147, 148, 149, 150, 151, 152, 153**
ACMNA123 Year 6	Select and apply efficient mental and written strategies and appropriate digital technologies to solve problems involving all four operations with whole numbers	**150, 151, 152, 153, 154, 155, 156, 157**
ACMNA130 Year 6	Multiply and divide decimals by powers of 10	**156, 157**
ACMNA134 Year 6	Explore the use of brackets and order of operations to write number sentences	**158, 159**

Australian Curriculum Correlations continued

5. Division — PAGES 166–187

ACARA CODE	Content Description	Pages
ACMNA076 Year 4	Develop efficient mental and written strategies, and use appropriate digital technologies for multiplication and for division where there is no remainder	168, 169, 170, 171
ACMNA100 Year 5	Solve problems involving multiplication of large numbers by one- or two-digit numbers using efficient mental, written strategies and appropriate digital technologies	176, 177
ACMNA101 Year 5	Solve problems involving division by a one-digit number, including those that result in a remainder	166, 167, 172, 173, 174, 175, 176, 177, 178, 179, 180, 181
ACMNA291 Year 5	Use efficient mental and written strategies and apply appropriate digital technologies to solve problems	180, 181, 182, 183

6. Fractions — PAGES 188–211

ACARA CODE	Content Description	Pages
ACMNA058 Year 3	Model and represent unit fractions including $\frac{1}{2}, \frac{1}{4}, \frac{1}{3}, \frac{1}{5}$ and their multiples to a complete whole	192, 193
ACMNA077 Year 4	Investigate equivalent fractions used in contexts	194, 195, 204, 205
ACMNA078 Year 4	Count by quarters, halves and thirds, including with mixed numerals. Locate and represent these fractions on a number line	200, 201
ACMNA102 Year 5	Compare and order common unit fractions and locate and represent them on a number line	196, 197
ACMNA103 Year 5	Investigate strategies to solve problems involving addition and subtraction of fractions with the same denominator	198, 199, 202, 203
ACMNA125 Year 6	Compare fractions with related denominators and locate and represent them on a number line	196
ACMNA126 Year 6	Solve problems involving addition and subtraction of fractions with the same or related denominators	204, 205

7. Decimals — PAGES 212–239

ACARA CODE	Content Description	Pages
ACMNA077 Year 4	Investigate equivalent fractions used in contexts	230, 231
ACMNA079 Year 4	Recognise that the place value system can be extended to tenths and hundredths. Make connections between fractions and decimal notation	212, 213, 216, 217, 228, 229
ACMNA104 Year 5	Recognise that the place value system can be extended beyond hundredths	214, 215, 216, 217
ACMNA127 Year 6	Find a simple fraction of a quantity where the result is a whole number, with and without digital technologies	228, 229, 232, 233
ACMNA128 Year 6	Add and subtract decimals, with and without digital technologies, and use estimation and rounding to check the reasonableness of answers	219, 220, 221
ACMNA129 Year 6	Multiply decimals by whole numbers and perform divisions by non-zero whole numbers where the results are terminating decimals, with and without digital technologies	222, 223, 234, 235, 236, 237
ACMNA130 Year 6	Multiply and divide decimals by powers of 10	226, 227
ACMNA131 Year 6	Make connections between equivalent fractions, decimals and percentages	228, 229, 230, 231
ACMNA132 Year 6	Investigate and calculate percentage discounts of 10%, 25% and 50% on sale items, with and without digital technologies	232, 233

8. Patterns & Algebra — PAGES 240–257

ACARA CODE	Content Description	Pages
ACMNA060 Year 3	Describe, continue, and create number patterns resulting from performing addition or subtraction	240, 241
ACMNA081 Year 4	Explore and describe number patterns resulting from performing multiplication	240, 241
ACMNA107 Year 5	Describe, continue and create patterns with fractions, decimals and whole numbers resulting from addition and subtraction	246, 247, 248, 249
ACMNA121 Year 5	Find unknown quantities in number sentences involving multiplication and division and identify equivalent number sentences involving multiplication and division	244, 245, 248
ACMNA123 Year 6	Select and apply efficient mental and written strategies and appropriate digital technologies to solve problems involving all four operations with whole numbers	248
ACMMG143 Year 6	Introduce the Cartesian coordinate system using all four quadrants	250, 251

9. Chance — PAGES 258–271

ACARA CODE	Content Description	Pages
ACMSP067 Year 3	Conduct chance experiments, identify and describe possible outcomes and recognise variation in results	260, 261
ACMSP092 Year 4	Describe possible everyday events and order their chances of occurring	258, 259
ACMSP116 Year 5	List outcomes of chance experiments involving equally likely outcomes and represent probabilities of those outcomes using fractions	262, 263, 264, 265
ACMSP117 Year 5	Recognise that probabilities range from 0 to 1	258, 259
ACMSP144 Year 6	Describe probabilities using fractions, decimals and percentages	266, 267

10. Data — PAGES 272–301

ACARA CODE	Content Description	Pages
ACMSP068 Year 3	Identify questions or issues for categorical variables. Identify data sources and plan methods of data collection and recording	272, 273, 274
ACMSP069 Year 3	Collect data, organise into categories and create displays using lists, tables, picture graphs and simple column graphs, with and without the use of digital technologies	278, 279, 280, 281, 282, 283
ACMSP096 Year 4	Construct suitable data displays, with and without the use of digital technologies, from given or collected data. Include tables, column graphs and picture graphs where one picture can represent many data values	272, 275, 278, 281, 282, 283, 286, 289, 292
ACMSP118 Year 5	Pose questions and collect categorical or numerical data by observation or survey	272, 273, 274
ACMSP119 Year 5	Construct displays, including column graphs, dot plots and tables, appropriate for data type, with and without the use of digital technologies	281, 282, 283, 284, 285, 286, 287, 288, 289, 290, 291
ACMSP120 Year 5	Describe and interpret different data sets in context	294, 295, 296
ACMSP147 Year 6	Interpret and compare a range of data displays, including side-by-side column graphs for two categorical variables	275, 276

THREE-DIGIT, FOUR-DIGIT AND FIVE-DIGIT NUMBERS

A three-digit number has 3 digits. Numbers from 100 to 999 are 3-digit numbers. These numbers are in the hundreds.
Examples: 625, 109, 834

A four-digit number has 4 digits. Numbers from 1000 to 9999 are 4-digit numbers. These numbers are in the thousands.
Examples: 2493, 1795, 9461

A five-digit number has 5 digits. Numbers from 10 000 to 99 999 are 5-digit numbers. These numbers are in the ten thousands.
Examples: 53 289, 42 536, 14 003

Example 1:

437
7 in the units place
3 in the tens place
4 in the hundreds place

Example 2: Complete.

5823
3 in the _______ place
2 in the _______ place
8 in the _______________ place
5 in the thousands place

Example 3: Complete.

23 170
2 in the ten thousands place
0 in the _______ place
7 in the _______ place
1 in the _______________ place
3 in the _______________ place

Use blue to circle the three-digit numbers, red to circle the four-digit numbers and green to circle the five-digit numbers.

● (326) 58 320 94 371 2496 1359 453
1002 103 57 195 42 858 954 9420

SELF CHECK Tick how you feel
Got it! ☐ Need help... ☐ I don't get it ☐

Check your answers
How many did you get correct? ☐

  ISBN: 9781925726183

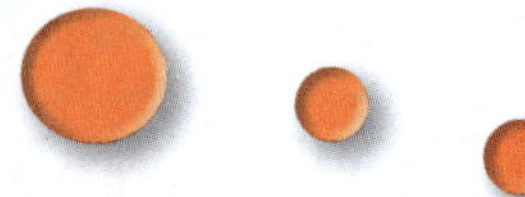

PRACTICE

1 Write these numbers in words.

- 3205 three thousand, two hundred and five
- **a** 349 ______
- **b** 428 ______
- **c** 503 ______
- **d** 1937 ______
- **e** 5642 ______
- **f** 7030 ______
- **g** 24 346 ______
- **h** 41 930 ______
- **i** 57 005 ______

2 Fill in the table.

	Number	Ten Thousands	Thousands	Hundreds	Tens	Ones
●	253	0	0	2	5	3
a	749					
b	903					
c	6245					
d	7142					
e	29 307					
f	99 390					

3 Match the numbers to the correct label.

140 359 3008 41 580 1439 82 493 2413

3 digits 4 digits 5 digits

71 345 157 23 205 752 203 2893 9981 50 000

CATCH UP MATHS YEAR 6 BOOK A © PASCAL PRESS ISBN: 9781925726183

PLACE VALUE

THREE-DIGIT, FOUR-DIGIT AND FIVE-DIGIT NUMBERS

Place value is the position of a digit in a number.

SCAN to watch video

A **three-digit number** has a hundreds place, a tens place and a ones place.

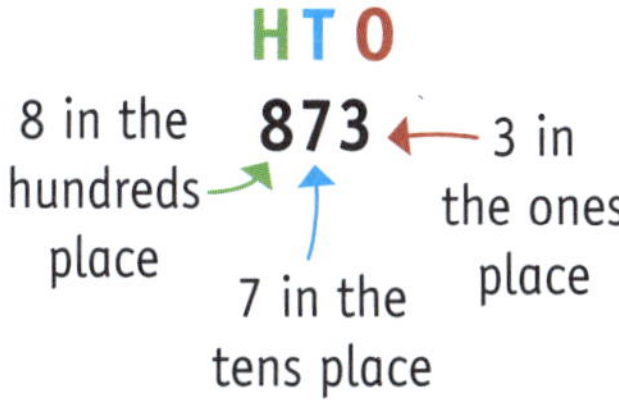

Example 1:

284 has:

2 in the hundreds place
8 in the tens place
4 in the ones place

Example 2:

432 has:

__ in the hundreds place
__ in the tens place
2 in the ones place

A **four-digit number** has a thousands place, a hundreds place, a tens place and a ones place.

Th H T O

2631

2 in the thousands place
1 in the ones place
6 in the hundreds place
3 in the tens place

Example 3:

4375 has:

4 in the thousands place
3 in the hundreds place
7 in the tens place
5 in the ones place

Example 4:

5820 has:

5 in the thousands place
__ in the hundreds place
__ in the tens place
0 in the ones place

A **five-digit number** has a ten thousands place, a thousands place, a hundreds place, a tens place and a ones place.

Example 5:

26 150 has:

2 in the ten thousands place
6 in the thousands place
1 in the hundreds place
5 in the tens place
0 in the ones place

Example 6:

53 894 has:

__ in the ten thousands place
3 in the thousands place
__ in the hundreds place
__ in the tens place
__ in the ones place

Place value = where is the digit?

Your turn

Circle the ten thousands in orange, thousands in purple, hundreds in green, tens in blue and ones in red.

● 32 584	**c** 4150	**f** 5836	**i** 177
a 352	**d** 41 384	**g** 1429	**j** 5389
b 93 526	**e** 580	**h** 409	**k** 83 715

SELF CHECK Tick how you feel

Got it!	Need help...	I don't get it
☐	☐	☐

Check your answers

How many did you get correct? ☐

PRACTICE

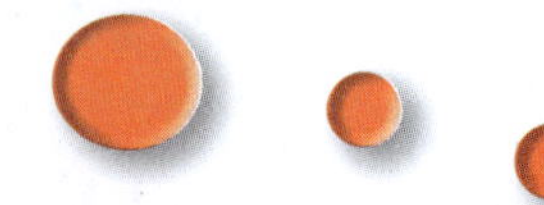

1 Write the place value of the underlined digit.

- 73 498 ten thousands
- a 1493
- b 392
- c 7230
- d 886
- e 2937
- f 59 827
- g 4493
- h 84 246
- i 15 281
- j 24 386
- k 53 815

2 Use orange to colour the digits in the ten thousands, use purple for thousands, green for hundreds, blue for tens and red for units.

- 1 3 7 2 5
- a 6 5 9 2 3
- b 5 1 4
- c 7 1 4 9 0
- d 4 8 3 0
- e 9 5 2 4 3
- f 7 8 1
- g 4 0 8 5 4
- h 5 8 2 1

3 Write the place value of the 4.

- 34 372 thousands
- a 6425
- b 43 821
- c 534
- d 3942
- e 5384
- f 65 941
- g 497

4 Write the numbers.

- 8 thousands, 2 ones, 3 tens 8032
- a 6 tens, 3 ones, 2 hundreds, 3 thousands
- b 8 ones, 9 hundreds, 4 tens
- c 7 ten thousands, 6 ones, 4 tens
- d 3 hundreds, 5 tens, 1 ones, 9 ten thousands, 4 thousands
- e 5 ten thousands, 3 ones
- f 6 thousands, 7 ones, 4 hundreds, 3 tens

 CATCH UP MATHS YEAR 6 BOOK A © PASCAL PRESS ISBN: 9781925726183

VALUE

THREE-DIGIT, FOUR-DIGIT AND FIVE-DIGIT NUMBERS

The value of a number is how much a number is worth.
To find the value of a digit, look at its place in a number.

For example, what is the value of the 3 in 23 000?
The 3 is in the thousands place, so the value of 3 is 3000.

SCAN to watch video

3-digit numbers

Example 1:
In the number 623,
the value of 6 is 600,
the value of 2 is 20
and the value of 3 is 3.

Example 2:
In the number 459,
the value of 4 is ____,
the value of 5 is 50
and the value of 9 is ___.

4-digit numbers

Example 3:
In the number 7438,
the value of 7 is 7000,
the value of 4 is 400,
the value of 3 is 30
and the value of 8 is 8.

Example 4:
In the number 6247,
the value of 6 is _____,
the value of 2 is ____,
the value of ___ is 40
and the value of 7 is 7.

5-digit numbers

Example 5:
In the number 25 681,
the value of 2 is 20 000,
the value of 5 is 5000,
the value of 6 is 600,
the value of 8 is 80
and the value of 1 is 1.

Example 6:
In the number 34 592,
the value of 3 is ________,
the value of 4 is 4000,
the value of 5 is ____,
the value of ___ is 90
and the value of ___ is 2.

Remember, the place value is where the digit is in a number.

The value is how much the digit is worth.

Your turn

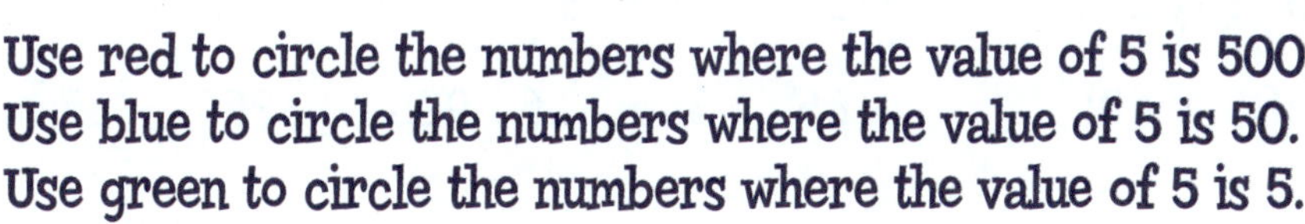

Use red to circle the numbers where the value of 5 is 500.
Use blue to circle the numbers where the value of 5 is 50.
Use green to circle the numbers where the value of 5 is 5.

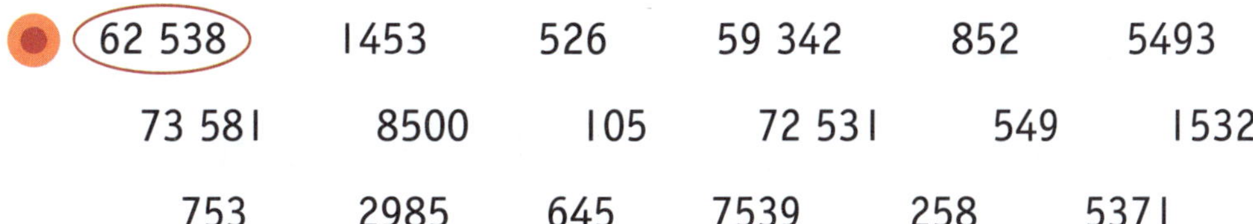

62 538   1453   526   59 342   852   5493

73 581   8500   105   72 531   549   1532

753   2985   645   7539   258   5371

SELF CHECK Tick how you feel

Got it!	Need help...	I don't get it
☐	☐	☐

Check your answers
How many did you get correct? ☐

PRACTICE

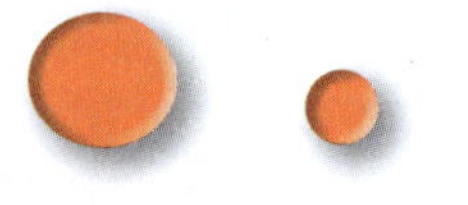

1 What is the value of 6 in these numbers?

- ● 2463 60
- a 362 ______
- b 6534 ______
- c 2436 ______
- d 64 281 ______
- e 4639 ______
- f 16 382 ______
- g 5624 ______
- h 3760 ______
- i 7246 ______
- j 6 ______
- k 60 ______

2 Write a number that has a 7 with the given value.

- ● 70 79
- a 700 ______
- b 70 000 ______
- c 7000 ______
- d 70 ______
- e 7 ______

3 Circle the digit with the most value and underline the digit with the least value.

- ● (7)3 82_1_
- a 439
- b 1384
- c 593
- d 879
- e 41 589
- f 2491
- g 34 285
- h 1509
- i 16 073
- j 97 329
- k 42 444
- l 555
- m 9009
- n 24 823
- o 60 399

4 Circle the numbers with the matching value.

- ● The value of 3 is 30: (37) 483 342 (5037) (64 736) 3491
- a The value of 5 is 500: 359 562 6593 51 34 531 5632
- b The value of 7 is 7000: 736 57 111 3792 7000 47 328 74 385
- c The value of 9 is 90 000: 93 246 49 371 90 331 893 914 98 363
- d The value of 2 is 2: 5236 24 593 61 592 125 59 472 1582
- e The value of 6 is 60: 4165 346 64 928 3265 37 469 16 431

5 Write numbers that match the descriptions.

- ● 3 is worth the most 34 172
- a 3 is worth the least __ ___
- b 4 is worth the most ___
- c 4 is worth the least ___
- d 7 is worth the most __ ___
- e 7 is worth the least __ ___
- f 9 is worth the most ____
- g 9 is worth the least ____
- h 8 is worth the most ___
- i 8 is worth the least ___

CATCH UP MATHS YEAR 6 BOOK A © PASCAL PRESS ISBN: 9781925726183

EXPANDED NUMBERS

THREE-DIGIT, FOUR-DIGIT AND FIVE-DIGIT NUMBERS

We expand a number by writing the value of each digit.

Example 1: 273 in expanded form = 200 + 70 + 3

Example 2: 706 in expanded form = 700 + 6

Example 3: 8491 in expanded form = 8000 + 400 + 90 + 1

Example 4: 9082 in expanded form = 9000 + 80 + 2

Example 5: 42 183 in expanded form = 40 000 + 2000 + 100 + 80 + 3

Example 6: 40 324 in expanded form = 40 000 + 300 + 20 + 4

Example 7: 435 in expanded form = 400 + ___ + 5

Example 8: 5907 in expanded form = 5000 + ____ + __

Example 9: 37 381 in expanded form = 30 000 + ______ + 300 + 80 + __

Write these numbers in expanded form.

- 24 387 = 20 000 + 4000 + 300 + 80 + 7

a 7351 = ______ + _____ + ___ + __

b 501 = ____ + __

c 56 204 = ________ + ______ + _____ + __

d 293 = ____ + ___ + __

e 1482 = ______ + _____ + ___ + __

Check your answers
How many did you get correct?

  ISBN: 9781925726183

PRACTICE

1 Expand the numbers.

27 683 = 20 000 + 7000 + 600 + 80 + 3

a 76 941 = ______ + ______ + ____ + ___ + __

b 1536 = ______ + ____ + ___ + __

c 30 245 = ______ + ______ + ____ + ___ + __

d 370 = ____ + ___ + __

e 539 = ____ + ___ + __

f 7065 = ______ + ____ + ___ + __

g 93 250 = ______ + ______ + ____ + ___ + __

h 4358 = ______ + ____ + ___ + __

i 160 = ____ + ___ + __

2 Write these numbers in expanded form.

42 736 = 40 000 + 2000 + 700 + 30 + 6

a 24 623 = ____________________

b 15 094 = ____________________

c 3343 = ____________________

d 1750 = ____________________

e 249 = ____________________

f 503 = ____________________

3 Write the place value and value of the 4.

	Number	Place value	Value
●	73 428	hundreds	400
a	249		
b	64 731		
c	49 387		
d	62 354		

	Number	Place value	Value
e	104		
f	4321		
g	43 877		
h	473		
i	34 920		

CATCH UP MATHS YEAR 6 BOOK A © PASCAL PRESS ISBN: 9781925726183

ORDERING NUMBERS

THREE-DIGIT, FOUR-DIGIT AND FIVE-DIGIT NUMBERS

When numbers are ordered from smallest to largest, it is called ascending order.

These numbers are in ascending order: 325, 1246, 3874, 15 928.

When numbers are ordered from largest to smallest, it is called descending order.

These numbers are in descending order: 1592, 387, 142, 132.

SCAN to watch video

Example 1:

Write these numbers in ascending order.

153	297	184	493

Ascending order:

153	184	297	493

smallest number first — largest number last

Example 2:

Write these numbers in ascending order:
243, 137, 1492, 3457.

137, ______, ______, ______

Example 3:

Write these numbers in descending order.

3411	2523	1098	4381

Descending order:

4381	3411	2523	1098

largest number first — smallest number last

Example 4:

Write these numbers in descending order:
3815, 4934, 732, 541.

4934, ______, ______, ______

Write ascending or descending to describe how the numbers are ordered.

- ● 153, 792, 3518 — ascending
- **a** 523, 532, 1503 — ______
- **b** 9843, 1623, 403 — ______
- **c** 158, 297, 438 — ______
- **d** 5411, 4511, 1541 — ______
- **e** 89 437, 84 739, 83 759 — ______
- **f** 12 456, 14 265, 16 542 — ______
- **g** 72 491, 74 923, 79 001 — ______

Check your answers
How many did you get correct?

PRACTICE

1 Write each set of numbers in ascending order.

● 7394, 3497, 9374, 4239

3497, 4239, 7394, 9374

a 126, 621, 162, 612, 261

______

b 743, 734, 437, 347, 374

______

c 1526, 6251, 2651, 5261, 6512

______

d 5934, 4395, 5349, 4539, 9453

______

e 13 426, 62 431, 31 462, 41 623

______

f 83 560, 85 630, 30 856, 80 653

______

g 62 378, 26 783, 73 862, 32 867

______

2 Write each set of numbers in descending order.

● 247, 723, 413, 652, 490 — 723, 652, 490, 413, 247

a 472, 384, 646, 909, 800 ______

b 803, 642, 246, 380, 890 ______

c 1623, 2361, 3162, 1236, 6132 ______

d 4382, 1459, 2357, 6841, 3981 ______

e 13 826, 62 813, 26 138, 83 621, 32 183

______

f 72 415, 15 472, 24 715, 45 147, 71 524

______

3 Order these states according to the money raised from smallest (1) to largest (6).

	State	Money raised	Order
a	New South Wales	$ 72 549	
b	Victoria	$ 56 206	
c	Queensland	$ 29 689	
d	Tasmania	$ 13 421	
e	Western Australia	$ 97 370	
 f	South Australia	$ 5 639	1

CATCH UP MATHS YEAR 6 BOOK A © PASCAL PRESS ISBN: 9781925726183

ROUNDING TO THE NEAREST 100, 1000 AND 10 000

Rounding is useful when you need to estimate an answer.

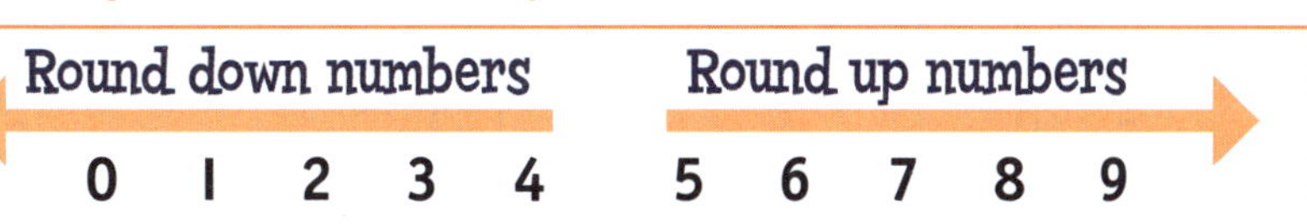

Rounding to 100s

1 Go to the hundreds place.
2 Write the round up number above the hundreds.
3 Circle the number in the tens place.
4 Is it a round up or round down number?
5 Round the number.

Example 1: Round to the nearest 100.

6
5(3)6 ↺ down

500

Example 2: Round to the nearest 100.

8 ← up
7(5)3

______

Rounding to 1000s

1 Go to the thousands place.
2 Write the round up number above the thousands.
3 Circle the number in the hundreds place.
4 Is it a round up or round down number?
5 Round the number.

Example 3: Round to the nearest 1000.

5 ← up
4(9)34

5000

Example 4: Round to the nearest 1000.

9
8(4)25 ↺ down

______

Rounding to 10 000s

1 Go to the ten thousands place.
2 Write the round up number above the ten thousands.
3 Circle the number in the thousands place.
4 Is it a round up or round down number?
5 Round the number.

Example 5: Round to the nearest 10 000.

7
6(3) 282 ↺ down

60 000

Example 6: Round to the nearest 10 000.

9 ← up
8(9) 256

______

Round these numbers to the nearest 100.

● 6
35(4)2 ↺ down
3500

a 7239 ______

b 385 ______

c 49 563 ______

d 489 ______

e 63 428 ______

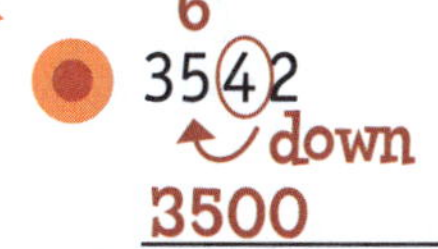

Check your answers
How many did you get correct?

1 Round these numbers to the nearest 100.

- ● 126 498 — 126 500
- a 1526 ______
- b 8739 ______
- c 403 ______
- d 15 438 ______
- e 152 493 ______
- f 1 573 249 ______
- g 7 594 ______

2 Round these numbers to the nearest 1000.

- ● 4 593 628 — 4 594 000
- a 1928 ______
- b 92 635 ______
- c 17 268 ______
- d 124 253 ______
- e 74 629 ______
- f 3427 ______
- g 6 372 871 ______

3 Round these numbers to the nearest 10 000.

- ● 53 497 — 50 000
- a 42 651 ______
- b 274 389 ______
- c 15 491 ______
- d 3 262 950 ______
- e 15 583 ______
- f 7 458 218 ______
- g 81 818 ______

4 Complete the table.

	Nearest 100	Nearest 1000	Nearest 10 000
● 53 852	53 900	54 000	50 000
a 34 568			
b 59 731			
c 54 836			
d 97 425			
e 580 263			
f 742 589			
g 1 429 632			
h 5 643 859			

CATCH UP MATHS YEAR 6 BOOK A © PASCAL PRESS ISBN: 9781925726183

PLACE VALUE
SIX-DIGIT AND SEVEN-DIGIT NUMBERS

Place value is the position of a digit in a number.

SCAN to watch video

A six-digit number has 6 digits. Numbers from 100 000 to 999 999 are 6-digit numbers. These numbers are in the hundred thousands.

Examples:
124 382, 674 950, 987 203.

Example 1: 583 497 has

5 in the hundred thousands place
8 in the ten thousands place
3 in the thousands place
4 in the hundreds place
9 in the tens place
7 in the ones place

In words:
Five hundred and eighty-three thousand, four hundred and ninety-seven

Example 2: 641 209 has

6 in the hundred thousands place
__ in the ten thousands place
1 in the thousands place
2 in the hundreds place
__ in the tens place
__ in the ones place

A seven-digit number has 7 digits. Numbers from 1 000 000 to 9 999 999 are 7-digit numbers. These numbers are in the millions.

Examples:
1 347 423, 3 248 903, 7 343 151.

Example 3: 6 841 295 has

6 in the millions place
8 in the hundred thousands place
4 in the ten thousands place
1 in the thousands place
2 in the hundreds place
9 in the tens place
5 in the units place

In words:
Six million, eight hundred and forty-one thousand, two hundred and ninety-five

Example 4: 7 436 281 has

__ in the millions place
__ in the hundred thousands place
3 in the ten thousands place
6 in the thousands place
2 in the hundreds place
8 in the tens place
1 in the units place

Your turn

Use yellow to circle the six-digit numbers and black to circle the seven-digit numbers.

(3 435 624) 120 036 950 133 236 158

8 542 187 1 248 970 5 341 286 3 436 219

415 273 847 311 7 436 281 641 209

Check your answers
How many did you get correct?

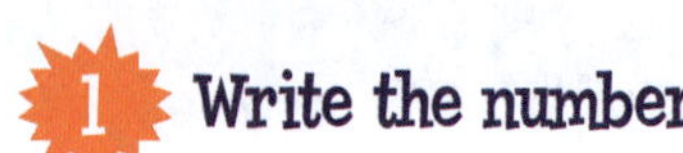

PRACTICE

1 **Write the number.**

- seven hundred and thirty-six thousand, two hundred and nineteen — 736 219

a six hundred and ninety-four thousand, three hundred and eighty-seven ______

b nine million, three hundred and seventy-seven thousand, four hundred and twenty-five ______

2 **Write these numbers in words.**

- 432 971 — four hundred and thirty-two thousand, nine hundred and seventy-one

a 307 281 ______

b 8 702 900 ______

3 **Write the place value of the underlined digit.**

- <u>3</u> 268 149 — millions

a 4 93<u>7</u> 937 ______

b <u>8</u>42 154 ______

c 3 <u>8</u>59 268 ______

d 842 59<u>1</u> ______

e 4 8<u>5</u>6 319 ______

4 **Write the place value of the 8.**

- 248 361 — thousands

a 1 843 262 ______

b 315 849 ______

c 4 893 215 ______

d 784 150 ______

e 3 924 181 ______

5 **Write the numbers.**

- 6 thousands, 7 ten thousands, 6 hundred thousands, 5 tens, 7 ones, 8 hundreds — 676 857

a 5 ones, 3 hundreds, 8 ten thousands, 6 millions, 4 hundred thousands, 9 thousands ______

b 2 hundred thousands, 8 thousands, 4 millions ______

c 9 hundreds, 4 thousands, 7 hundred thousands ______

CATCH UP MATHS YEAR 6 BOOK A © PASCAL PRESS ISBN: 9781925726183

VALUE
SIX-DIGIT AND SEVEN-DIGIT NUMBERS

The value of a number is how much a number is worth.
To find the value of a digit, look at where it is in a number.

6-digit numbers

Example 1:
In the number 526 391,
the value of 5 is 500 000
the value of 2 is 20 000
the value of 6 is 6000
the value of 3 is 300
the value of 9 is 90
the value of 1 is 1.

Example 2:
In the number 742 186,
the value of 7 is __________
the value of 4 is 40 000
the value of 2 is _______
the value of 1 is ____
the value of 8 is 80
the value of 6 is 6.

7-digit numbers

Example 3:
In the number 6 384 173,
the value of 6 is 6 000 000
the value of 3 is 300 000
the value of 8 is 80 000
the value of 4 is 4000
the value of 1 is 100
the value of 7 is 70
the value of 3 is 3.

Example 4:
In the number 5 872 493,
the value of 5 is ____________
the value of 8 is 800 000
the value of 7 is 70 000
the value of 2 is _______
the value of 4 is 400
the value of 9 is ___
the value of 3 is __.

Remember, the place value is where the digit is in a number. The value is how much the digit is worth.

Your turn

Use yellow to circle the numbers where 7 has a value of 700 000.
Use black to circle the numbers where 4 has a value of 4 000 000.

● 725 435   4 373 362   736 489   4 759 328   5 768 430

9 759 015   4 873 581   6 700 000   4 573 000

SELF CHECK Tick how you feel

Got it!	Need help...	I don't get it
☐	☐	☐

Check your answers
How many did you get correct? ☐

PRACTICE

1 What is the value of 8 in these numbers?

- 8 052 436 — 8 000 000
- a 3 842 431 ________
- b 842 376 ________
- c 8 645 590 ________
- d 136 852 ________
- e 4 382 915 ________
- f 5 493 183 ________
- g 7 841 320 ________
- h 1 573 108 ________
- i 583 462 ________
- j 642 837 ________
- k 149 383 ________

2 Complete the information for 7 324 951.

- 7 324 951 ones
- a ________ tens + __ ones
- b ______ hundreds + __ tens + __ ones
- c _____ thousands + __ hundreds + __ tens + __ ones
- d ____ ten thousands + __ thousands + __ hundreds + __ tens + __ ones
- e ___ hundred thousands + __ ten thousands + __ thousands + __ hundreds + __ tens + __ units
- f __ millions + __ hundred thousands + __ ten thousands + __ thousands + __ hundreds + __ tens + __ ones

3 Complete the information for 583 284.

- 583 284 ones
- a ______ tens + __ ones
- b _____ hundreds + __ tens + __ units
- c ____ thousands + __ hundreds + __ tens + __ ones
- d ___ ten thousands + __ thousands + __ hundreds + __ tens + __ ones
- e __ hundred thousands + __ ten thousands + __ thousands + __ hundreds + __ tens + __ ones

4 Write three numbers that fit the description.

- a 4 has a value of 400 000 ________ ________ ________
- b 6 has a value of 6 000 000 ________ ________ ________
- c 3 has a value of 30 000 ________ ________ ________
- d 7 has a value of 7000 ________ ________ ________

EXPANDED NUMBERS

SIX-DIGIT AND SEVEN-DIGIT NUMBERS

We expand a number by writing the value of each digit.
A number written this way is in expanded form.

SCAN to watch video

Example 1: Write 839 241 in expanded form.

800 000 + 30 000 + 9000 + 200 + 40 + 1

Example 2: Write 7 563 498 in expanded form.

7 000 000 + 500 000 + 60 000 + 3000 + 400 + 90 + 8

Example 3: Write 839 241 using powers of 10.

$8 \times 10^5 + 3 \times 10^4 + 9 \times 10^3 + 2 \times 10^2 + 4 \times 10 + 1 \times 1$

Example 4: Write 7 563 498 using powers of 10.

$7 \times 10^6 + 5 \times 10^5 + 6 \times 10^4 + 3 \times 10^3 + 4 \times 10^2 + 9 \times 10 + 8 \times 1$

Example 5: Write 523 964 in expanded form.

500 000 + ______ + 3000 + ____ + ___ + 4

Example 6: Write 3 524 378 using powers of 10.

$3 \times 10^6 + 5 \times 10^{\_} + 2 \times 10^4 + 4 \times 10^{\_} + 3 \times 10^2 + 7 \times \_\_ + 8 \times 1$

Compare the power of 10 with the number of zeroes: $10^3 = 1000$

$10^1 = 10$
$10^0 = 1$

Write these numbers in expanded form and using powers of 10.

● 3 429 715

= 3 000 000 + 400 000 + 20 000 + 9000 + 700 + 10 + 5

$= \underline{3} \times 10^{\underline{6}} + \underline{4} \times 10^{\underline{5}} + \underline{2} \times 10^{\underline{4}} + \underline{9} \times 10^{\underline{3}} + \underline{7} \times 10^{\underline{2}} + \underline{1} \times 10 + \underline{5} \times 1$

a 4 378 231

= ________ + ________ + ______ + _____ + ____ + ___ + __

$= \_\_ \times 10^{\_} + \_\_ \times 10^{\_} + \_\_ \times 10^{\_} + \_\_ \times 10^{\_} + \_\_ \times 10^{\_} + \_\_ \times 10 + \_\_ \times 1$

b 536 928

= ________ + ______ + _____ + ____ + ___ + __

$= \_\_ \times 10^{\_} + \_\_ \times 10^{\_} + \_\_ \times 10^{\_} + \_\_ \times 10^{\_} + \_\_ \times 10 + \_\_ \times 1$

SELF CHECK Tick how you feel

Got it!	Need help...	I don't get it
 ☐	 ☐	☐

Check your answers
How many did you get correct? ☐

PRACTICE

1 Expand the numbers.

- 436 281 = 400 000 + 30 000 + 6000 + 200 + 80 + 1

a 1 372 489 = ______ + ______ + ______ + ______ + ______ + ______ + ______

b 839 204 = ______ + ______ + ______ + ______ + ______

c 5 682 473 = ______ + ______ + ______ + ______ + ______ + ______ + ______

d 569 320 = ______ + ______ + ______ + ______ + ______

2 Write the numbers that have been expanded.

- 1 000 000 + 500 000 + 30 000 + 6000 + 400 + 20 + 1 = 1 536 421

a 3 000 000 + 400 000 + 20 000 + 4000 + 200 + 10 + 9 = ______

b 500 000 + 30 000 + 7000 + 600 + 30 + 7 = ______

c 4 000 000 + 200 000 + 300 + 50 + 2 = ______

d 800 000 + 7000 + 20 = ______

3 Write using powers of 10.

- 5 863 124 = $5 \times 10^{6} + 8 \times 10^{5} + 6 \times 10^{4} + 3 \times 10^{3} + 1 \times 10^{2} + 2 \times 10 + 4 \times 1$

a 243 659 = $\_\_ \times 10^{\_} + \_\_ \times 10^{\_} + \_\_ \times 10^{\_} + \_\_ \times 10^{\_} + \_\_ \times 10 + \_\_ \times 1$

b 1 478 831 = $\_\_ \times 10^{\_} + \_\_ \times 10^{\_} + \_\_ \times 10^{\_} + \_\_ \times 10^{\_} + \_\_ \times 10^{\_} + \_\_ \times 10 + \_\_ \times 1$

c 4 036 528 = $\_\_ \times 10^{\_} + \_\_ \times 10^{\_} + \_\_ \times 10^{\_} + \_\_ \times 10^{\_} + \_\_ \times 10 + \_\_ \times 1$

d 758 361 = $\_\_ \times 10^{\_} + \_\_ \times 10^{\_} + \_\_ \times 10^{\_} + \_\_ \times 10^{\_} + \_\_ \times 10 + \_\_ \times 1$

4 Write the numbers.

- $4 \times 10^6 + 5 \times 10^5 + 3 \times 10^4 + 8 \times 10^3 + 2 \times 10^2 + 7 \times 10 + 3 \times 1$ = 4 538 273

a $5 \times 10^5 + 6 \times 10^4 + 7 \times 10^3 + 2 \times 10^2 + 1 \times 1$ = ______

b $8 \times 10^6 + 4 \times 10^5 + 7 \times 10^4 + 9 \times 10^3 + 1 \times 10^2 + 5 \times 10 + 4 \times 1$ = ______

c $9 \times 10^5 + 4 \times 10^4 + 3 \times 10^2 + 8 \times 1$ = ______

d $3 \times 10^5 + 3 \times 10^4 + 2 \times 10^3 + 1 \times 10^2 + 3 \times 10 + 3 \times 1$ = ______

e $6 \times 10^6 + 4 \times 10^3 + 6 \times 10^2 + 5 \times 10 + 9 \times 1$ = ______

ORDERING NUMBERS
SIX-DIGIT AND SEVEN-DIGIT NUMBERS

Ascending order means numbers are ordered from smallest to largest.
These numbers are in ascending order:
135 426, 243 857, 392 450, 487 920.

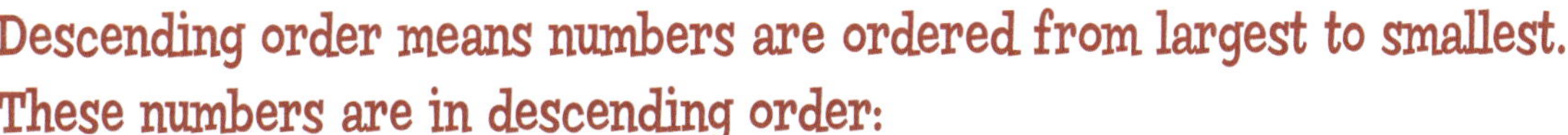
Descending order means numbers are ordered from largest to smallest.
These numbers are in descending order:
1 582 343, 1 329 470, 1 295 341, 1 124 381.

Example 1:
These numbers are in ascending order: 681 267, 748 532, 1 182 637, 4 431 972

smallest number first

largest number last

Example 2:
These numbers are in descending order: 8 352 461, 2 412 733, 736 281, 381 432

largest number first

smallest number last

Example 3: Write in ascending order: 573 219, 415 219, 793 291, 329 468

329 468, ________, ________, 793 291

smallest number first

largest number last

Example 4: Write in descending order: 4 430 215, 483 134, 5 419 993, 919 993

5 419 993, ________, ________, ________

largest number first

smallest number last

Is each group of numbers in ascending order or descending order?

- ● 3 294 137, 2 398 736, 1 497 298 descending
- **a** 534 240, 435 042, 403 524 ________
- **b** 1 425 362, 1 523 264, 1 643 225 ________
- **c** 5 943 273, 3 594 327, 2 334 954 ________
- **d** 849 274, 983 421, 1 090 352 ________
- **e** 4 432 138, 5 493 140, 6 539 415 ________

Check your answers
How many did you get correct?

PRACTICE

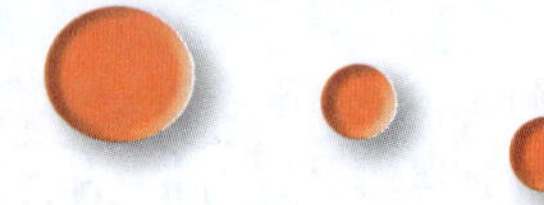

1 Write these numbers in ascending order.

- 242 361, 539 247, 635 149, 140 036

  140 036, 242 361, 539 247, 635 149

a 1 384 236, 632 483, 843 362, 233 846

____

b 2 469 137, 4 269 713, 3 716 924, 6 719 423

____

c 438 120, 102 483, 832 014, 410 283

____

d 109 413, 901 341, 471 013, 104 731

____

e 5 436 923, 3 296 345, 4 932 523, 9 234 352

____

2 Order these numbers from largest (1) to smallest (5).

●	5	124 373	3	337 421	2	412 733	4	142 337	1	734 243
a		736 281		182 637		763 812		681 267		276 186
b		1 437 249		9 427 341		4 729 314		4 431 972		7 944 213
c		5 382 416		6 142 835		8 352 461		6 412 583		3 851 436
d		4 324 928		4 426 829		9 824 264		6 944 826		8 944 263

3 Order the money from smallest (1) to largest (5).

●	3	$524 378	2	$428 735	5	$847 325	4	$748 532	1	$253 874
a		$1 359 267		$9 531 726		$6 217 539		$5 624 134		$5 462 431
b		$632 148		$236 814		$381 432		$184 332		$433 281
c		$1 470 431		$4 413 071		$7 144 310		$4 403 117		$3 407 114
d		$1 936 054		$6 315 409		$5 316 490		$9 613 045		$9 531 640

  ISBN: 9781925726183

ODD AND EVEN NUMBERS

These numbers are even:

525 630   3724   4 236 438   1840   53 666   984 432   10

They end in 0, 2, 4, 6 or 8 and can be divided by 2.

These numbers are odd:

796 153   23 351   1 446 249   7253   4267   3   937 395

They end in 1, 3, 5, 7 or 9 and cannot be divided by 2.

It doesn't matter how big the number is – always look at the ones place to find out if it is odd or even.

Example 1:
Write a 5-digit odd number.
89 421

Example 2:
Write a 7-digit even number.
1 896 472

Example 3:
Write a 3-digit odd number.
______

Example 4:
Write a ones digit that makes these numbers even.
15<u>0</u>
2 850 12__

Example 5:
Write a units digit that makes these numbers odd.
7 638 28<u>9</u>
9__

Use red to circle the odd numbers and green to circle the even numbers.

- 3240   989 113   42   143 744   4 386 420

54 982   14 243   5945   1115   654 321

2 459 998   73 248   1 473 846   274 447

1 824 739   937 937   7   876 534   120

Check your answers
How many did you get correct?

  ISBN: 9781925726183

PRACTICE

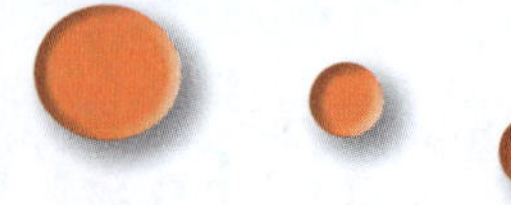

1 Write the numbers in the correct box.

● ~~13 565~~ 2 456 702 24 893 1591 24 983 1460
434 362 3579 287 369 708 1 475 672 13 504
85 342 1 493 246 44 467 131

Odd Numbers	Even Numbers
13 565	

2 Write the next three numbers.

● 2 332 462, 2 332 464, 2 332 466, 2 332 468, 2 332 470

a 149 351, 149 353, ______, ______, ______

b 3 762 449, 3 762 451, ______, ______, ______

c 289 400, 289 402, ______, ______, ______

d 138, 140, ______, ______, ______

e 6846, 6848, ______, ______, ______

f 37 237, 37 239, ______, ______, ______

g 4 037 499, 4 037 501, ______, ______, ______

3 Cross out the odd number in each set of numbers.

● 238 424, ~~238 517~~, 238 456, 238 422

a 1 424 328, 1 532 436, 1 652 140, 1 783 143

b 583 252, 584 253, 584 258, 583 260

c 6 581 370, 6 831 421, 6 632 482, 6 624 444

d 673 541, 637 152, 625 434, 683 488

e 3 871 435, 3 736 248, 3 439 256, 3 562 544

f 736 532, 764 155, 783 460, 739 862

g 8 437 281, 8 473 184, 8 743 368, 8 592 452

  ISBN: 9781925726183

GREATER THAN, LESS THAN, EQUAL TO

> greater than   **=** equal to   **<** less than

Example 1: 2493 > 1082

This statement is read as: 2493 is greater than 1082.
It is true.

Example 2: 64 391 < 89 839

This statement is read as: 64 391 is less than 89 839.
It is true.

Using the symbols for greater than, less than and equal to is a way to compare two values.

Example 3: 1 425 378 = 1 425 378

This statement is read as: 1 425 378 is equal to 1 425 378.
It is true.

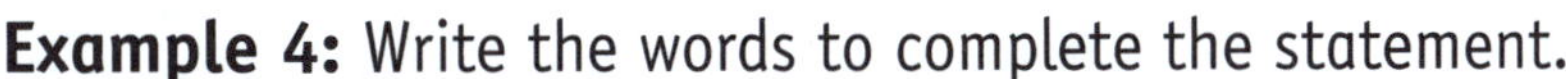

Example 4: Write the words to complete the statement.

3 482 364 ___is equal to___ 3 482 364.

Example 5: Write the words to complete the statement.

436 483 ________________ 844 394.

Example 6: Write the words to complete the statement.

149 ________________ 108.

Write T for true or F for false.

- ● [T] 29 375 < 34 539
- **a** [ ] 59 362 > 49 820
- **b** [ ] 146 801 = 146 801
- **c** [ ] 419 387 < 391 258
- **d** [ ] 1 498 326 > 1 098 683
- **e** [ ] 4 397 481 > 4 279 184
- **f** [ ] 9894 < 9496
- **g** [ ] 100 320 = 100 230
- **h** [ ] 8439 = 8439
- **i** [ ] 68 495 < 89 995
- **j** [ ] 486 248 > 468 842
- **k** [ ] 1097 > 1930
- **l** [ ] 5 638 246 > 5 361 289
- **m** [ ] 843 436 < 942 638

Check your answers
How many did you get correct?

  ISBN: 9781925726183

PRACTICE

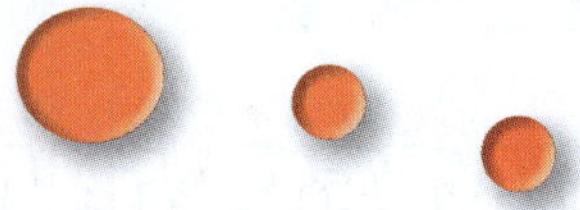

1 Are these statements correct? Write yes or no.

- 382 < 483 yes
- a 1263 = 1362 ____
- b 58 245 < 58 246 ____
- c 1 357 920 > 1 742 321 ____
- d 89 743 < 98 473 ____
- e 24 690 = 24 690 ____
- f 471 = 471 ____
- g 643 410 < 634 104 ____
- h 915 214 < 916 412 ____
- i 4 124 638 > 3 436 159 ____
- j 843 286 < 814 415 ____
- k 64 382 > 46 283 ____
- l 5861 < 8516 ____
- m 98 283 > 89 389 ____
- n 614 384 < 583 215 ____
- o 32 495 < 24 359 ____

2 Write >, < or = to make the statements true.

- 4579 [<] 5462
- a 7184 [ ] 4843
- b 15 384 [ ] 27 249
- c 3 249 147 [ ] 3 152 743
- d 45 539 [ ] 45 539
- e 389 [ ] 983
- f 59 453 [ ] 95 637
- g 159 743 [ ] 519 347
- h 8 874 581 [ ] 6 439 284
- i 1498 [ ] 8491
- j 587 432 [ ] 587 432
- k 8459 [ ] 8594
- l 4 386 410 [ ] 4 683 104
- m 5 409 328 [ ] 5 940 823
- n 26 495 [ ] 21 543
- o 9 372 495 [ ] 9 372 495

3 Cross out the incorrect words.

- 723 241 is greater than / ~~is less than~~ / ~~is equal to~~ 327 241
- a 36 409 is greater than / is less than / is equal to 36 409
- b 1 489 370 is greater than / is less than / is equal to 1 984 730
- c 248 324 is greater than / is less than / is equal to 248 324
- d 4 738 298 is greater than / is less than / is equal to 7 438 982
- e 349 is greater than / is less than / is equal to 394
- f 1932 is greater than / is less than / is equal to 1543

  ISBN: 9781925726183

LARGEST AND SMALLEST NUMBERS

How would you move these numbers around to make the smallest number possible and the largest number possible?

3 digits: 3 0 2

Smallest: 2 0 3

Largest: 3 2 0

Look for the digit with the smallest value and the digit with the greatest value.

The smallest number is not 023 as this is 23, a two-digit number.

Example 1:

4 digits:	2	5	1	9
Smallest	1	2	5	9
Largest	9	5	2	1

Example 2:

5 digits:	6	9	2	0	3
Smallest	2	0	3	6	9
Largest	9	6	3	2	0

Example 3:

6 digits:	9	1	7	2	6	4
Smallest	1	2	4	6	7	9
Largest	9	7			2	1

Example 4:

7 digits:	1	5	9	8	2	3	1
Smallest	1	1					9
Largest	9						

Check your answer on the video!

Write the smallest number and the largest number.

● 2 5 9 3

Smallest: 2359

Largest: 9532

a 5 0 3 0 9

Smallest: ________

Largest: ________

b 6 2 0 5 1 3

Smallest: ________

Largest: ________

c 2 4 9 3 8 6 4

Smallest: ________

Largest: ________

SELF CHECK Tick how you feel

Got it!	Need help...	I don't get it
☐	☐	☐

Check your answers

How many did you get correct? ☐

  ISBN: 9781925726183

PRACTICE

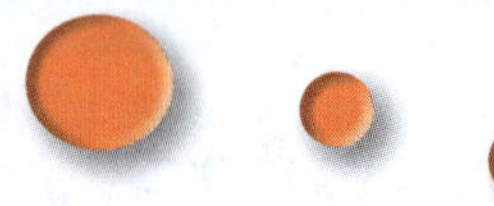

1 Write the smallest and largest numbers using the digits.

	Digits	Smallest number	Largest number
●	7 3 2 8 9 5	235 789	987 532
a	4 9 0 3		
b	5 8 7 4 6 2		
c	0 0 3 9 5		
d	1 1 4 6		
e	3 8 8		
f	1 3 9 9 8 7 7		
g	8 2 6 4 8 5 9		
h	5 1 4 9 8 3		
i	6 8 4 9 3 7		
j	9 9 9 3 8 6 2		

2 Here is a test Tran did where he crossed out the number that is not the largest or smallest number that could be made with the digits. Mark Tran's test.

	Digits	Numbers	Mark
●	3 2 6 4 2 3	~~223 436~~, 223 346, 643 322	✓
a	1 8 9	189, 981, ~~891~~	
b	2 8 9 3	~~9832~~, 2839, 2389	
c	2 0 5 3 8	~~02 358~~, 85 320, 20 358	
d	1 9 3 2 1 0	101 239, ~~110 239~~, 932 110	
e	8 4 9 7 4 6	896 447, 987 644, ~~446 789~~	
f	7 6 1	761, 176, ~~167~~	
g	4 9 3 0 0 1	~~010 394~~, 943 100, 100 349	
h	5 0 3 1	5310, 1035, ~~3051~~	
i	6 0 1 0 1	61 100, 10 016, ~~10 106~~	
j	3 7 2 2 0 3 8	8 733 220, 2 023 378, ~~2 203 378~~	

CATCH UP MATHS YEAR 6 BOOK A © PASCAL PRESS ISBN: 9781925726183

ROUNDING TO 100 000 AND 1 000 000

Rounding is useful when you need to estimate an answer.

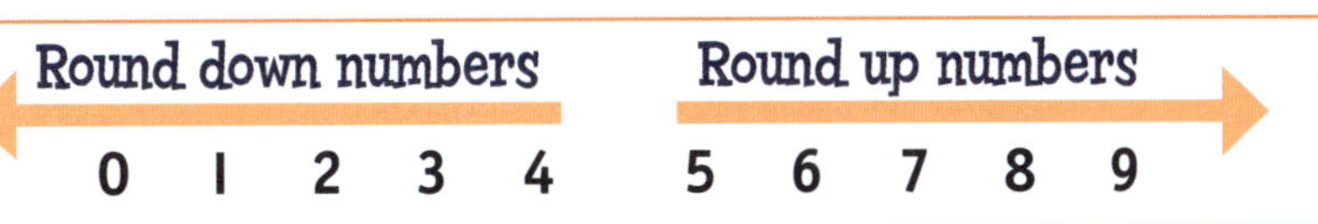

SCAN to watch video

Rounding to 100 000

1 Go to the hundred thousands place.
2 Write the round up number above the hundred thousands.
3 Circle the number in the ten thousands place.
4 Is it a round up or round down number?
5 Round the number.

Example 1: Round to the nearest 100 000.

7
6(3)5 927 down

600 000

Example 2: Round to the nearest hundred thousand.

5
1 4(7)2 985

____

Rounding to 1 000 000

1 Go to the millions place.
2 Write the round up number above the millions.
3 Circle the number in the hundred thousands place.
4 Is it a round up or round down number?
5 Round the number.

Example 3: Round to the nearest million.

9 up
8(9)93 645

9 000 000

Check your answer on the video!

Example 4: Round to the nearest 1 000 000.

6
5(6)74 829

____

1 Round to the nearest 100 000.

4 900 000

a 3 695 411 __________

b 526 439 __________

2 Round to the nearest 1 000 000.

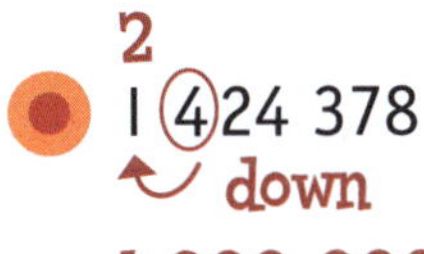

a 4 739 246 __________

b 8 471 115 __________

SELF CHECK	Tick how you feel	
Got it! ☐	Need help... ☐	I don't get it ☐

Check your answers
How many did you get correct? ☐

PRACTICE

1 Round these numbers to the nearest 100 000.

● 1 529 612 (6, down) → 1 500 000	c 316 489	f 743 836	i 429 382
a 6 573 248	d 5 974 389	g 533 906	j 630 490
b 536 212	e 9 873 283	h 899 374	k 8 884 486

2 Round these numbers to the nearest 1 000 000.

● 8 524 949 (9, up) → 9 000 000	c 6 918 369	f 5 643 294	i 2 673 899
a 9 537 735	d 4 845 100	g 7 473 095	j 1 529 216
b 8 210 005	e 1 000 005	h 3 650 038	k 1 977 150

3 Complete the table.

	Round to nearest 100 000	Round to nearest 1 000 000
● 3 424 382	3 400 000	3 000 000
a 5 243 781		
b 9 501 331		
c 8 525 887		
d 6 303 525		
e 3 116 952		
f 5 266 123		
g 8 714 251		
h 9 896 244		

CATCH UP MATHS YEAR 6 BOOK A © PASCAL PRESS ISBN: 9781925726183

THE ROLE OF ZERO

Zero (0) is a symbol used to describe a place holder for a value when no other number is in that place.

Example 1: 509

The zero is holding the tens place.
Otherwise, the number would be 59.

Why don't we use zero at the start of a number?

Example 2: 6085

The zero is holding the hundreds place.
Otherwise, the number would be 685.

Example 3: 103 552

The zero is holding the ten thousands place.
Otherwise, the number would be 13 552.

Example 4: 390

The zero is holding the ________ place.

Otherwise, the number would be ___________.

Example 5: 501 236

The zero is holding the ______________________ place.

Otherwise, the number would be ___________.

Example 6: 6 053 281

The zero is holding the ______________________ place.

Otherwise, the number would be ___________.

Write the place that zero is holding.

- 40 365 ______________________
- **a** 705 341 ______________________
- **b** 1 563 032 ______________________
- **c** 10 436 ______________________
- **d** 530 928 ______________________

SELF CHECK Tick how you feel		
Got it! ☐	Need help... ☐	I don't get it ☐

Check your answers
How many did you get correct? ☐

PRACTICE

1 What place is zero holding in each number below?

- 4 850 493 — thousands

a 124 506 ____

b 10 563 ____

c 9 036 521 ____

d 524 903 ____

e 1 598 420 ____

f 6 420 124 ____

g 430 281 ____

h 3 850 132 ____

i 240 ____

j 1056 ____

k 3 082 153 ____

l 120 ____

m 5 243 109 ____

n 524 302 ____

o 6302 ____

p 7 306 249 ____

q 4087 ____

2 Write five numbers that match each description.

- four-digit numbers with a zero in the hundreds place

4023, 9074, 3095, 1000, 7037

a five-digit numbers with a zero in the tens place

____, ____, ____, ____, ____

b six-digit numbers with a zero in the thousands place

____, ____, ____, ____, ____

c seven-digit numbers with a zero in the hundred thousands place

____, ____, ____, ____, ____

d three-digit numbers with a zero in the ones place

____, ____, ____, ____, ____

e four-digit numbers with a zero in the tens place

____, ____, ____, ____, ____

f six-digit numbers with a zero in the ten thousands place

____, ____, ____, ____, ____

 CATCH UP MATHS YEAR 6 BOOK A © PASCAL PRESS ISBN: 9781925726183

ABBREVIATIONS OF LARGE NUMBERS

You can use the letter K, the word 'thousand' or the abbreviation 'thous.' instead of writing three zeros at the end of a number.

250 000 = 250K = 250 thous.

SCAN to watch video

Example 1:

Use the word 'thousand' to write these numbers.

4000 = 4 thousand

21 000 = 21 thousand

999 000 = ____________________

Example 2:

Use K to write these numbers.

1000 = 1K

3 000 000 = 3000K

53 000 = ______

Example 3:

Use 'thous.' to write these numbers.

28 000 = 28 thous.

76 000 = 76 thous.

5000 = ________________

Example 4:

Use 3 zeroes to write these numbers.

3140 thous. = 3 140 000

8K = 8000

64 thousand = _______

1 Use K to write these numbers.

- 242 thousand = 242K
- a 615 000 = _________
- b 14 000 = _________
- c 824 thousand = _________

There are lots of different ways that you can show numbers.

2 Use thous. to write these numbers.

- 51 000 = 51 thous.
- a 716 000 = ____________________
- b 179 thousand = ____________________
- c 623 thousand = ____________________

SELF CHECK Tick how you feel

Got it!	Need help...	I don't get it
☐	☐	☐

Check your answers

How many did you get correct? ☐

PRACTICE

1 Write these numbers using K for thousand.

- 462 000 = 462K
- a 197 000 = ______
- b 27 thousand = ______
- c 503 000 = ______
- d 6000 = ______
- e 90 thousand = ______
- f 810 000 = ______
- g 403 000 = ______
- h 20 thousand = ______
- i 38 thousand = ______
- j 524 000 = ______
- k 100 thousand = ______

2 Write these numbers using thous. for thousand.

- 247 000 = 247 thous.
- a 23 thousand = ________
- b 17 thousand = ________
- c 840 000 = ________
- d 99 000 = ________
- e 103 000 = ________
- f 711 thousand = ________
- g 616 thousand = ________
- h 8 thousand = ________
- i 400 thousand = ________
- j 430 000 = ________
- k 595 000 = ________

3 Match the numbers.

23 000	248 thous.
a 162 000	670K
b 89 000	5K
c 511 000	23K
d 670 000	4 thous.
e 403 000	425 thous.
f 820 000	162K
g 4000	511 thous.
h 309 000	309K
i 5000	403 thous.
j 248 000	89 thous.
k 425 000	820K

CATCH UP MATHS YEAR 6 BOOK A © PASCAL PRESS ISBN: 9781925726183

FACTORS

A factor is a number that multiplies with another number to get a product.

5 × 2 = 10 ← product

factors

Remember – the product is the answer you get when you multiply.

Example 1: What are the factors of 18?

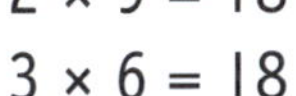

1 × 18 = 18
2 × 9 = 18
3 × 6 = 18

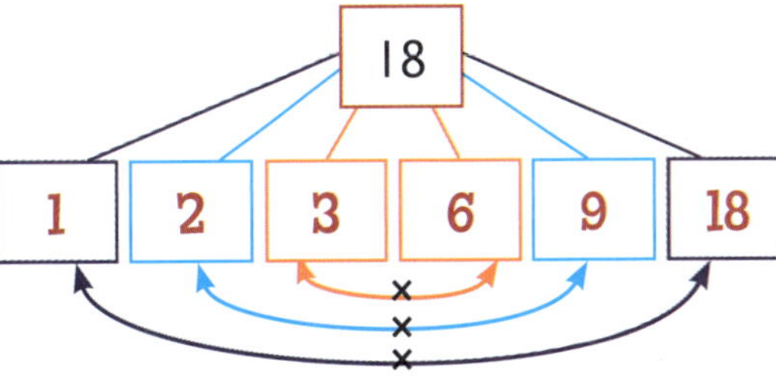

1, 2, 3, 6, 9 and 18 are the factors of 18.

Example 2: What are the factors of 12?

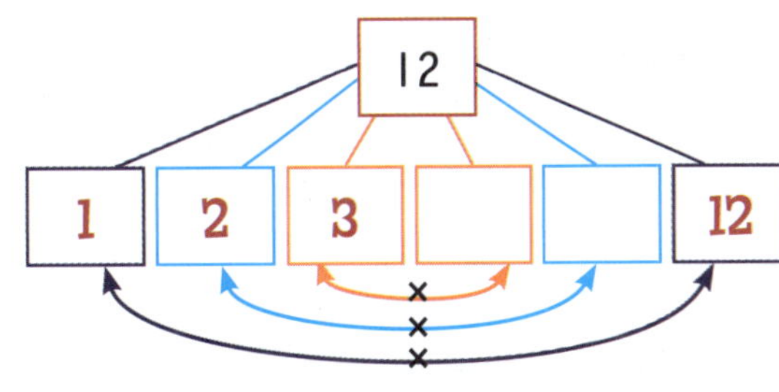

1, 2, 3, ___, ___ and 12 are the factors of 12.

Example 3: What are the factors of 8?

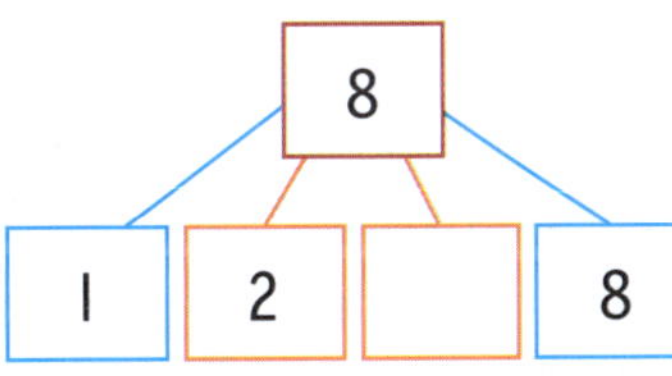

1, 2, ___ and 8 are the factors of 8.

Cross out the number that is NOT a factor.

- 6: 1, 2, 3, ~~4~~, 6
- **a** 10: 1, 2, 3, 5, 10
- **b** 5: 1, 2, 5
- **c** 15: 1, 3, 4, 5, 15
- **d** 20: 1, 2, 4, 5, 6, 10, 20
- **e** 21: 1, 2, 3, 7, 21

Check your answers
How many did you get correct?

PRACTICE

1 Circle the number that is NOT a factor.

- 10: 1, 2, (4), 5, 10

a 14: 1, 2, 3, 7, 14

b 16: 1, 2, 3, 4, 8, 16

c 30: 1, 2, 3, 4, 5, 6, 10, 30

d 32: 1, 2, 3, 4, 8, 16, 32

e 36: 1, 2, 3, 4, 5, 6, 9, 12, 18, 36

f 20: 1, 2, 4, 5, 6, 10, 20

g 22: 1, 2, 4, 11, 22

2 Write all the factors of each number.

- 21: 1, 3, 7, 21

a 27: ______________________

b 40: ______________________

c 100: ______________________

d 18: ______________________

e 24: ______________________

f 25: ______________________

g 28: ______________________

h 19: ______________________

i 144: ______________________

j 64: ______________________

k 7: ______________________

l 50: ______________________

m 56: ______________________

n 26: ______________________

  ISBN: 9781925726183

HIGHEST COMMON FACTOR (HCF)

SCAN to watch video

The Highest Common Factor (HCF) is the highest number that is a factor of two other numbers.

Example 1:

What is the Highest Common Factor (HCF) of 8 and 12?

8: 1, 2, (4), 8

12: 1, 2, 3, (4), 6, 12

- Write all the factors of each number.
- Look for the highest number that is in both lists.

4 is the Highest Common Factor of 8 and 12.

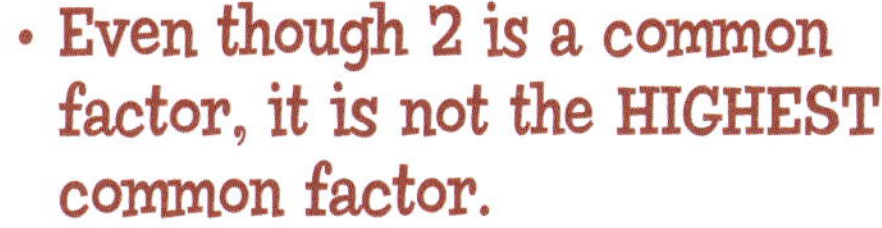

- Even though 2 is a common factor, it is not the HIGHEST common factor.

Example 2:

What is the Highest Common Factor (HCF) of 30 and 45?

30: 1, 2, 3, 5, 6, 10, (15), 30

45: 1, 3, 5, (15), 45

15 is the HCF of 30 and 45.

Example 3:

What is the Highest Common Factor (HCF) of 9 and 27?

9: 1, 3, 9

27: 1, 3, 9, 27

__ is the HCF of 9 and 27.

Example 4:

What is the Highest Common Factor (HCF) of 35 and 25?

35: ___, ___, ___, ___

25: ___, ___, ___

__ is the HCF of 35 and 25.

Circle the Highest Common Factor (HCF) for each pair of numbers.

● **8:** 1, 2, 4, 8
  10: 1, 2, 5, 10

a **12:** 1, 2, 3, 4, 6, 12
  15: 1, 3, 5, 15

b **30:** 1, 2, 3, 5, 6, 10, 15, 30
  60: 1, 2, 3, 4, 5, 6, 10, 12, 15, 20, 30, 60

Check your answers
How many did you get correct?

PRACTICE

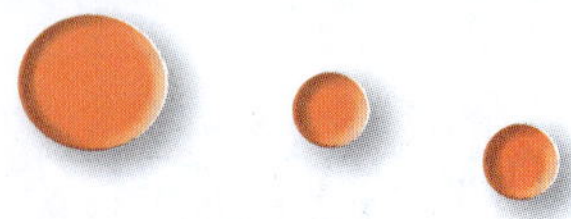

1 Work out the Highest Common Factor (HCF).

- 12: 1, 2, 3, (4), 6, 12
  16: 1, (4), 16
  4 is the HCF of 12 and 16.

a 16: ____
  20: ____
  ☐ is the HCF of 16 and 20.

b 5: ____
  10: ____
  ☐ is the HCF of 5 and 10.

c 21: ____
  24: ____
  ☐ is the HCF of 21 and 24.

d 10: ____
  24: ____
  ☐ is the HCF of 10 and 24.

e 20: ____
  36: ____
  ☐ is the HCF of 20 and 36.

f 18: ____
  36: ____
  ☐ is the HCF of 18 and 36.

g 33: ____
  55: ____
  ☐ is the HCF of 33 and 55.

h 60: ____
  42: ____
  ☐ is the HCF of 60 and 42.

i 50: ____
  110: ____
  ☐ is the HCF of 50 and 110.

j 91: ____
  65: ____
  ☐ is the HCF of 91 and 65.

  ISBN: 9781925726183

MULTIPLES

A multiple is the number you get when you multiply two factors together.

18 is a multiple of 3 and 6 because 3 × 6 = 18

factor factor multiple

18 is also a multiple of 1 and 18 because 1 × 18 = 18

factor factor multiple

It is also a multiple of 2 and 9 because 2 × 9 = 18

Every number has both 1 and itself as factors.

Example 1: What are the first five multiples of 4?

The multiples of 4 are the numbers you get when you multiply 4 by another number.

The first five multiples of 4 are **4**, **8**, **12**, **16** and **20**.

4 × 1 = **4**
4 × 2 = **8**
4 × 3 = **12**
4 × 4 = **16**
4 × 5 = **20**

Example 2: What are the first five multiples of 6?

The multiples of 6 are the numbers you get when you multiply 6 by another number.

The first five multiples of 6 are ___, ___, ___, ___ and ___.

6 × 1 = 
6 × 2 = ___
6 × 3 = ___
6 × 4 = ___
6 × 5 = ___

Write the first four multiples.

a 3: ___, ___, ___, ___

b 5: ___, ___, ___, ___

c 10: ___, ___, ___, ___

Check your answers
How many did you get correct?

PRACTICE

1 Write the first eight multiples.

- 4: 4, 8, 12, 16, 20, 24, 28, 32

a 12: ____, ____, ____, ____, ____, ____, ____, ____

b 6: ____, ____, ____, ____, ____, ____, ____, ____

c 1: ____, ____, ____, ____, ____, ____, ____, ____

d 3: ____, ____, ____, ____, ____, ____, ____, ____

e 5: ____, ____, ____, ____, ____, ____, ____, ____

f 8: ____, ____, ____, ____, ____, ____, ____, ____

g 10: ____, ____, ____, ____, ____, ____, ____, ____

h 2: ____, ____, ____, ____, ____, ____, ____, ____

i 7: ____, ____, ____, ____, ____, ____, ____, ____

2 Write the next four multiples.

- …, 20, 24, 28, 32, 36, 40, 44

a …, 14, 21, 28, ____, ____, ____, ____

b …, 25, 30, 35, ____, ____, ____, ____

c …, 48, 56, 64, ____, ____, ____, ____

d …, 45, 54, 63, ____, ____, ____, ____

3 Cross out the number that is not a multiple.

- 5: 10, 15, 20, 23

a 6: 24, 30, 37, 42

b 11: 11, 12, 22, 33

c 2: 24, 26, 28, 29

d 7: 56, 63, 70, 78

e 3: 15, 18, 20, 24

f 4: 24, 28, 33, 36

g 8: 40, 48, 58, 64

CATCH UP MATHS YEAR 6 BOOK A © PASCAL PRESS ISBN: 9781925726183

LOWEST COMMON MULTIPLE (LCM)

The Lowest Common Multiple (LCM) of two numbers is the smallest number that is a multiple of both numbers.

Example 1:

What is the Lowest Common Multiple of 8 and 12?

Multiples of **8**: 8, 16, (24), 32 ...

Multiples of **12**: 12, (24), 36, 48 ...

The LCM of 8 and 12 is **24**.

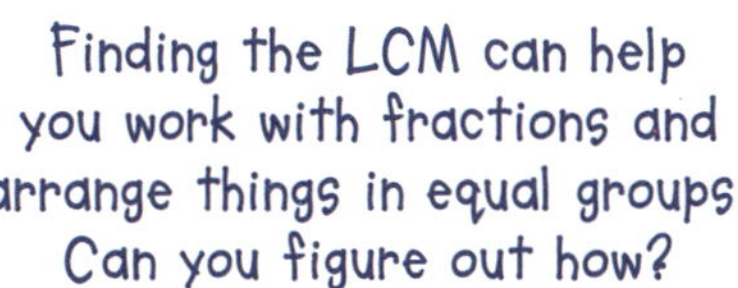

Example 2:

What is the Lowest Common Multiple of 3 and 5?

Multiples of **3**: 3, 6, 9, 12, 15, 18 ...

Multiples of **5**: 5, 10, 15, 20, 25 ...

The LCM of 3 and 5 is ___.

Example 3:

What is the Lowest Common Multiple of 4 and 6?

Multiples of **4**: 4, ___, ___, ___, ___ ...

Multiples of **6**: 6, ___, ___, ___, ___ ...

The LCM of 4 and 6 is ___.

What is the Lowest Common Multiple (LCM)?

● **2:** 2, 4, 6, (8), 10
8: (8), 16, 24, 32, 40

[8] is the Lowest Common Multiple of 2 and 8

a **3:** 3, 6, 9, 12, 15
4: 4, 8, 12, 16, 20

[ ] is the Lowest Common Multiple of 3 and 4

b **3:** 3, 6, 9, 12, 15
9: 9, 18, 27, 36, 45

[ ] is the Lowest Common Multiple of 3 and 9

SELF CHECK Tick how you feel

Got it!	Need help...	I don't get it
☐	☐	☐

Check your answers
How many did you get correct? ☐

PRACTICE

1 Work out the Lowest Common Multiple (LCM).

- 2: 2, 4, 6, 8, (10)
  5: 5, (10), 15, 20, 25
  [10] is the LCM of 2 and 5

a 4: ____
5: ____
[ ] is the LCM of 4 and 5

b 4: ____
10: ____
[ ] is the LCM of 4 and 10

c 2: ____
6: ____
[ ] is the LCM of 2 and 6

d 1: ____
5: ____
[ ] is the LCM of 1 and 5

e 5: ____
10: ____
[ ] is the LCM of 5 and 10

f 4: ____
8: ____
[ ] is the LCM of 4 and 8

g 2: ____
8: ____
[ ] is the LCM of 2 and 8

h 2: ____
10: ____
[ ] is the LCM of 2 and 10

i 3: ____
4: ____
[ ] is the LCM of 3 and 4

j 7: ____
5: ____
[ ] is the LCM of 7 and 5

CATCH UP MATHS YEAR 6 BOOK A © PASCAL PRESS ISBN: 9781925726183

INTEGERS

Integers are the positive and negative whole numbers on each side of zero on a number line. Fractions and decimals are NOT integers.

Backwards from 0 | Forwards from 0

Negative Integers | Positive Integers

−5 −4 −3 −2 −1 0 1 2 3 4 5

5 = +5

Negative numbers are less than zero. They are written with a minus sign in front of them.

Positive numbers are greater than zero. They can be written with or without a plus sign.

Example 1: Write the missing numbers.

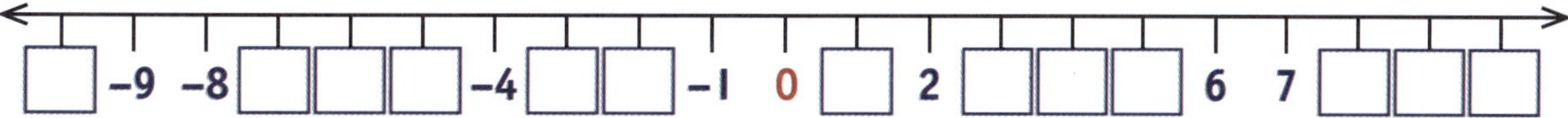

Example 2: Follow the directions on the number line and write the number you end on.

Start at −3
Go back 2
Go forwards 5
End on: ___

Remember – forwards is the positive direction and backwards is the negative direction.

1 Follow the directions on the number line above, and write the number you end on.

- Start at + 3, go forwards 2, back 1. End on: 4
- **a** Start at −5, go forwards 5, back 4. End on: ___
- **b** Start at 2, go forwards 1, back 6. End on: ___
- **c** Start at 4, go back 5, back 1. End on: ___

2 Circle the integers.

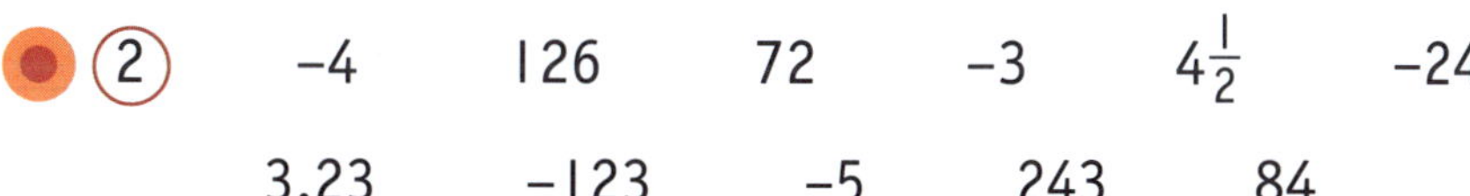

Check your answers
How many did you get correct?

  ISBN: 9781925726183

PRACTICE

1 The opposite integer of –3 is +3. Match the opposite integers.

3 4 5 2 1 7 6 8

–8 –5 –3 –6 –1 –4 –7 –2

2 Order the integers from smallest to greatest.

- 5, –7, 7, –2, 0 –7, –2, 0, 5, 7
- **a** 5, –3, –8, 1, 9 ____________
- **b** –32, 16, 32, –6, –19 ____________
- **c** 12, –24, –100, 15, 29 ____________
- **d** 53, 27, –140, 1, –32 ____________

3 Use the diagram to help you find the difference in temperature.

- –15 °C and 5 °C 20 °C
- **a** 20 °C and 5 °C ______
- **b** 25 °C and –15 °C ______
- **c** 0 °C and –10 °C ______
- **d** 15 °C and 10 °C ______
- **e** 10 °C and –10 °C ______
- **f** 25 °C and 0 °C ______
- **g** –10 °C and 5 °C ______
- **h** –5 °C and 10 °C ______
- **i** –15 °C and 25 °C ______

25 °C
20 °C
15 °C
10 °C
5 °C
0 °C
-5 °C
-10 °C
-15 °C

4 Follow the directions by drawing arrows on the number line.
Then write the number you end on. – means negative, backwards or left.
\+ means positive, forwards or right.

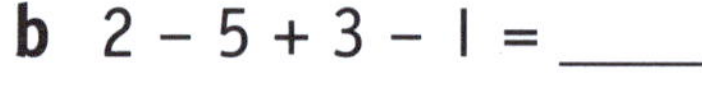

- 2 + 2 – 3 – 2 = –1

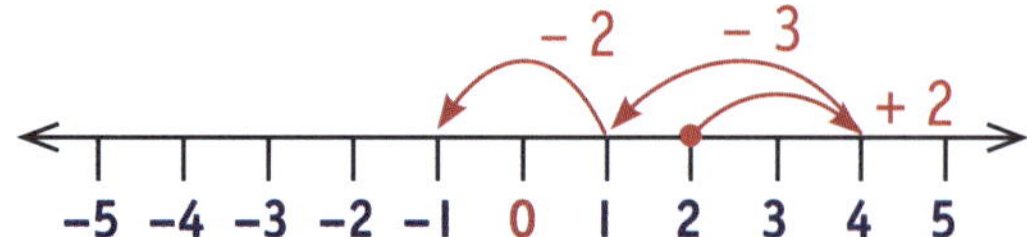

- **b** 2 – 5 + 3 – 1 = ____

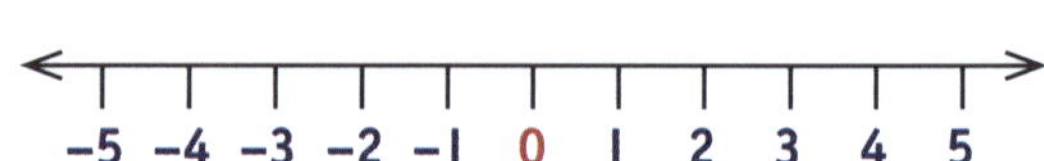

- **a** – 5 + 3 – 2 + 3 = ____

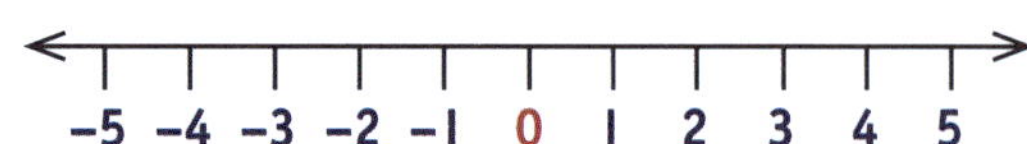

- **c** – 4 – 1 + 3 + 2 = ____

CATCH UP MATHS YEAR 6 BOOK A © PASCAL PRESS ISBN: 9781925726183

PRIME AND COMPOSITE NUMBERS

Prime numbers have only themselves and 1 as factors.
3 is a prime number because its factors are 1 and 3.

Composite numbers have more than two factors.
4 is a composite number because its factors are 1, 2 and 4.

Example 1:
8 has factors of 1, 2, 4 and 8, so 8 is a composite number.

Example 2:
7 has factors of 1 and 7, so 7 is a prime number.

Example 3:
2 has factors of 1 and 2, so 2 is a ________________ number.

Example 4:
6 has factors of 1, __, __ and __, so 6 is a ________________ number.

Complete.

● List the factors of 5: 1, 5

5 is a prime ~~composite~~ number.

a List the factors of 12: ________________

12 is a prime / composite number.

b List the factors of 21: ________________

21 is a prime / composite number.

c List the factors of 13: ________________

13 is a prime / composite number.

SELF CHECK Tick how you feel

Got it!	Need help...	I don't get it
☐	☐	☐

Check your answers
How many did you get correct?

PRACTICE

1 Use red to circle the prime numbers and blue to circle the composite numbers.

- (3) 16 17 39 9 14 56 30 82
- 121 24 11 101 36 51 7

2 Write all the composite numbers between these numbers.

- 20 and 30 21, 22, 24, 25, 26, 27, 28

a 0 and 20 ______

b 50 and 75 ______

c 125 and 150 ______

3 Write all the prime numbers between these numbers.

- 20 and 30 23, 29

a 0 and 20 ______

b 50 and 100 ______

c 110 and 150 ______

4 Underline the greatest prime number.

- 3 <u>151</u> 153 43

a 78 34 23 67

b 29 54 47 168

c 139 147 155 129

5 Underline the smallest composite number.

- 7 <u>14</u> 22 3

a 5 8 13 4

b 113 101 112 140

c 87 125 137 157

6 Write the next prime number after each of these numbers.

- 9 11

a 15 ____ **b** 27 ____ **c** 153 ____

7 Write the next composite number after each of these numbers.

- 18 20

a 29 ____ **b** 38 ____ **c** 149 ____

CATCH UP MATHS YEAR 6 BOOK A © PASCAL PRESS ISBN: 9781925726183

WHOLE NUMBERS REVIEW

1 Circle all the three-digit numbers.

248 24 6263 82 431 585 493 821 215

1543 385 15 378 2462 16 147 3490

2 Circle all the four-digit numbers.

824 42 6632 13 842 855 394 128 305

3541 835 37 851 6422 61 7410 9043

3 Circle all the five-digit numbers.

4136 950 133 9163 653 1436 40 052 78

52 437 13 959 920 452 82 634 15 3498

4 Write these numbers in words.

a 253 ______________________________

b 820 ______________________________

c 705 ______________________________

d 509 ______________________________

e 1536 ______________________________

f 8290 ______________________________

g 4053 ______________________________

h 2002 ______________________________

i 89 216 ______________________________

j 70 014 ______________________________

k 60 050 ______________________________

l 80 927 ______________________________

m 50 005 ______________________________

n 22 465 ______________________________

o 19 240 ______________________________

p 92 300 ______________________________

REVIEW

5 **Fill in the table.**

	Number	Ten Thousands	Thousands	Hundreds	Tens	Ones
a	503					
b	857					
c	960					
d	1795					
e	2803					
f	5036					
g	6100					
h	37 389					
i	34 090					
j	26 000					

6 **Write the place value of the underlined digit.**

a 24 30$\underline{6}$ ____________________

b $\underline{2}$3 ____________________

c 1$\underline{5}$32 ____________________

d 5$\underline{9}$6 ____________________

e 15 37$\underline{3}$ ____________________

f $\underline{6}$598 ____________________

7 **Write the place value of the 3.**

a 236 ____________________

b 3415 ____________________

c 13 ____________________

d 23 020 ____________________

e 36 421 ____________________

f 347 ____________________

8 **Write the numbers.**

a 6 thousands, 2 hundreds, 5 tens, 8 ones ____________

b 9 hundreds, 4 ones, 7 tens ____________

c 5 ten thousands, 9 ones ____________

d 3 tens, 4 ones, 5 hundreds, 6 thousands, 1 ten thousand ____________

e 7 hundreds, 3 ones, 6 tens, 3 thousands ____________

f 5 ones, 4 tens, 8 ten thousands ____________

CATCH UP MATHS YEAR 6 BOOK A © PASCAL PRESS ISBN: 9781925726183

9 What is the value of 2 in these numbers?

a 362 ________
b 6234 ________
c 2436 ________
d 64 281 ________
e 3260 ________
f 72 ________
g 53 702 ________
h 27 389 ________
i 49 325 ________
j 42 153 ________
k 287 ________
l 53 321 ________

10 Write a three-digit number that matches the description.

a 4 is worth the most ________
b 4 is worth the least ________
c 2 is worth the most ________
d 2 is worth the least ________
e 9 is worth the most ________
f 9 is worth the least ________

11 Write a four-digit number that matches the description.

a 5 is worth the most ________
b 5 is worth the least ________
c 7 is worth the most ________
d 7 is worth the least ________
e 1 is worth the most ________
f 1 is worth the least ________

12 Write a five-digit number that matches the description.

a 3 is worth the most ________
b 3 is worth the least ________
c 6 is worth the most ________
d 6 is worth the least ________
e 4 is worth the most ________
f 4 is worth the least ________

13 Write these numbers in expanded form.

a 325 = ________________
b 130 = ________________
c 589 = ________________
d 1542 = ________________
e 6970 = ________________
f 9053 = ________________
g 24 905 = ________________
h 15 419 = ________________
i 60 326 = ________________

REVIEW

14 Write the place value and value of 7 in each number.

	Number	Place Value of 7	Value of 7
a	947		
b	704		
c	3873		
d	78 326		
e	4738		
f	17 365		
g	2789		
h	117		
i	8724		
j	8673		
k	52 784		
l	74 123		

15 Write each set of numbers in ascending order.

a 105, 510, 501, 515, 551

____

b 1536, 6315, 5136, 3165, 1653

____

c 34 790, 79 430, 40 973, 94 307, 74 390

____

16 Write each set of numbers in descending order.

a 734, 374, 437, 743, 347

____

b 1029, 9201, 1209, 9102, 2190

____

c 36 413, 63 431, 33 461, 13 346, 64 313

____

CATCH UP MATHS YEAR 6 BOOK A © PASCAL PRESS ISBN: 9781925726183

17 Write these numbers in words.

a 562 431 ______________________________

b 703 235 ______________________________

c 498 107 ______________________________

d 5 090 500 ______________________________

e 7 006 250 ______________________________

f 8 429 621 ______________________________

18 Circle all the six-digit numbers.

395 24 538 5 264 73 495 285 362

624 398 2 568 243 10 368 3 875 000 402 581

19 Circle all the seven-digit numbers.

204 518 59 437 2564 2 846 443 589

582 623 5 738 246 80 631 4 254 124 136 286

20 Fill in the table.

	Number	Millions	Hund. Thous.	Ten Thous.	Thous.	Hundreds	Tens	Ones
a	583 296							
b	145 370							
c	209 460							
d	602 000							
e	1 672 438							
f	4 053 203							
g	5 374 002							
h	3 006 060							

REVIEW

21 Write the place value of the underlined digit.

a <u>8</u>84 523 ________

b 6 413 28<u>0</u> ________

c 373 <u>4</u>89 ________

d <u>5</u> 420 015 ________

e 529 6<u>1</u>2 ________

f 4 <u>5</u>15 116 ________

g 746 4<u>6</u>0 ________

h <u>2</u> 125 764 ________

22 Write the place value of the 6.

a 524 637 ________

b 6 154 291 ________

c 647 315 ________

d 5 461 593 ________

e 950 635 ________

f 4 652 419 ________

g 163 487 ________

h 3 492 136 ________

23 Write the number.

a 3 hundred thousands, 4 million, 2 hundreds, 1 ten, 5 ones, 7 ten thousands ________

b 6 ten thousands, 4 tens, 3 millions, 7 ones ________

c 8 millions, 5 ten thousands, 8 ones, 6 tens ________

24 Write the number.

a 6 ones, 5 ten thousands, 7 tens, 6 hundred thousands, 4 thousands, 2 hundreds ________

b 8 hundred thousands, 3 ones, 5 tens, 7 hundreds, 5 thousands, 9 ten thousands ________

c 2 tens, 3 ten thousands, 5 hundred thousands ________

25 What is the value of 1 in each of these numbers?

a 341 273 ________

b 1 436 287 ________

c 581 325 ________

d 149 857 ________

e 613 429 ________

f 284 381 ________

g 513 208 ________

h 1 826 253 ________

26 Write four numbers that match each description.

a 5 has a value of 500 000 ________ ________ ________ ________

b 3 has a value of 300 000 ________ ________ ________ ________

c 7 has a value of 7 000 000 ________ ________ ________ ________

d 9 has a value of 9 000 000 ________ ________ ________ ________

27 Write in expanded form.

a 238 209 = ____________________

b 147 283 = ____________________

c 590 451 = ____________________

d 2 486 281 = ____________________

e 3 015 215 = ____________________

f 7 000 009 = ____________________

g 9 320 553 = ____________________

28 Write using powers of 10.

a 632 497

= ____________________

b 249 352

= ____________________

c 451 643

= ____________________

d 5 347 243

= ____________________

e 6 842 150

= ____________________

f 2 950 243

= ____________________

g 3 202 536

= ____________________

REVIEW

29 Write the numbers.

a 3 000 000 + 400 000 + 20 000 + 5000 + 300 + 20 + 5 ______________

b 7 000 000 + 100 000 + 50 000 + 8000 + 200 + 10 + 6 ______________

c 300 000 + 50 000 + 7000 + 300 + 70 ______________

d 400 000 + 5000 + 600 + 2 ______________

e 6 000 000 + 200 000 + 5000 + 400 + 10 + 7 ______________

f 500 000 + 10 000 + 3 000 + 100 + 50 + 3 ______________

g 1 000 000 + 80 000 + 600 000 + 700 + 4000 + 20 + 5 ______________

h 8 + 30 + 2000 + 40 000 + 5 000 000 + 200 000 + 700 ______________

i 6000 + 800 + 30 000 + 4 000 000 + 100 000 + 7 + 50 ______________

30 Write the numbers into the place value chart below.

- three million and twenty-six thousand

a two million, four hundred and sixteen thousand, eight hundred and two

b nine hundred and forty-two thousand, four hundred and sixty-three

c eight hundred and eleven thousand, seven hundred and fifty

d one million, two hundred and twelve thousand, four hundred and thirty-seven

e five hundred and fifty-six thousand, eight hundred and three

f one million and two

	Millions	Hund. Thous.	Ten Thous.	Thous.	Hundreds	Tens	Ones
	3		2	6			
a							
b							
c							
d							
e							
f							

CATCH UP MATHS YEAR 6 BOOK A © PASCAL PRESS ISBN: 9781925726183

31 Write these numbers in words.

a 3 427 093 ____________________

____

b 2 581 320 ____________________

____

c 842 501 ____________________

____

d 92 689 ____________________

____

32 Circle the digit in the hundred thousands place and underline the digit in the millions place.

a 5 973 181

b 7 349 138

c 5 282 112

d 6 493 651

e 4 382 400

f 7 059 368

33 Write a six-digit number that matches each description.

a 3 hundreds ____________

b 5 thousands ____________

c 7 ten thousands ____________

c 9 hundred thousands ____________

34 Write a seven-digit number that matches each description.

a 5 ten thousands ____________

b 8 hundreds ____________

c 6 millions ____________

d 4 hundred thousands ____________

e 3 tens ____________

f 9 thousands ____________

35 Write these numbers in ascending order.

a 524 378, 243 875, 234 785, 874 532, 735 842

____

b 943 286, 629 438, 98 643, 698 432, 489 326

____

c 4 682 359, 5 469 283, 6 459 382, 2 389 546, 3 596 849

____

REVIEW

36 Use red to circle the odd numbers and green to circle the even numbers.

324 1527 43 824 947 23 465 324 190 4374

125 8009 7342 78 475 641 382 9 435 152

37 Write the next three numbers.

a 247 124, 247 126, 247 128, ____________, ____________, ____________

b 502 137, 502 139, 502 141, ____________, ____________, ____________

c 3 553 415, 3 553 417, 3 553 419, ____________, ____________, ____________

d 8 409 514, 8 409 516, 8 409 518, ____________, ____________, ____________

38 Write T for true or F for false.

a ☐ 641 108 < 146 801

b ☐ 249 327 > 219723

c ☐ 3 437 204 > 3 437 204

d ☐ 7 432 135 < 7 243 531

e ☐ 743 829 = 743 829

f ☐ 913 452 < 513 954

g ☐ 5 632 008 > 5 236 800

h ☐ 4 311 137 < 4113731

39 Write the smallest and largest numbers you can make with these digits.

	Digits	Smallest number	Largest number
a	3 4 2 9 8 4		
b	1 0 5 3 2 1		
c	6 4 0 1 5 7 5		
d	9 4 5 3 8 6 1		

40 Round to the nearest 100.

a 327 ____________

b 489 ____________

c 3562 ____________

d 5943 ____________

e 591 274 ____________

f 1 471 328 ____________

41 Round to the nearest 1000.

a 4275 ____________

b 2859 ____________

c 7093 ____________

d 11 495 ____________

e 482 157 ____________

f 2 493 581 ____________

CATCH UP MATHS YEAR 6 BOOK A © PASCAL PRESS ISBN: 9781925726183

42 Round to the nearest 10 000.

a 42 475 ______

b 54 912 ______

c 68 739 ______

d 491 283 ______

e 675 297 ______

f 849 483 ______

g 6 847 920 ______

h 8 782 421 ______

43 Round to the nearest 100 000.

a 629 373 ______

b 784 292 ______

c 3 427 281 ______

d 1 493 284 ______

e 8 813 287 ______

f 389 455 ______

g 4 297 389 ______

h 7 138 212 ______

44 Round to the nearest 1 000 000.

a 6 493 287 ______

b 2 738 898 ______

c 1 237 831 ______

d 3 723 419 ______

e 4 935 284 ______

f 7 324 437 ______

g 9 899 956 ______

h 629 382 ______

45 Complete the table.

	Number	Round to the nearest				
		100	1000	10 000	100 000	1 000 000
a	2 836 129					
b	3 439 312					
c	7 632 790					
d	5 153 251					
e	6 758 469					
f	9 529 612					
g	8 135 293					
h	4 689 424					

REVIEW

46 What place value is zero holding?

a 5 302 137 ______________________

b 7 436 120 ______________________

c 4 710 316 ______________________

d 703 243 ______________________

e 580 963 ______________________

f 230 468 ______________________

g 302 ______________________

h 504 253 ______________________

i 30 492 ______________________

j 6 537 130 ______________________

k 4 215 307 ______________________

l 283 059 ______________________

47 Write four numbers that match each description.

a three-digit numbers with zero in the tens place

______________, ______________, ______________, ______________

b four-digit numbers with zero in hundreds place

______________, ______________, ______________, ______________

c six-digit numbers with zero in the ten thousands place

______________, ______________, ______________, ______________

d seven-digit numbers with zero in the hundred thousands place

______________, ______________, ______________, ______________

e five-digit numbers with zero in the thousands place

______________, ______________, ______________, ______________

f seven-digit numbers with zero in the ones place

______________, ______________, ______________, ______________

48 Write these numbers using the abbreviation K.

a 2 000 = ____________

b 51 thousand= ____________

c 27 000 = ____________

d 243 thousand= ____________

e 49 000 = ____________

f 103 thousands = ____________

CATCH UP MATHS YEAR 6 BOOK A © PASCAL PRESS ISBN: 9781925726183

49 Write these numbers using the abbreviation thous.

a 16 000 = ____________

b 215 thousand = ____________

c 84 000 = ____________

d 929 000 = ____________

e 102 thousand = ____________

f 92 000 = ____________

50 Write all the factors of the numbers below.

a 12 ____________________

b 27 ____________________

c 16 ____________________

d 18 ____________________

e 24 ____________________

f 50 ____________________

51 What is the Highest Common Factor (HCF) of each pair of numbers?

a 10: ____________________

12: ____________________

☐ is the HCF of 10 and 12.

b 2: ____________________

3: ____________________

☐ is the HCF of 2 and 3.

c 20: ____________________

24: ____________________

☐ is the HCF of 20 and 24.

d 15: ____________________

20: ____________________

☐ is the HCF of 15 and 20.

e 12: ____________________

15: ____________________

☐ is the HCF of 12 and 15.

52 Write the first 5 multiples.

a 5: ___, ___, ___, ___, ___

b 11: ___, ___, ___, ___, ___

c 3: ___, ___, ___, ___, ___

d 10: ___, ___, ___, ___, ___

53 Write the next three multiples.

a ..., 28, 32, 36, ___, ___, ___

b ..., 40, 45, 50, ___, ___, ___

c ..., 15, 18, 21, ___, ___, ___

d ..., 30, 36, 42, ___, ___, ___

REVIEW

54 What is the Lowest Common Multiple (LCM) in each pair?

a 3: ____________________
5: ____________________ ☐ is the LCM of 3 and 5.

b 4: ____________________
8: ____________________ ☐ is the LCM of 4 and 8.

c 3: ____________________
8: ____________________ ☐ is the LCM of 3 and 8.

d 3: ____________________
4: ____________________ ☐ is the LCM of 3 and 4.

e 7: ____________________
4: ____________________ ☐ is the LCM of 7 and 4.

55 Write the opposite integer.

a +3 ___	c −4 ___	e −10 ___	g −1 ___	i −6 ___
b −7 ___	d +5 ___	f −8 ___	h +2 ___	j +9 ___

56 Order the integers from smallest to greatest.

a −5, 6, −8, −1, 0, 2 ___, ___, ___, ___, ___, ___

b +4, −2, −3, +6, 14, 1 ___, ___, ___, ___, ___, ___

c −32, 30, 18, −24, −11, 2 ___, ___, ___, ___, ___, ___

d −3, 7, 5, −10, 2, −26 ___, ___, ___, ___, ___, ___

e −8, 2, 0, 3, −7, −18 ___, ___, ___, ___, ___, ___

f −50, −24, 20, 16, −36, 3 ___, ___, ___, ___, ___, ___

57 Cross out the numbers that are NOT integers.

a 3, −7, $\frac{1}{2}$, 4.3, $-1\frac{1}{2}$, $3\frac{1}{3}$

b −4, −7, 3, 8, $-3\frac{1}{4}$, 3.18

c −20, −20.8, $15\frac{1}{2}$, −12, 4.9

d 17, $14\frac{1}{2}$, 8.4, $6\frac{1}{4}$, 15, 19

e +42, −17, $14\frac{1}{4}$, −83, +51, 9.4

f −32, 2.384, $2\frac{1}{2}$, $+3\frac{1}{4}$, $-2\frac{1}{2}$, 5

CATCH UP MATHS YEAR 6 BOOK A © PASCAL PRESS ISBN: 9781925726183

58 Write the difference in temperature.

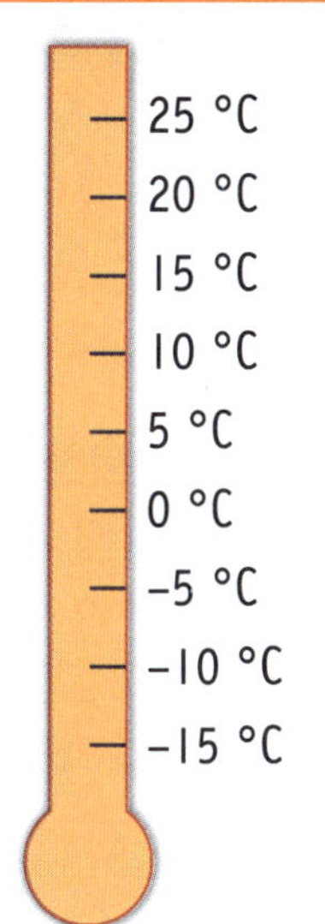

a −10 °C and 0 °C ______

b 20 °C and 10 °C ______

c −15 °C and 5 °C ______

d 0 °C and 20 °C ______

e −20 °C and 5 °C ______

f −5 °C and −20 °C ______

g 10 °C and −5 °C ______

h −10 °C and 10 °C ______

i 5 °C and 15 °C ______

j −5 °C and −10 °C ______

59 Follow the directions by drawing arrows on the number line. Then write the number you end on. – means negative, backwards or left. + means positive, forwards or right.

a 3 + 2 − 4 − 3 = ___

−5 −4 −3 −2 −1 0 1 2 3 4 5

b − 2 + 4 − 5 + 1 = ___

−5 −4 −3 −2 −1 0 1 2 3 4 5

c − 5 + 3 − 2 − 1 = ___

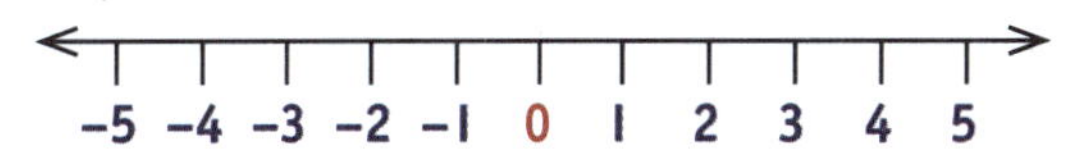

d − 4 − 1 + 3 + 2 = ___

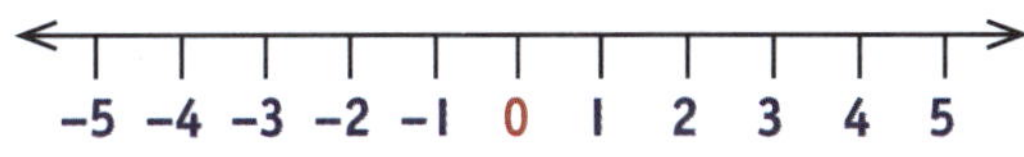

60 Complete.

a List the factors of 6: ____________________ 6 is a prime / composite number.

b List the factors of 14: ____________________ 14 is a prime / composite number.

c List the factors of 23: ____________________ 23 is a prime / composite number.

61 Use red to circle the prime numbers and blue to circle the composite numbers.

101 32 3 67 153 27 151

78 14 150 24 23 168

15 100 30 7 54 28

ADDITION WITHOUT TRADING

TWO-DIGIT AND THREE-DIGIT NUMBERS

Addition is finding the total, or sum, of two or more numbers.

SCAN to watch video

Add 2-digit numbers

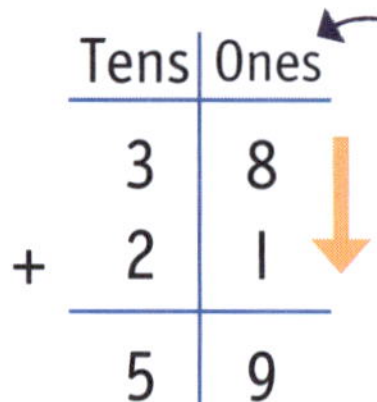

	Tens	Ones
	3	8
+	2	1
	5	9

Start at the ones column and add downwards.

Add the numbers in the same column. Write the total at the bottom.

	T	O
	4	3
+	2	5
	6	8

Add a 3-digit and a 2-digit number

	H	T	O
	6	2	3
+		3	4
	6	5	7

	H	T	O
	5	1	7
+		3	2
	5	4	9

Add 3-digit numbers

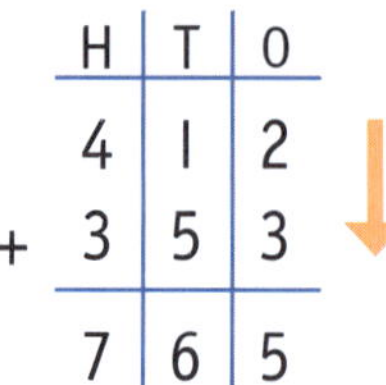

	H	T	O
	4	1	2
+	3	5	3
	7	6	5

	H	T	O
	3	2	5
+	6	4	3
	9	6	8

Example 1:

	T	O
	3	2
+	4	6
		8

Example 2:

	H	T	O
	4	3	5
+		2	2
			7

Example 3:

	H	T	O
	5	6	1
+	3	0	5
		6	

Check your answer on the video!

Your turn

Complete the additions.

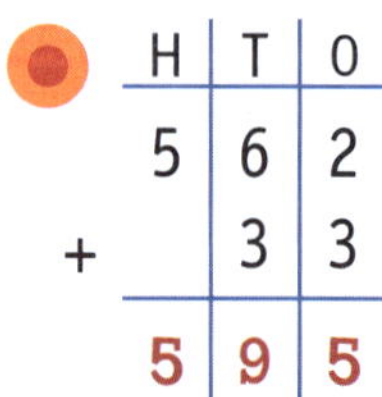

	H	T	O
	5	6	2
+		3	3
	5	9	5

a

	H	T	O
	5	8	4
+	1	0	2

b

	H	T	O
	3	7	2
+		2	6

c

	H	T	O
		3	4
+		4	3

d

	H	T	O
	6	6	4
+	1	2	5

e

	H	T	O
		4	9
+		3	0

SELF CHECK Tick how you feel

Got it!	Need help...	I don't get it
☐	☐	☐

Check your answers

How many did you get correct? ☐

CATCH UP MATHS YEAR 6 BOOK A © PASCAL PRESS ISBN: 9781925726183

1 Complete these additions.

	●	a	b	c	d
	T O	T O	T O	T O	T O
	3 4	5 2	3 5	6 1	5 3
+	1 5	4 1	3 4	2 7	4 3
	4 9				

2 Now complete these additions.

	●	a	b	c	d
	8 2	3 4	8 1	2 7	1 9
+	1 4	1 3	1 6	5 2	5 0
	9 6				

3 Answer.

	●	a	b	c	d
	H T O	H T O	H T O	H T O	H T O
	1 2 3	4 0 2	7 4 0	5 4 8	2 5 7
+	3 2	3 5	4 5	3 1	4 1
	1 5 5				

4 Complete these additions.

	●	a	b	c	d
	5 2 1	7 2 7	4 3 2	8 0 3	1 5 6
+	1 5	2 1	5 5	7 4	4 3
	5 3 6				

5 Solve these additions.

	●	a	b	c	d
	H T O	H T O	H T O	H T O	H T O
	6 3 2	5 2 4	3 5 0	7 1 1	1 4 3
+	3 4 1	1 7 3	2 3 9	2 5 8	7 3 6
	9 7 3				

6 Add

	●	a	b	c	d
	8 5 2	2 3 8	1 5 8	5 0 3	7 3 2
+	1 1 6	3 4 1	6 4 1	4 0 4	1 5 6
	9 6 8				

ADDITION WITH TRADING

TWO-DIGIT AND THREE-DIGIT NUMBERS

Sometimes when we add, the sum of the two digits in a place value column is more than 9. Then we trade 10 ones for 1 ten.

Add 2-digit numbers

	H	T	O
		$^{1}5$	3
+		4	8
	1	0	1

	H	T	O
		$^{1}3$	8
+		2	3
		6	1

Add a 3-digit and a 2-digit number

We traded twice here.

	H	T	O
	$^{1}3$	$^{1}5$	4
+		4	9
	4	0	3

	H	T	O
	$^{1}6$	$^{1}9$	6
+		8	7
	7	8	3

Add 3-digit numbers

	H	T	O
	6	$^{1}3$	2
+	2	4	8
	8	8	0

	H	T	O
	2	$^{1}3$	7
+	7	3	5
	9	7	2

Example 1:

	H	T	O
		$^{1}5$	3
+		1	8
			1

Example 2:

	H	T	O
	$^{1}2$	$^{1}6$	3
+		3	9
			2

Example 3:

	H	T	O
	3	$^{1}6$	9
+	3	1	4
			3

Check your answer on the video!

Complete the additions.

●

	H	T	O
	7	$^{1}3$	2
+	2	1	9
	9	5	1

a

	H	T	O
		5	3
+		4	7

b

	H	T	O
	5	1	9
+	2	3	7

c

	H	T	O
	6	8	1
+		5	9

SELF CHECK Tick how you feel

Got it!	Need help...	I don't get it
☐	☐	☐

Check your answers

How many did you get correct? ☐

CATCH UP MATHS YEAR 6 BOOK A © PASCAL PRESS ISBN: 9781925726183

PRACTICE

1 Add.

	T	O
	[1]5	9
+	1	2
	7	1

a

	T	O
	7	3
+	2	8

b

	T	O
	8	8
+	2	7

c

	T	O
	4	9
+	3	6

2 Solve these additions.

$$\begin{array}{r} {}^{1}4\ 8 \\ +\ 3\ 7 \\ \hline 8\ 5 \end{array}$$

a $\begin{array}{r} 8\ 3 \\ +\ 1\ 9 \\ \hline \end{array}$

b $\begin{array}{r} 5\ 2 \\ +\ 2\ 9 \\ \hline \end{array}$

c $\begin{array}{r} 8\ 7 \\ +\ 1\ 8 \\ \hline \end{array}$

3 Complete these additions.

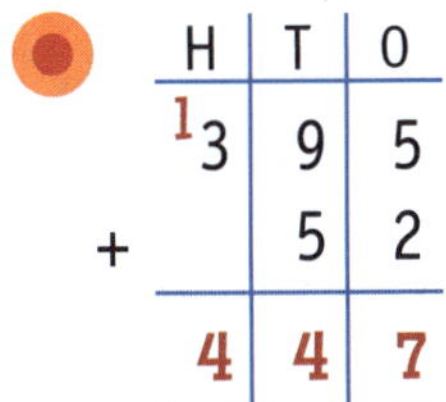

a

	H	T	O
	4	0	9
+		9	5

b

	H	T	O
	7	3	7
+		9	1

c

	H	T	O
	8	5	2
+		9	9

4 Add these numbers.

$$\begin{array}{r} {}^{1}5\ 7\ 2 \\ +\ \ \ 5\ 1 \\ \hline 6\ 2\ 3 \end{array}$$

a $\begin{array}{r} 6\ 4\ 6 \\ +\ \ \ 6\ 8 \\ \hline \end{array}$

b $\begin{array}{r} 3\ 0\ 7 \\ +\ \ \ 4\ 4 \\ \hline \end{array}$

c $\begin{array}{r} 4\ 7\ 1 \\ +\ \ \ 4\ 8 \\ \hline \end{array}$

5 Solve.

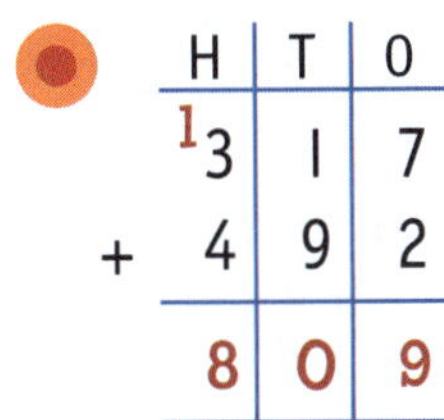

a

	H	T	O
	5	3	1
+	3	8	7

b

	H	T	O
	6	1	4
+	2	9	2

c

	H	T	O
	1	3	8
+	4	2	9

6 Complete these additions.

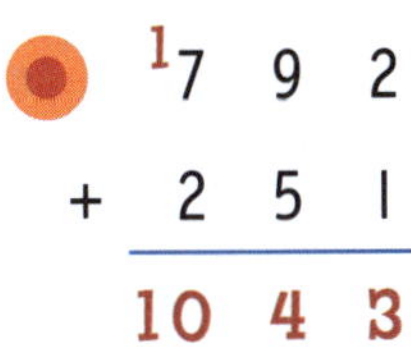

a $\begin{array}{r} 5\ 4\ 3 \\ +\ 8\ 4\ 7 \\ \hline \end{array}$

b $\begin{array}{r} 2\ 4\ 3 \\ +\ 2\ 3\ 9 \\ \hline \end{array}$

c $\begin{array}{r} 1\ 9\ 5 \\ +\ 6\ 8\ 2 \\ \hline \end{array}$

ADDITION WITHOUT TRADING
FOUR-DIGIT AND FIVE-DIGIT NUMBERS

Here are examples of adding two-digit, three-digit, four-digit and five-digit numbers to four-digit and five-digit numbers.

4-digit + 2-digit

	Th	H	T	O
	7	3	2	5
+			4	1
	7	3	6	6

4-digit + 3-digit

	Th	H	T	O
	6	4	1	5
+		3	4	2
	6	7	5	7

4-digit + 4-digit

	Th	H	T	O
	4	1	3	2
+	2	3	5	4
	6	4	8	6

5-digit + 2-digit

	TT	Th	H	T	O
	1	4	3	7	4
+				2	5
	1	4	3	9	9

5-digit + 3-digit

	TT	Th	H	T	O
	5	3	2	7	1
+			4	1	3
	5	3	6	8	4

5-digit + 4-digit

	TT	Th	H	T	O
	4	5	6	3	8
+		3	2	4	1
	4	8	8	7	9

5-digit + 5-digit

	TT	Th	H	T	O
	8	3	4	0	6
+	1	2	5	9	3
	9	5	9	9	9

Example 1:

	Th	H	T	O
	2	4	3	2
+			2	5
			5	

Example 2:

	Th	H	T	O
	5	3	2	4
+	1	3	7	3
	6			7

Example 3:

	TT	Th	H	T	O
	6	7	3	9	1
+			3	0	2
	6				3

Example 4:

	TT	Th	H	T	O
	4	1	3	8	5
+		4	3	1	4
		5			9

Check your answer on the video!

Complete these additions.

	TT	Th	H	T	O
	1	2	7	3	5
+		4	1	6	4
	1	6	8	9	9

a

	Th	H	T	O
	5	3	8	2
+		3	1	4

b

	TT	Th	H	T	O
	6	7	3	4	1
+	2	1	3	3	5

c

	Th	H	T	O
	5	8	3	0
+	2	1	4	0

SELF CHECK Tick how you feel

Got it!	Need help...	I don't get it
 ☐	 ☐	 ☐

Check your answers
How many did you get correct? ☐

CATCH UP MATHS YEAR 6 BOOK A © PASCAL PRESS ISBN: 9781925726183

PRACTICE

1 Add

Example:

	Th	H	T	O
	4	3	7	1
+			1	3
	4	3	8	4

a

	Th	H	T	O
	5	8	3	6
+		1	6	2

b

	Th	H	T	O
	6	4	1	5
+		3	8	3

c

	Th	H	T	O
	7	1	0	3
+			2	5

2 Solve these additions.

Example: 5381 + 12 = 5393

a 7415 + 23 = ______

b 3814 + 185 = ______

c 5678 + 321 = ______

3 Complete these additions.

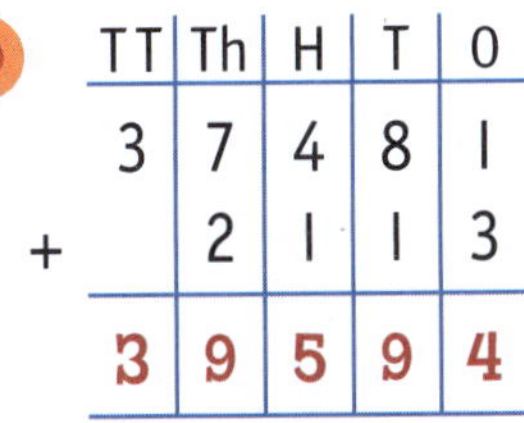

	TT	Th	H	T	O
	3	7	4	8	1
+		2	1	1	3
	3	9	5	9	4

a

	Th	H	T	O
	6	3	7	3
+	2	1	2	4

b

	Th	H	T	O
	5	8	0	9
+	1	1	3	0

c

	TT	Th	H	T	O
	7	2	4	5	0
+		3	2	3	8

d

	TT	Th	H	T	O
	6	1	7	4	1
+	2	3	1	5	0

e

	TT	Th	H	T	O
	1	4	3	5	1
+	5	3	4	2	4

4 Add these numbers.

Example: 68241 + 458 = 68699

a 97324 + 235 = ______

b 54321 + 21542 = ______

c 43212 + 43111 = ______

d 64953 + 13024 = ______

e 17132 + 31243 = ______

ADDITION WITH TRADING

FOUR-DIGIT AND FIVE-DIGIT NUMBERS

Here are some examples of adding four-digit and five-digit numbers with trading.

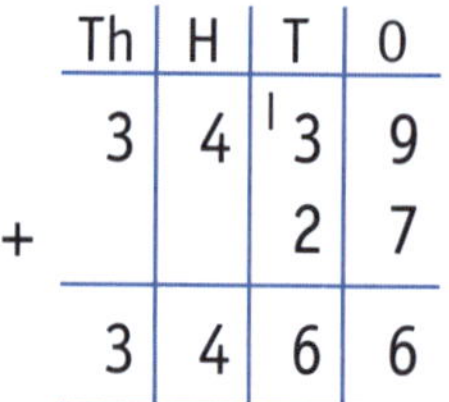

	Th	H	T	O
	3	4	$^{1}3$	9
+			2	7
	3	4	6	6

	TT	Th	H	T	O
	$^{1}5$	8	$^{1}3$	7	2
+		5	4	3	1
	6	3	8	0	3

	Th	H	T	O
	$^{1}7$	$^{1}3$	$^{1}5$	9
+		9	7	4
	8	3	3	3

	TT	Th	H	T	O
	7	$^{1}4$	3	8	1
+	1	4	9	1	5
	8	9	2	9	6

Example 1:

	TT	Th	H	T	O
	3	$^{1}4$	6	$^{1}2$	5
+			5	3	7
					2

Example 2:

	Th	H	T	O
	6	$^{1}2$	8	5
+		4	9	3
				8

Example 3:

	TT	Th	H	T	O
	5	3	7	2	4
+				3	9

Check your answer on the video!

Complete these additions.

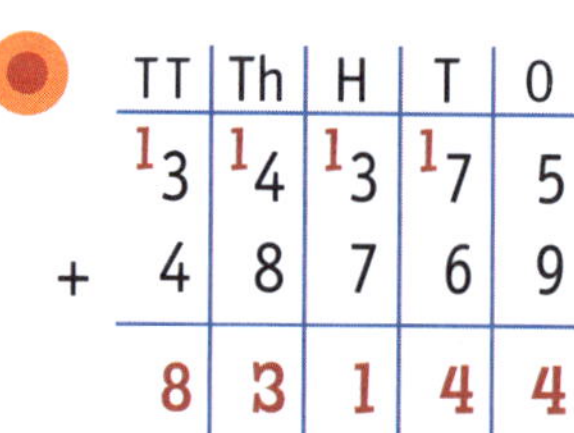

	TT	Th	H	T	O
	$^{1}3$	$^{1}4$	$^{1}3$	$^{1}7$	5
+	4	8	7	6	9
	8	3	1	4	4

a

	Th	H	T	O
	7	4	9	7
+	1	5	8	3

b

	TT	Th	H	T	O
	6	3	2	9	5
+			3	5	9

SELF CHECK	Tick how you feel	
Got it! ☐	Need help... ☐	I don't get it ☐

Check your answers

How many did you get correct? ☐

CATCH UP MATHS YEAR 6 BOOK A © PASCAL PRESS ISBN: 9781925726183

1 Add.

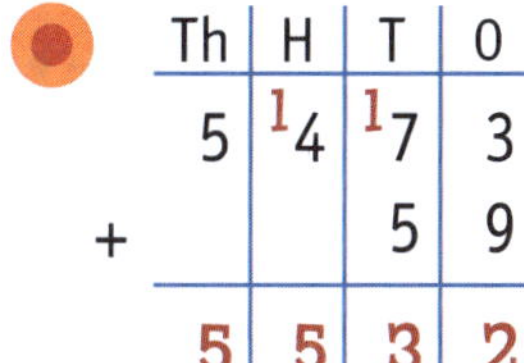

	Th	H	T	O
	5	[1]4	[1]7	3
+			5	9
	5	**5**	**3**	**2**

a

	Th	H	T	O
	8	5	9	2
+		8	7	5

b

	Th	H	T	O
	8	1	3	2
+	4	5	9	3

c

	TT	Th	H	T	O
	7	4	9	3	5
+				9	2

d

	TT	Th	H	T	O
	4	8	2	1	1
+		4	6	8	9

e

	TT	Th	H	T	O
	6	5	9	2	3
+	3	7	8	9	4

2 Solve these additions.

Example:
5 [1]6 [1]9 3
+ 2 9
5 7 2 2

a 2 4 6 7 + 5 3 = ______

b 5 8 2 3 + 5 4 7 = ______

c 7 8 9 5 + 6 3 2 = ______

d 6 3 9 5 + 4 5 8 3 = ______

e 8 5 3 6 + 2 5 2 = ______

f 7 6 9 9 + 1 7 9 3 = ______

g 7 8 4 2 3 + 5 4 9 = ______

h 6 3 2 7 8 + 4 5 7 9 = ______

i 3 8 4 3 7 + 1 5 7 8 = ______

j 2 4 6 8 0 + 2 9 = ______

k 8 3 0 9 8 + 1 4 9 3 7 = ______

l 4 5 6 3 8 + 3 2 7 8 9 = ______

m 3 2 4 9 3 + 4 7 7 3 8 = ______

n 7 4 9 6 8 + 2 3 8 4 9 = ______

ROUNDING TO ESTIMATE ANSWERS

We can use rounding to get an answer that is close to the actual answer.

An estimate is near the answer, but it isn't perfectly accurate.

Round to the nearest 10

Example 1:

141 + 26 is about 170

140 30

Example 2:

249 + 32 is about 280

250 30

Example 3:

Round to the nearest 100

Example 4:

359 + 247 is about 600

400 200

Example 5:

142 + 834 is about 900

100 800

Example 6:

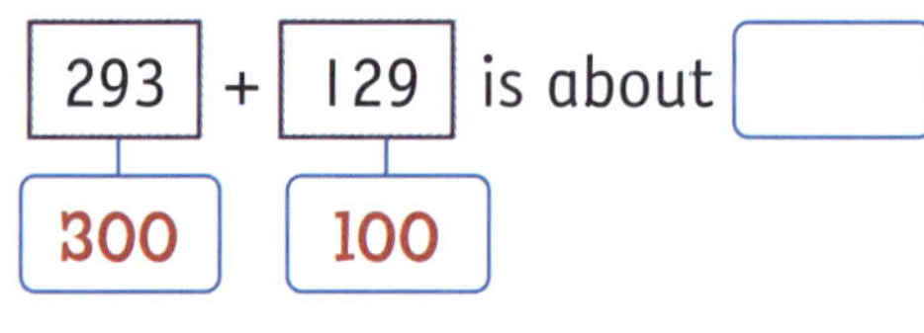

Check your answer on the video!

1 Round to the nearest 10 and add.

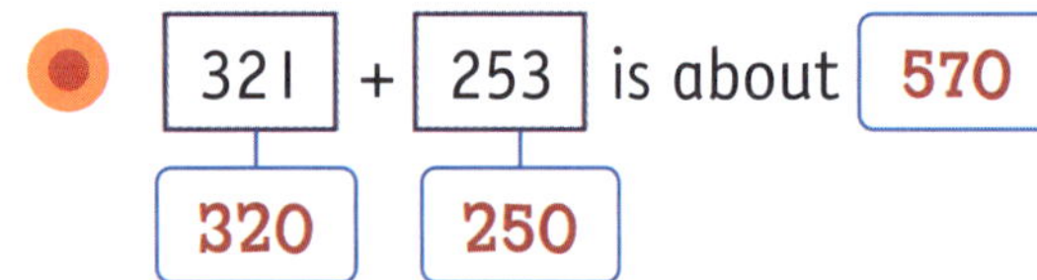

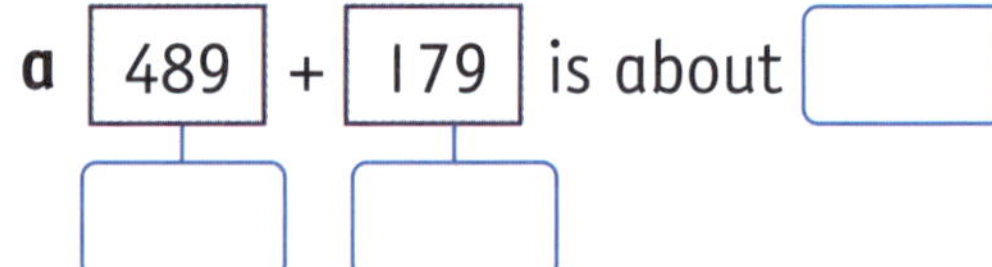

2 Round to the nearest 100 and add.

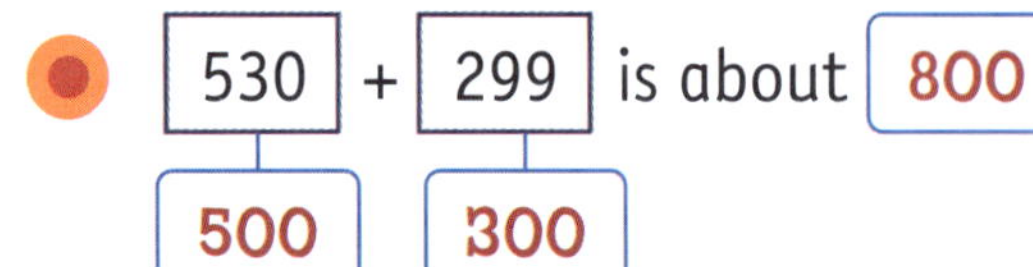

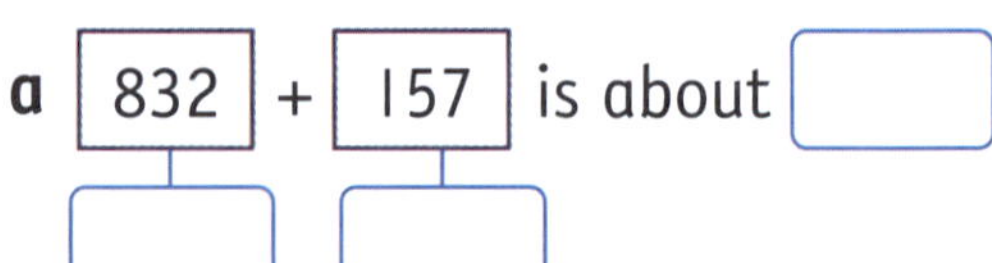

Check your answers

How many did you get correct?

CATCH UP MATHS YEAR 6 BOOK A © PASCAL PRESS ISBN: 9781925726183

PRACTICE

1 Round each number to the nearest 10 and add to estimate the answer.

●	143 + 124 is about 160	140	120	
a	236 + 189 is about ☐	☐	☐	
b	737 + 163 is about ☐	☐	☐	
c	451 + 215 is about ☐	☐	☐	
d	193 + 534 is about ☐	☐	☐	
e	621 + 736 is about ☐	☐	☐	
f	181 + 275 is about ☐	☐	☐	
g	117 + 532 is about ☐	☐	☐	
h	239 + 544 is about ☐	☐	☐	
i	333 + 407 is about ☐	☐	☐	

2 Round each number to the nearest 100 and add to estimate the answer.

●	157 + 319 is about 500	200	300	
a	291 + 152 is about ☐	☐	☐	
b	536 + 299 is about ☐	☐	☐	
c	409 + 391 is about ☐	☐	☐	
d	737 + 301 is about ☐	☐	☐	
e	370 + 151 is about ☐	☐	☐	
f	835 + 137 is about ☐	☐	☐	
g	294 + 285 is about ☐	☐	☐	

USING THE JUMP STRATEGY TO ADD
TWO-DIGIT AND THREE-DIGIT NUMBERS

We can use a number line and the jump strategy to solve addition problems.

SCAN to watch video

Example 1: 72 + 24 = 96

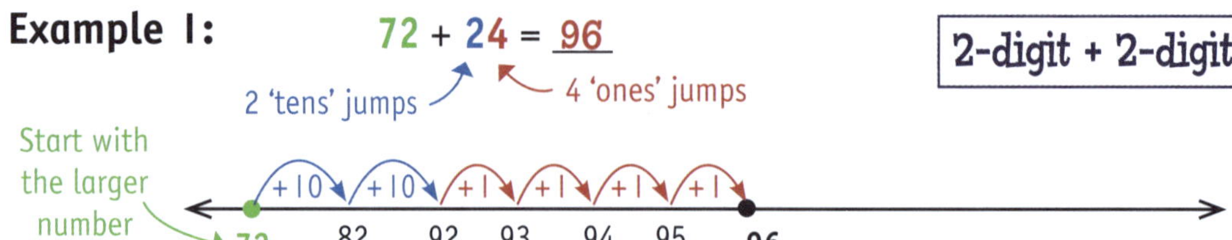

Example 2: 432 + 35 = 467

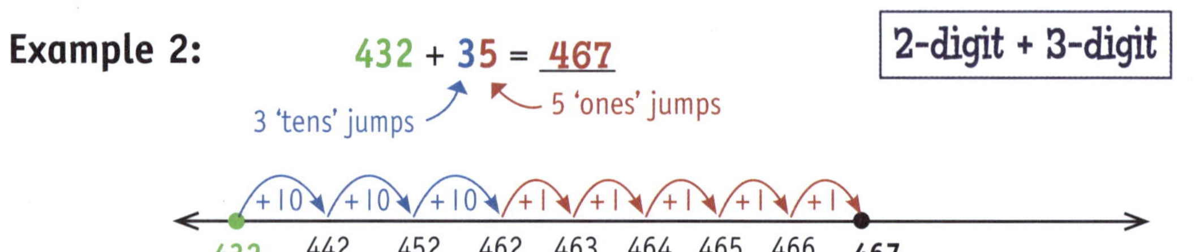

Example 3: 24 + 13 = ____

+10 +1 +1 +1

24

Example 4: 136 + 34 = ____

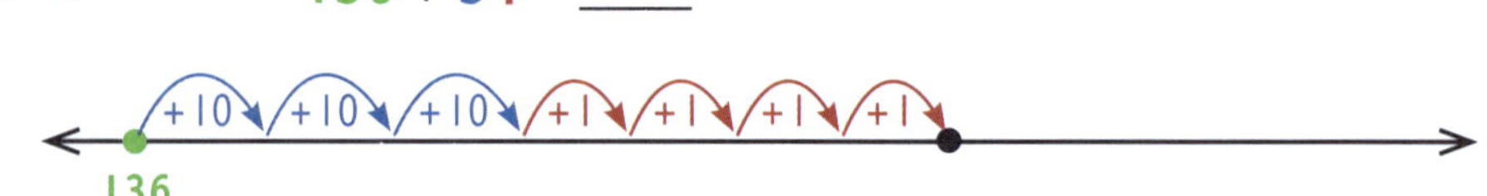

Solve using the number lines and the jump strategy.

379 + 21 = 400

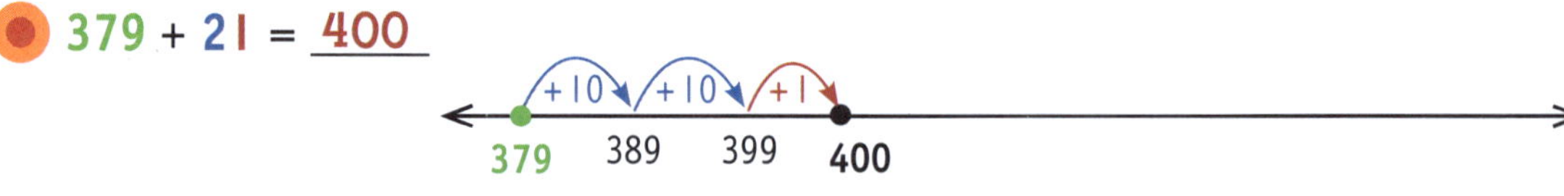

a 58 + 24 = ____

58

b 526 + 72 = ____

526

Check your answers
How many did you get correct?

CATCH UP MATHS YEAR 6 BOOK A © PASCAL PRESS ISBN: 9781925726183

1 Solve using the jump strategy and the number lines.

63 + 24 = 87

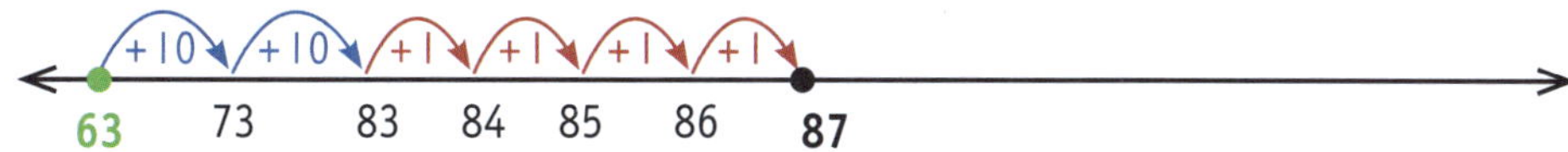

a 33 + 62 = ____

b 21 + 15 = ____

c 43 + 22 = ____

d 654 + 23 = ____

e 407 + 41 = ____

f 230 + 53 = ____

g 846 + 15 = ____

h 709 + 40 = ____

USING THE JUMP STRATEGY TO ADD
THREE-DIGIT AND FOUR-DIGIT NUMBERS

We can use the jump strategy to solve addition problems with numbers in the hundreds and thousands.

SCAN to watch video

Example 1: 316 + 432 = 748 **3-digit + 3-digit**

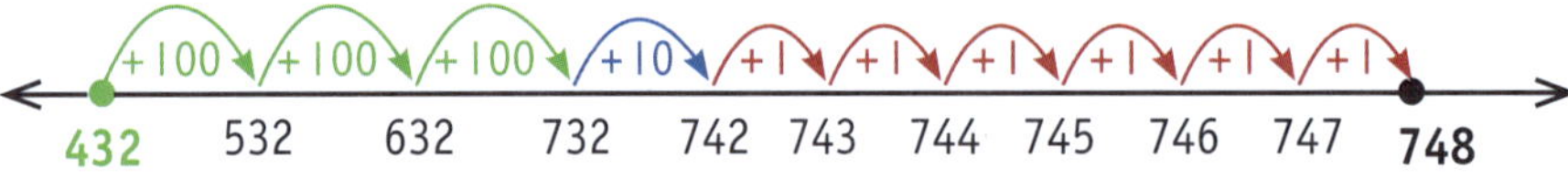

Start with the bigger number.

Example 2: 4283 + 152 = 4435 **4-digit + 3-digit**

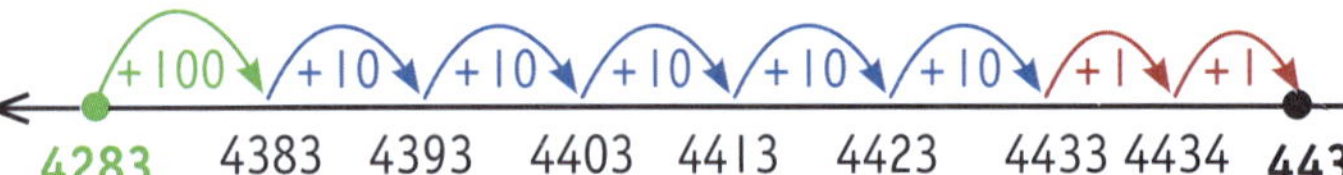

Example 3: 3415 + 5636 = 9051 **4-digit + 4-digit**

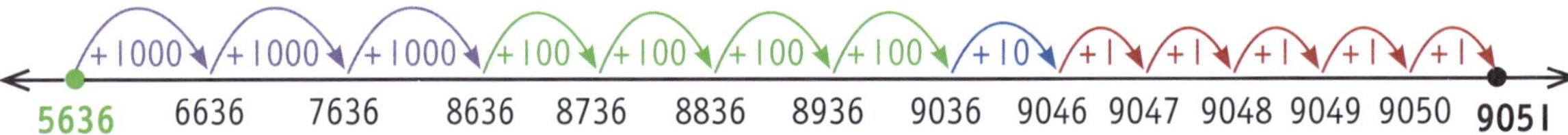

Example 4: 345 + 1254 = _____

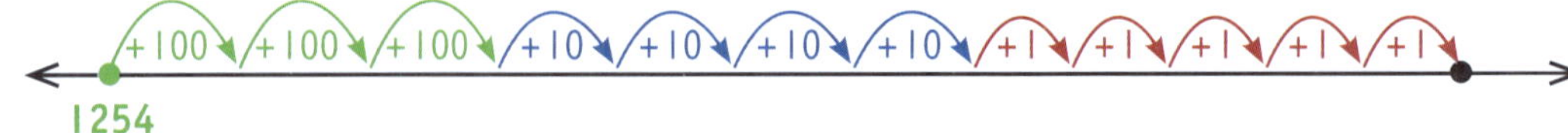

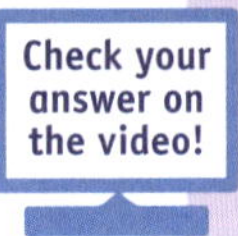

Your turn

Solve using the number lines and the jump strategy.

● 896 + 210 = 1106

a 351 + 504 = _____

b 8295 + 1430 = _____

Check your answers
How many did you get correct?

CATCH UP MATHS YEAR 6 BOOK A © PASCAL PRESS ISBN: 9781925726183

1 Solve using the jump strategy and the number lines.

513 + 749 = 1262

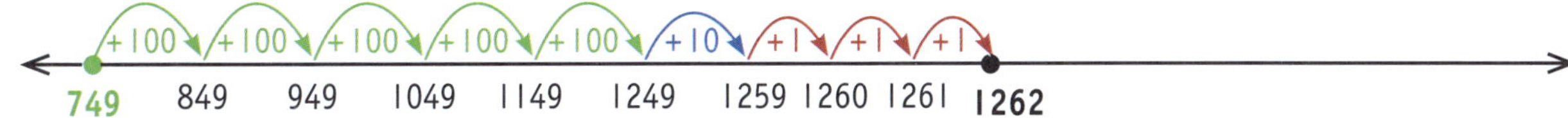

a 983 + 142 = ________

b 711 + 936 = ________

c 512 + 220 = ________

d 1843 + 501 = ________

e 6849 + 2132 = ________

f 4897 + 1050 = ________

g 3132 + 9741 = ________

h 5000 + 3120 = ________

USING THE SPLIT STRATEGY TO ADD
TWO-DIGIT AND THREE-DIGIT NUMBERS

With the split strategy, you split the numbers to show their value, and then add.

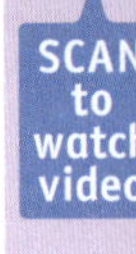

Add 2-digit numbers

Example 1:

55 + 24
50 + 20 = 70
5 + 4 = 9
70 + 9
= 79

Example 2:

72 + 15
70 + 10 = ____
2 + 5 = __
____ + __
= ____

Add a 3-digit and a 2-digit number

Example 3:

214 + 54
200
10 + 50 = 60
4 + 4 = 8
200 + 60 + 8
= 268

Example 4:

424 + 25
____
20 + 20 = ____
4 + 5 = __
____ + ____ + __
= ____

Add 3-digit numbers

Example 5:

523 + 316
500 + 300 = 800
20 + 10 = 30
3 + 6 = 9
800 + 30 + 9
= 839

Example 6:

224 + 542
200 + 500 = ____
20 + 40 = ____
4 + 2 = __
____ + ____ + __
= ____

Check your answer on the video!

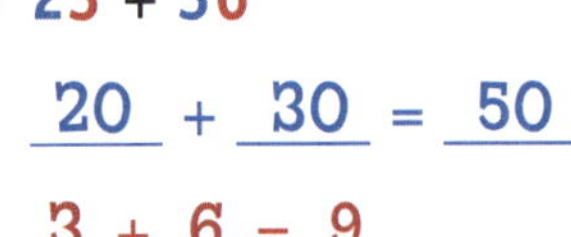

Your turn

Solve using the split strategy.

● 23 + 36
20 + 30 = 50
3 + 6 = 9
50 + 9
= 59

a 41 + 13
____ + ____ = ____
__ + __ = __
____ + __
= ____

b 133 + 56
____
____ + ____ = ____
__ + __ = __
____ + ____ + __
= ____

SELF CHECK Tick how you feel		
Got it! ☐	Need help... ☐	I don't get it ☐

Check your answers
How many did you get correct? ☐

CATCH UP MATHS YEAR 6 BOOK A © PASCAL PRESS ISBN: 9781925726183

PRACTICE

1 Use the split strategy to solve these additions.

924 + 132

900 + 100 = 1000

20 + 30 = 50

4 + 2 = 6

1000 + 50 + 6 = 1056

a 37 + 21

____ + ____ = ____

___ + ___ = ___

____ + ___ = ____

b 48 + 30

____ + ____ = ____

___ + ___ = ___

____ + ___ = ____

c 85 + 14

____ + ____ = ____

___ + ___ = ___

____ + ___ = ____

d 253 + 36

_____ + _____ = _____

____ + ____ = ____

___ + ___ = ___

______ + ____ + ___ = ______

e 441 + 23

_____ + _____ = _____

____ + ____ = ____

___ + ___ = ___

______ + ____ + ___ = ______

f 802 + 47

_____ + _____ = _____

____ + ____ = ____

___ + ___ = ___

______ + ____ + ___ = ______

g 154 + 532

_____ + _____ = _____

____ + ____ = ____

___ + ___ = ___

______ + ____ + ___ = ______

h 837 + 240

_____ + _____ = _____

____ + ____ = ____

___ + ___ = ___

______ + ____ + ___ = ______

i 556 + 243

_____ + _____ = _____

____ + ____ = ____

___ + ___ = ___

______ + ____ + ___ = ______

j 905 + 642

_____ + _____ = _____

____ + ____ = ____

___ + ___ = ___

______ + ____ + ___ = ______

USING THE SPLIT STRATEGY TO ADD
THREE-DIGIT AND FOUR-DIGIT NUMBERS

You can use the split strategy to solve addition with larger numbers, in the hundreds and thousands.

3-digit + 3-digit

Example 1:

413 + 246

400 + 200 = 600

10 + 40 = 50

3 + 6 = 9

600 + 50 + 9

= 659

Example 2:

327 + 452

300 + 400 = ____

20 + 50 = ____

7 + 2 = __

____ + ____ + __

= ____

3-digit + 4-digit

Example 3:

5123 + 743

5000

100 + 700 = 800

20 + 40 = 60

3 + 3 = 6

5000 + 800 + 60 + 6

= 5866

Example 4:

4153 + 542

____

100 + 500 = ____

50 + 40 = ____

3 + 2 = __

____ + ____ + ____ + __

= ____

4-digit + 4-digit

Example 5:

6814 + 2155

6000 + 2000 = 8000

800 + 100 = 900

10 + 50 = 60

4 + 5 = 9

8000 + 900 + 60 + 9

= 8969

Example 6:

7271 + 1516

____ + ____ = ____

200 + 500 = ____

70 + 10 = ____

1 + 6 = __

____ + ____ + ____ + __

= ____

Solve using split strategy.

- 425 + 123

  400 + 100 = 500

  20 + 20 = 40

  5 + 3 = 8

  500 + 40 + 8

  = 548

a 7271 + 1516

____ + ____ = ____

____ + ____ = ____

___ + ___ = ___

__ + __ = __

____ + ____ + ___ + __

= ____

SELF CHECK Tick how you feel

Got it!	Need help...	I don't get it

Check your answers

How many did you get correct?

CATCH UP MATHS YEAR 6 BOOK A © PASCAL PRESS ISBN: 9781925726183

1 Use the split strategy to find the sum.

514 + 234

500 + 200 = 700

10 + 30 = 40

4 + 4 = 8

700 + 40 + 8 = 748

a 261 + 317

____ + ____ = ____

___ + ___ = ___

__ + __ = __

____ + ___ + __ = ____

b 143 + 352

____ + ____ = ____

___ + ___ = ___

__ + __ = __

____ + ___ + __ = ____

c 741 + 258

____ + ____ = ____

___ + ___ = ___

__ + __ = __

____ + ___ + __ = ____

d 1435 + 362

_____ + _____ = _____

____ + ____ = ____

___ + ___ = ___

__ + __ = __

_____ + ____ + ___ + __ = _____

e 2538 + 420

_____ + _____ = _____

____ + ____ = ____

___ + ___ = ___

__ + __ = __

_____ + ____ + ___ + __ = _____

f 6543 + 2454

_____ + _____ = _____

____ + ____ = ____

___ + ___ = ___

__ + __ = __

_____ + ____ + ___ + __ = _____

g 7153 + 2633

_____ + _____ = _____

____ + ____ = ____

___ + ___ = ___

__ + __ = __

_____ + ____ + ___ + __ = _____

h 4938 + 2710

_____ + _____ = _____

____ + ____ = ____

___ + ___ = ___

__ + __ = __

_____ + ____ + ___ + __ = _____

i 6592 + 103

_____ + _____ = _____

____ + ____ = ____

___ + ___ = ___

__ + __ = __

_____ + ____ + ___ + __ = _____

ROUNDING TO THE NEAREST 5 CENTS

When we shop, the totals often end in numbers that are not 0 or 5, like \$8.23, \$4.62 or \$22.64.

We can't use Australian coins to make 1 cent, 2 cents, 3 cents, 4 cents, 6 cents, 7 cents, 8 cents or 9 cents, so we round the cents to 0 or 5.

When you are rounding, look at the number in the hundredths place:

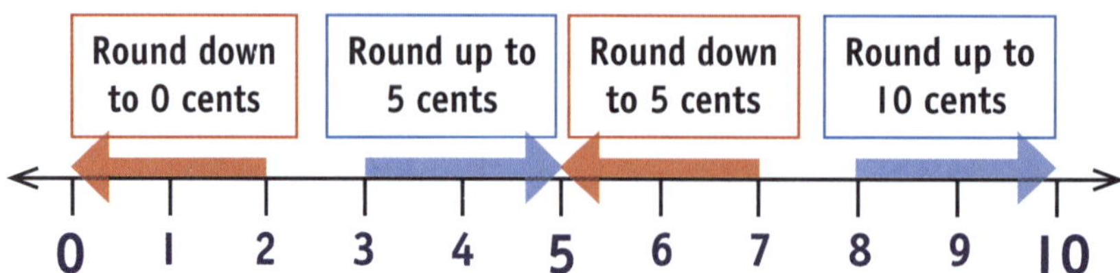

Example 1: Round \$2.34.

4 cents rounds up to 5 cents.

So \$2.34 rounded to the nearest 5 cents is \$2.35.

Example 2: Round \$5.49.

9 cents rounds up to 10 cents.

So \$5.49 rounded to the nearest 5 cents is \$5.50.

Example 3: Round \$10.51.

1 cent rounds down to 0 cents.

So \$10.51 rounded to the nearest 5 cents is \$10.50.

Example 4: Round \$3.27.

7 cents rounds down to __ cents.

So \$3.27 rounded to the nearest 5 cents is \$__.____.

Example 5: Round \$6.48.

8 cents rounds up to ___ cents.

So \$6.48 rounded to the nearest 5 cents is \$__.____.

Example 6: Round \$8.52.

2 cents rounds down to __ cents.

So \$8.52 rounded to the nearest 5 cents is \$__.____.

Round to the nearest 5 cents.

● \$1.31   \$1.30

a \$2.44 ________

b \$8.58 ________

c \$10.63 ________

d \$20.57 ________

e \$140.22 ________

Check your answers
How many did you get correct?

CATCH UP MATHS YEAR 6 BOOK A © PASCAL PRESS ISBN: 9781925726183

PRACTICE

1 Round to the nearest 5 cents by circling the correct answer.

●	$2.86	$2.80	($2.85)	$2.90
a	$6.47	$6.40	$6.45	$6.50
b	$17.52	$17.50	$17.55	$17.60
c	$42.63	$42.60	$42.65	$42.70
d	$21.48	$21.40	$21.45	$21.50
e	$89.44	$89.40	$89.45	$89.50
f	$1.01	$1.00	$1.05	$1.10
g	$66.66	$66.60	$66.65	$66.70
h	$15.58	$15.50	$15.55	$15.60
i	$0.94	$0.90	$0.95	$1.00
j	$121.11	$121.10	$121.15	$121.20
k	$0.13	$0.10	$0.15	$0.20
l	$19.42	$19.40	$19.45	$19.50
m	$12.06	$12.00	$12.05	$12.10
n	$38.56	$38.50	$38.55	$38.60
o	$19.97	$19.90	$19.95	$20.00

2 Round to the nearest 5 cents. Write the answer in the box.

●	$12.43	$12.45	h	$3.56	
a	$0.17		i	$24.01	
b	$65.73		j	$99.99	
c	$39.39		k	$57.62	
d	$8.29		l	$39.81	
e	$66.13		m	$104.04	
f	$5.54		n	$209.09	
g	$149.02		o	$49.28	

WORKING OUT CHANGE

When you use cash to buy something, you can work out the change by counting on.

SCAN to watch video

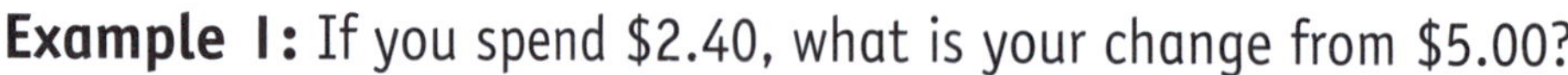

Example 1: If you spend $2.40, what is your change from $5.00?

up to the next dollar

$2.40 → $3.00 → $5.00

60c | $2.00 — The change is $2.60.

Example 2: If you have $10.00 and spend $4.35, what is your change?

up to the next dollar

$4.35 → $5.00 → $10.00

65c | $5.00 — The change is $5.65.

Example 3: If you spend $3.20, what is your change from $5.00?

$3.20 → $4.00 → $5.00

___c | $______ — The change is _______.

Example 4: If you spend $5.60, what is your change from $10.00?

$5.60 → $______ → $10.00

___c | $______ — The change is _______.

Work out the change.

	Spend		Amount given	Change
●	$1.25	$2.00	$10.00	$8.75
	75c	$8.00		
a	$2.75	$3.00	$5.00	
b	$11.55	$12.00	$20.00	

SELF CHECK Tick how you feel

Got it!	Need help...	I don't get it
☐	☐	☐

Check your answers
How many did you get correct? ☐

1 Work out the change by counting on to the nearest dollar and then counting on the dollars to the amount given.

	Spend	Next $	Amount given	Change
●	45c	$1.00	$2.00	$1.55
	55c		$1.00	

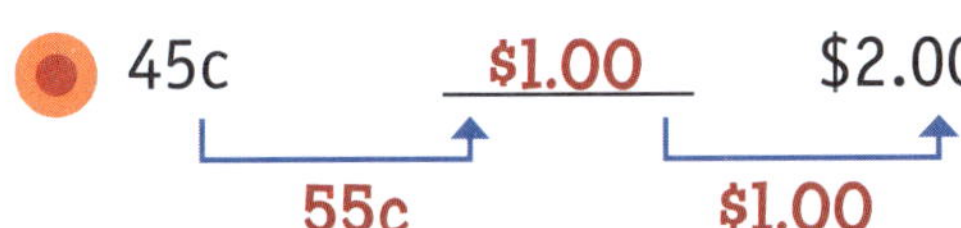

	Spend	Next $	Amount given	Change
a	$1.85	______	$5.00	
b	$2.20	______	$5.00	
c	$7.65	______	$10.00	
d	$1.25	______	$20.00	
e	$14.95	______	$20.00	
f	$27.40	______	$50.00	
g	$46.45	______	$50.00	
h	$72.15	______	$100.00	
i	$34.95	______	$50.00	
j	$16.40	______	$100.00	
k	$8.60	______	$50.00	

SPENDING AND SAVING MONEY

Expenses are the things that people spend money on.
Any money left over is called savings.

SCAN to watch video

Example 1:
Ngoc gets paid $150.20 per week. Here are her expenses:

Food $73.10 | Movies $18.00 | Clothes $25.00 | Gift $20.00

Ngoc spends $136.10:

	$ 73.10	Food
	$ 18.00	Movies
	$ 25.00	Clothes
+	$ 20.00	Gift
	$ 136.10	

Ngoc saves $14.10:

	$ 1~~5~~0.20 (4 1)
–	$ 136.10
	$ 14.10

Example 2: Add up Allan's expenses.

Food $22.20 | Clothes $15.00 | Medicine $20.00

	$ 22.20	Food
	$ 15.00	Clothes
+	$ 20.00	Medicine
	$.	

Allan earns $240.00 per week. How much can he save?

	$ 240.00
–	$ ______
	$ ______

When you put money into a bank account, you are making a deposit. When you take money out of a bank account, it is called a withdrawal.

Amin spends $22.50 on food and $20.00 on clothes. He earns $120.00.

● What are Amin's expenses?

	$ 22.50
+	$ 20.00
	$ 42.50

Amin spent $42.50.

a How much can Amin deposit into his bank account?

______

Check your answers
How many did you get correct?

CATCH UP MATHS YEAR 6 BOOK A © PASCAL PRESS ISBN: 9781925726183

1 Maria and Carmen are working out how much they spend (expenses) and how much they can save every week.

Maria

Payslip	$289.45	
Expenses	$20.20	Lunches
	$40.00	Movies
	$100.00	Food

Carmen

Payslip	$372.50	
Expenses	$10.00	Lunches
	$35.00	Medicine
	$200.00	Food

a Work out Maria and Carmen's expenses.

Maria: ________ Carmen: ________

b How much can Maria save? ________

c How much money can Carmen deposit? ________

d Who earns more money? ____________

e Who saves more money, Maria or Carmen? ____________

f With their savings every week, how long will it take Maria and Carmen to save for a kayak that costs $1600.00?

$1600 ÷ (________ + ________) = ___ weeks

2 If you put money into a bank account, you deposit the money. If you take money out of a bank account, it is called a withdrawal.

Here, Danny deposits $230.20 and withdraws $20:

Deposit	Withdraw	Balance
$230.20		$230.20
	$20.00	$210.20

Fill in the tables below to show Maria and Carmen depositing their savings and then withdrawing $20.

a Maria

Deposit	Withdraw	Balance
		$ 0.00
$		$
	$	$

b Carmen

Deposit	Withdraw	Balance
		$ 0.00
$		$
	$	$

Look at Paul's bank statement. A bank statement shows the date deposits and withdrawals are made and the bank balance (how much money is left).

Date	Deposit	Withdrawal	Balance
2/5	$100.00		$100.00
3/5	$210.00		$310.00
5/5		$50.00	$260.00
7/5	$500.00		$760.00
15/5		$210.50	$549.50
23/5	$975.00		$1524.50
27/5		$543.20	$981.30
30/5	$175.00		$1156.30

- What month is this bank statement from? May

a How many deposits did Paul make? ___

b What was the total amount of deposits made? ___________

c How many withdrawals were made? ___

d How much money did Paul withdraw this month? ___________

e What was the balance on 2/5? _________

f What was the balance on 7/5? _________

g What was the balance on 27/5? _________

h When was $175.00 deposited? ________

i When was $50.00 withdrawn? ________

j What was the highest balance? _________

k What was the largest withdrawal? _________

l What was the largest deposit? _________

CATCH UP MATHS YEAR 6 BOOK A © PASCAL PRESS ISBN: 9781925726183

Write the deposits and withdrawals in the correct columns on Jenai's bank statement. Don't forget to fill out the balance!

Date	Deposit	Withdrawal	Balance
5/12	$50.00		$50.00
8/12	$35.00		$85.00

Deposits and Withdrawals

- ~~5/12 Deposit $50.00~~
- ~~8/12 Deposit $35.00~~
- 10/12 Deposit $435.00
- 13/12 Withdraw $75.00
- 15/12 Withdraw $62.55
- 18/12 Deposit $189.20
- 21/12 Withdraw $325.00
- 27/12 Deposit $520.00
- 29/12 Withdraw $415.75
- 30/12 Deposit $120.20

a What was Jenai's bank balance on 10/12? __________

b What was Jenai's bank balance on 21/12? __________

c What was Jenai's bank balance on 27/12? __________

d What was Jenai's bank balance on 30/12? __________

e What was the largest deposit made? __________

f How many withdrawals were made? ______

g How many deposits were made? ______

h What was the total amount of withdrawals made? __________

i What was the total amount of deposits made? __________

j What month is this bank statement from? ____________________

k When was $35 deposited? ______

l When was $75 withdrawn? ______

m What was the largest balance of Jenai's bank account? __________

n What was the smallest balance of Jenai's bank account? __________

Write these items in the correct part of Tara's account.

Pay $987.45	New drum $50.00	Pay for chores $150.00	Food $240.00
Karate fees $140.00	Pay for walking dogs $80.00	Clothes $135.50	Movies $30.00

Tara's Account

Earnings		Expenses	
Item	**Amount**	**Item**	**Amount**
Total		Total	

a Tara's earnings total is ________.

b Tara can save ________.

c What is Tara's biggest expense? ____________________

d What does Tara get paid the least to do? ____________________

e What is the total of Tara's expenses? ________

f How much did Tara get paid altogether to do chores and walk dogs? ________

g If Tara started with $420.30 in her bank account and deposited all her savings, how much money does she have now?

____

CATCH UP MATHS YEAR 6 BOOK A © PASCAL PRESS ISBN: 9781925726183

Necessities are things that are necessary to live. Luxuries are things that are nice to have, but they are not essential.

Sort the items below as necessities or luxuries and then calculate the total costs.

Necessity	Amount
Total	

Luxury	Amount
Total	

a What two items would you choose not to buy if you were trying to save money?

____

b How much money was spent on toys? ________

c How much money was spent on food? ________

d Was more money spent on necessities or luxuries? ________________

e What was the difference in money spent on luxuries and necessities? ________

f If you only had $100 to spend, what would you buy, and why?

____

____

WORD PROBLEMS USING ADDITION

A word problem is when a maths problem is written in sentences instead of only numbers and symbols.

Follow these steps:

When you read word problems, look for key words that help you work out how to solve the problem.

1 **Read the question.**

2 **Think about the question.**
- Do you need to add, subtract, multiply or divide?
- Is there more than one step or operation?

3 **Highlight important information.**
- Circle the numbers.
- Underline what you need to find out.
- Drawing a picture might help too.

4 **Solve it!**

Here are some key words about addition:
altogether, and, both, combined, extra, increase, join, more, plus, sum, total

---

Example 1:
Bella had (24) mistakes and Gabby had (52) mistakes.
How many mistakes do they have altogether?

Working out

$$\begin{array}{r} 24 \\ +\ 52 \\ \hline 76 \end{array}$$

Example 2:
Deborah has (36) pencils and Allan has (23) pencils.
How many pencils do they both have?

Working out

$$\begin{array}{r} 36 \\ +\ 23 \\ \hline 59 \end{array}$$

Example 3:
Sam caught (47) fish on Monday and (51) fish on Wednesday.
What is the sum of fish Sam caught?

Working out

$$\begin{array}{r} 47 \\ +\ 51 \\ \hline \end{array}$$

Check your answer on the video!

Example 4:
In the bushfire, 26 koalas were rescued on Saturday and 13 on Sunday.
How many koalas were rescued altogether?

Working out

$$\begin{array}{r} \\ +\ \\ \hline \end{array}$$

CATCH UP MATHS YEAR 6 BOOK A © PASCAL PRESS ISBN: 9781925726183

Solve these word problems.

	Working out
● Antonia has 24 cats, 13 dogs and 22 birds at her rescue house for animals. How many animals has Antonia rescued altogether?	24 13 + 22 59
a Consecutive numbers follow one after the other. Find the sum of these consecutive numbers: 37, 38, 39	
b Find the total of these consecutive odd numbers: 47, 49, 51	
c Find the total sum of these consecutive even numbers: 62, 64, 66.	
d Benny watched 243 minutes of television on Saturday and 129 minutes on Sunday. How much television did Benny watch altogether? Give your answer in minutes.	

SELF CHECK Tick how you feel

Got it!	Need help...	I don't get it
☐	☐	☐

Check your answers
How many did you get correct? ☐

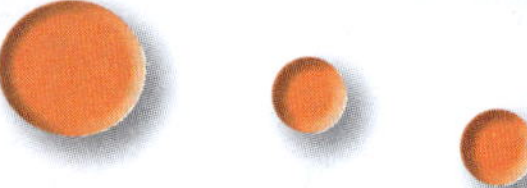

Solve these word problems.

	Working out
a The numbers of people watching movies at the cinema in two sessions on Saturday were 158 and 219. How many people altogether were in those two sessions?	
b A farmer sold 180 punnets of strawberries in September. He sold 93 more than this in August. How many punnets did the farmer sell in August and September?	
c Danny and Christine bought a house and added 29 square metres to the original house, which was 318 square metres. How many square metres is their house altogether?	
d Elise's bakery made 240 cupcakes on Friday, 184 on Saturday and 214 on Sunday. How many cupcakes were baked altogether on these three days?	
e Christian cycled 116 km on Tuesday, 88 km on Wednesday and 72 km on Thursday. What is total distance Christian covered?	
f Waterpipes were tested in 2375 homes in Canngbah, 3482 in Lilli Pilli and 1249 in Cronulla. How many homes were tested altogether?	

CATCH UP MATHS YEAR 6 BOOK A © PASCAL PRESS ISBN: 9781925726183

	Working out
g Beth's Art Shop had 3743 sheets of black paper, 968 sheets of pink paper and 4982 sheets of blue paper. How many sheets of paper did Beth have altogether?	
h Tiana read 252 pages of her book on Monday, 121 pages on Tuesday and 135 pages on Wednesday. How many more pages does Tiana have to read if her book has 635 pages?	
i A car dealer bought four cars. One car was \$42 356, the second car was \$37 605, the third was \$10 259 and the fourth was \$67 249. What was the purchase price of all the cars?	
j Zoe is a property investor. She bought a unit for \$725 636, a townhouse for \$936 242 and a house for \$1 357 282. How much did Zoe's properties cost altogether?	
k Nino loves travelling. He travelled 24 056 km in spring, 63 258 km in summer, 15 495 km in winter and 32 499 km in autumn. How many kilometres did Nino travel altogether?	
l Janet bought land. She bought 12 420 m^2 in Picton, 48 329 m^2 in Bathurst, 28 582 m^2 in Appin and 132 419 m^2 in Sussex Inlet. What is the total size of land that Janet bought?	

ADDITION REVIEW

1 Solve these additions without trading.

a $\begin{array}{r} 38 \\ +\ 31 \\ \hline \\ \hline \end{array}$

b $\begin{array}{r} 43 \\ +\ 24 \\ \hline \\ \hline \end{array}$

c $\begin{array}{r} 51 \\ +\ 18 \\ \hline \\ \hline \end{array}$

d $\begin{array}{r} 23 \\ +\ 52 \\ \hline \\ \hline \end{array}$

e $\begin{array}{r} 64 \\ +\ 25 \\ \hline \\ \hline \end{array}$

f $\begin{array}{r} 423 \\ +\ 41 \\ \hline \\ \hline \end{array}$

g $\begin{array}{r} 325 \\ +\ 34 \\ \hline \\ \hline \end{array}$

h $\begin{array}{r} 213 \\ +\ 55 \\ \hline \\ \hline \end{array}$

i $\begin{array}{r} 127 \\ +\ 41 \\ \hline \\ \hline \end{array}$

j $\begin{array}{r} 202 \\ +\ 93 \\ \hline \\ \hline \end{array}$

k $\begin{array}{r} 165 \\ +\ 231 \\ \hline \\ \hline \end{array}$

l $\begin{array}{r} 241 \\ +\ 305 \\ \hline \\ \hline \end{array}$

m $\begin{array}{r} 765 \\ +\ 234 \\ \hline \\ \hline \end{array}$

n $\begin{array}{r} 607 \\ +\ 382 \\ \hline \\ \hline \end{array}$

o $\begin{array}{r} 550 \\ +\ 349 \\ \hline \\ \hline \end{array}$

2 Solve these additions with trading.

a $\begin{array}{r} 29 \\ +\ 75 \\ \hline \\ \hline \end{array}$

b $\begin{array}{r} 83 \\ +\ 18 \\ \hline \\ \hline \end{array}$

c $\begin{array}{r} 93 \\ +\ 19 \\ \hline \\ \hline \end{array}$

d $\begin{array}{r} 57 \\ +\ 44 \\ \hline \\ \hline \end{array}$

e $\begin{array}{r} 72 \\ +\ 18 \\ \hline \\ \hline \end{array}$

f $\begin{array}{r} 123 \\ +\ 89 \\ \hline \\ \hline \end{array}$

g $\begin{array}{r} 764 \\ +\ 28 \\ \hline \\ \hline \end{array}$

h $\begin{array}{r} 524 \\ +\ 37 \\ \hline \\ \hline \end{array}$

i $\begin{array}{r} 819 \\ +\ 52 \\ \hline \\ \hline \end{array}$

j $\begin{array}{r} 707 \\ +\ 98 \\ \hline \\ \hline \end{array}$

k $\begin{array}{r} 347 \\ +\ 938 \\ \hline \\ \hline \end{array}$

l $\begin{array}{r} 502 \\ +\ 499 \\ \hline \\ \hline \end{array}$

m $\begin{array}{r} 711 \\ +\ 249 \\ \hline \\ \hline \end{array}$

n $\begin{array}{r} 189 \\ +\ 746 \\ \hline \\ \hline \end{array}$

o $\begin{array}{r} 637 \\ +\ 558 \\ \hline \\ \hline \end{array}$

3 Add.

a $\begin{array}{r} 8324 \\ +\ 32 \\ \hline \\ \hline \end{array}$

b $\begin{array}{r} 9238 \\ +\ 31 \\ \hline \\ \hline \end{array}$

c $\begin{array}{r} 3271 \\ +\ 26 \\ \hline \\ \hline \end{array}$

d $\begin{array}{r} 5632 \\ +\ 57 \\ \hline \\ \hline \end{array}$

CATCH UP MATHS YEAR 6 BOOK A © PASCAL PRESS ISBN: 9781925726183

e
$$\begin{array}{r} 7321 \\ +\ 134 \\ \hline \\ \hline \end{array}$$

f
$$\begin{array}{r} 8423 \\ +\ 402 \\ \hline \\ \hline \end{array}$$

g
$$\begin{array}{r} 6371 \\ +\ 518 \\ \hline \\ \hline \end{array}$$

h
$$\begin{array}{r} 4214 \\ +\ 550 \\ \hline \\ \hline \end{array}$$

i
$$\begin{array}{r} 5236 \\ +\ 3421 \\ \hline \\ \hline \end{array}$$

j
$$\begin{array}{r} 7267 \\ +\ 2520 \\ \hline \\ \hline \end{array}$$

k
$$\begin{array}{r} 8153 \\ +\ 1432 \\ \hline \\ \hline \end{array}$$

l
$$\begin{array}{r} 6381 \\ +\ 2203 \\ \hline \\ \hline \end{array}$$

4 Add.

a
$$\begin{array}{r} 14372 \\ +\ 15 \\ \hline \\ \hline \end{array}$$

b
$$\begin{array}{r} 24735 \\ +\ 43 \\ \hline \\ \hline \end{array}$$

c
$$\begin{array}{r} 71143 \\ +\ 50 \\ \hline \\ \hline \end{array}$$

d
$$\begin{array}{r} 56256 \\ +\ 412 \\ \hline \\ \hline \end{array}$$

e
$$\begin{array}{r} 27835 \\ +\ 103 \\ \hline \\ \hline \end{array}$$

f
$$\begin{array}{r} 85243 \\ +\ 243 \\ \hline \\ \hline \end{array}$$

g
$$\begin{array}{r} 47114 \\ +\ 1234 \\ \hline \\ \hline \end{array}$$

h
$$\begin{array}{r} 57329 \\ +\ 2510 \\ \hline \\ \hline \end{array}$$

i
$$\begin{array}{r} 74521 \\ +\ 4378 \\ \hline \\ \hline \end{array}$$

j
$$\begin{array}{r} 75243 \\ +\ 14351 \\ \hline \\ \hline \end{array}$$

k
$$\begin{array}{r} 64345 \\ +\ 24631 \\ \hline \\ \hline \end{array}$$

l
$$\begin{array}{r} 34579 \\ +\ 22420 \\ \hline \\ \hline \end{array}$$

5 Add. You will need to trade.

a
$$\begin{array}{r} 1347 \\ +\ 29 \\ \hline \\ \hline \end{array}$$

b
$$\begin{array}{r} 5263 \\ +\ 48 \\ \hline \\ \hline \end{array}$$

c
$$\begin{array}{r} 7342 \\ +\ 79 \\ \hline \\ \hline \end{array}$$

d
$$\begin{array}{r} 4364 \\ +\ 68 \\ \hline \\ \hline \end{array}$$

e
$$\begin{array}{r} 2382 \\ +\ 536 \\ \hline \\ \hline \end{array}$$

f
$$\begin{array}{r} 5869 \\ +\ 843 \\ \hline \\ \hline \end{array}$$

g
$$\begin{array}{r} 3125 \\ +\ 437 \\ \hline \\ \hline \end{array}$$

h
$$\begin{array}{r} 8690 \\ +\ 879 \\ \hline \\ \hline \end{array}$$

REVIEW

i $\begin{array}{r} 5965 \\ +\ 4375 \\ \hline \end{array}$

j $\begin{array}{r} 2479 \\ +\ 5945 \\ \hline \end{array}$

k $\begin{array}{r} 3815 \\ +\ 2719 \\ \hline \end{array}$

l $\begin{array}{r} 7234 \\ +\ 5096 \\ \hline \end{array}$

6 Complete these additions with trading.

a $\begin{array}{r} 12354 \\ +\ 59 \\ \hline \end{array}$

e $\begin{array}{r} 85371 \\ +\ 388 \\ \hline \end{array}$

i $\begin{array}{r} 58371 \\ +\ 1286 \\ \hline \end{array}$

b $\begin{array}{r} 24680 \\ +\ 63 \\ \hline \end{array}$

f $\begin{array}{r} 67249 \\ +\ 179 \\ \hline \end{array}$

j $\begin{array}{r} 39455 \\ +\ 87958 \\ \hline \end{array}$

c $\begin{array}{r} 13579 \\ +\ 82 \\ \hline \end{array}$

g $\begin{array}{r} 72346 \\ +\ 4937 \\ \hline \end{array}$

k $\begin{array}{r} 68245 \\ +\ 23081 \\ \hline \end{array}$

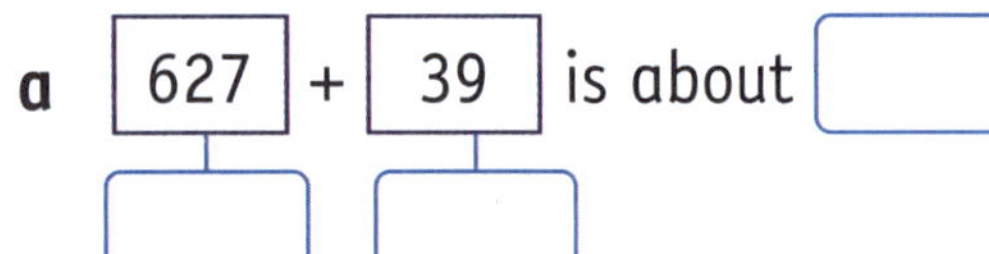

d $\begin{array}{r} 74328 \\ +\ 159 \\ \hline \end{array}$

h $\begin{array}{r} 92132 \\ +\ 5489 \\ \hline \end{array}$

l $\begin{array}{r} 75389 \\ +\ 32489 \\ \hline \end{array}$

7 Round each number to the nearest 10 and add to estimate the answer.

a 627 + 39 is about ☐

c 569 + 153 is about ☐

b 179 + 432 is about ☐

d 593 + 129 is about ☐

8 Round each number to the nearest 100 and add to estimate the answer.

a 179 + 294 is about ☐

b 657 + 421 is about ☐

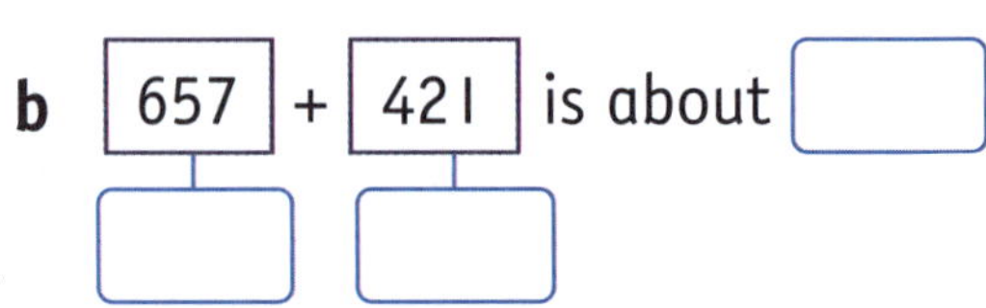

  ISBN: 9781925726183

c [114] + [229] is about [ ]

d [649] + [231] is about [ ]

9 **Solve using the number line and the jump strategy.**

a 89 + 26 = _____

b 153 + 47 = ______

c 109 + 52 = ______

d 526 + 343 = ______

e 603 + 907 = ______

f 4153 + 210 = _______

g 8965 + 1231 = _______

h 2030 + 9583 = _______

REVIEW

10 Use the split strategy to find the sum.

a 52 + 41

___ + ___ = ___

___ + ___ = ___

___ + ___ = ___

b 24 + 35

___ + ___ = ___

___ + ___ = ___

___ + ___ = ___

c 52 + 25

___ + ___ = ___

___ + ___ = ___

___ + ___ = ___

d 325 + 44

___ + ___ = ___

___ + ___ = ___

___ + ___ = ___

___ + ___ + ___ = ___

e 712 + 63

___ + ___ = ___

___ + ___ = ___

___ + ___ = ___

___ + ___ + ___ = ___

f 875 + 22

___ + ___ = ___

___ + ___ = ___

___ + ___ = ___

___ + ___ + ___ = ___

g 428 + 131

___ + ___ = ___

___ + ___ = ___

___ + ___ = ___

___ + ___ + ___ = ___

h 239 + 250

___ + ___ = ___

___ + ___ = ___

___ + ___ = ___

___ + ___ + ___ = ___

i 523 + 654

___ + ___ = ___

___ + ___ = ___

___ + ___ = ___

___ + ___ + ___ = ___

j 5243 + 712

___ + ___ = ___

___ + ___ = ___

___ + ___ = ___

___ + ___ = ___

___ + ___ + ___ + ___ = ___

k 2762 + 235

___ + ___ = ___

___ + ___ = ___

___ + ___ = ___

___ + ___ = ___

___ + ___ + ___ + ___ = ___

CATCH UP MATHS YEAR 6 BOOK A ISBN: 9781925726183

l 3824 + 143

______ + ______ = ______

_____ + _____ = _____

____ + ____ = ____

__ + __ = __

______ + _____ + ____ + __ = ______

m 4735 + 1244

______ + ______ = ______

_____ + _____ = _____

____ + ____ = ____

__ + __ = __

______ + _____ + ____ + __ = ______

n 6145 + 2743

______ + ______ = ______

_____ + _____ = _____

____ + ____ = ____

__ + __ = __

______ + _____ + ____ + __ = ______

o 5215 + 2463

______ + ______ = ______

_____ + _____ = _____

____ + ____ = ____

__ + __ = __

______ + _____ + ____ + __ = ______

11 Round to the nearest 5 cents.

a $2.13 ______	g $1.51 ______	m $0.19 ______	s $30.49 ______
b $7.52 ______	h $27.67 ______	n $0.56 ______	t $8.43 ______
c $8.09 ______	i $12.66 ______	o $0.37 ______	u $29.64 ______
d $16.49 ______	j $20.53 ______	p $0.02 ______	v $59.72 ______
e $10.11 ______	k $15.54 ______	q $30.23 ______	w $26.26 ______
f $4.28 ______	l $32.58 ______	r $25.74 ______	x $20.04 ______

REVIEW

12 Work out the change by counting on.

	Spend	Next $	Amount given	Change
a	$4.50	______	$10.00	
b	$3.55	______	$10.00	
c	$10.25	______	$20.00	
d	$19.20	______	$25.00	

13 Vlad gets paid $250.75 per week. This week he spent:

Food $120.00	Gift $50.00	Flowers $15.00	Movies $30.00	Lunches $20.00

a What do Vlad's expenses add up to? ________

b If Vlad saves the rest of his money, how much can he save? ________

c What was Vlad's biggest expense? ______________

d How much money did Vlad spend on food this week? ________

14 Antonia and Annie are working out how much they can save every week to buy a new mountain bike that costs $2500.

Antonia

Payslip	$755.00	
Expenses	$140.00	Groceries
	$20.00	Train ticket
	$10.00	Movies
	$40.00	Lunches

Annie

Payslip	$794.00	
Expenses	$225.00	Groceries
	$40.00	Petrol
	$10.00	Movies
	$50.00	Lunches

a Work out Antonia's and Annie's expenses.

Antonia: ________ Annie: ________

  ISBN: 9781925726183

b How much can Antonia save? ________

c How much can Annie deposit? ________

d Who earns more money? ____________

e Who saves more, Antonia or Annie? ____________

f Who spends more on food items every week? _________

g If Antonia and Annie save this much every week, how long will it take them to buy the mountain bike together? ____________

$2500 ÷ (________ + ________) = ___ weeks

15 Look at Max's bank statement and then answer the questions.

Date	Deposit	Withdrawal	Balance
3/6	$150.00		$150.00
5/6		$20.00	$130.00
8/6	$240.00		$370.00
8/6	$120.00		$490.00
10/6		$300.00	$190.00
15/6	$582.00		$772.00
28/6	$384.00		$1156.00
30/6		$100.00	$1055.50

a What month is this bank statement from? ___________

b How many withdrawals did Max make? ____

c How much money was deposited into Max's account? ________

d How much money did Max withdraw? ________

e What was the balance of Max's account at the end of the month? ________

f What was Max's balance on 5/6? ________

g What was Max's balance on 8/6? ________

h What was Max's balance on 28/6? ________

i What was Max's balance on 30/6? ________

j What was the largest balance? ________

k What was the largest deposit? ________

l The largest withdrawal was ________ and was made on ________.

m When was $582.00 deposited? ________

n When was $300.00 withdrawn? ________

REVIEW

16 Write the deposits and withdrawals in the correct columns on Xavier's bank statement. Don't forget to fill out the balance!

Date	Deposit	Withdrawal	Balance
4/11	$40.00		$40.00

Deposits and Withdrawals

- ~~4/11 Deposit $40.00~~
- 5/11 Deposit $50.00
- 7/11 Deposit $175.00
- 7/11 Deposit $120.00
- 9/11 Withdraw $70.00
- 11/11 Withdraw $115.50
- 13/11 Deposit $65.00
- 24/11 Deposit $72.00
- 28/11 Withdraw $179.00
- 30/11 Deposit $526.50

a What was Xavier's bank balance on 5/11? ________

b What was Xavier's bank balance on 24/11? ________

c What was Xavier's bank balance on 11/11? ________

d What was Xavier's bank balance on 28/11? ________

e What was the largest deposit made? ________

f How many deposits were made? ____

g Xavier made ____ withdrawals.

h This statement is for the month of ____________________.

i The total amount of deposits made was ________.

j The total amount of withdrawals made was ________.

k When was $65 deposited? ______

l When was $70 withdrawn? ______

m What was the largest balance of Xavier's bank account? ________

n What was the smallest balance of Xavier's bank account? ________

CATCH UP MATHS YEAR 6 BOOK A © PASCAL PRESS ISBN: 9781925726183

17 Write the following items in the correct column of Hassan's account.

- Pay $1943
- Food $325
- Pay for chores $150
- Pay for piano lessons $60
- Movies $35
- Golf $25
- Medicine $40
- Take away $30

Hassan's Account

Earnings		Expenses	
Item	**Amount**	**Item**	**Amount**
piano lessons			
Total		Total	

a Hassan's earnings total _____

b Hassan can save _______

c What was the total of Hassan's expenses? ________

d How much money would be in Hassan's account if he deposited all of his savings and he already had $515.20 in the account? ___________

18 Circle the necessities green and the luxuries red.

food   take away   clothes   medicine   toys   bike

19 Answer these word problems.

Working out

a Len had 34 mistakes and Kylie had 63.
How many mistakes do they have altogether?

____

b Ashleigh had 23 ducks, 44 goats and 17 horses on her farm.
How many animals does Ashleigh have altogether?

____

Working out

c Find the total of these consecutive numbers: 58, 59, 60

____

d Roisin watched Netflix for 532 minutes on Saturday and for 128 minutes on Sunday.
How much Netflix did Roisin watch altogether?

____

USING THE JUMP STRATEGY

TWO-DIGIT AND THREE-DIGIT NUMBERS

You can use the jump strategy on a number line to solve subtraction.

Example 1: Start number → 46 – 32 = 14

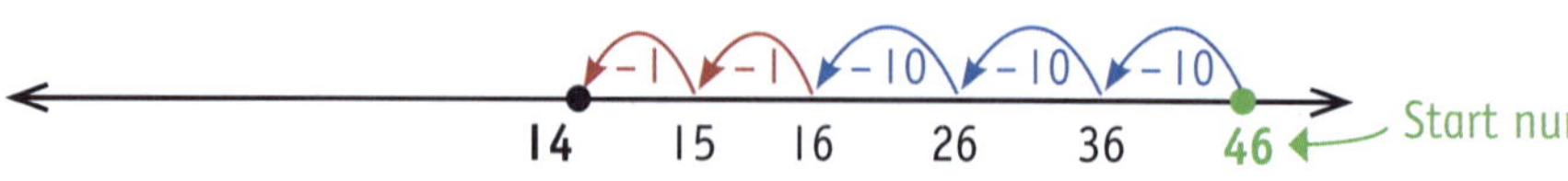

Example 2: 249 – 35 = 214

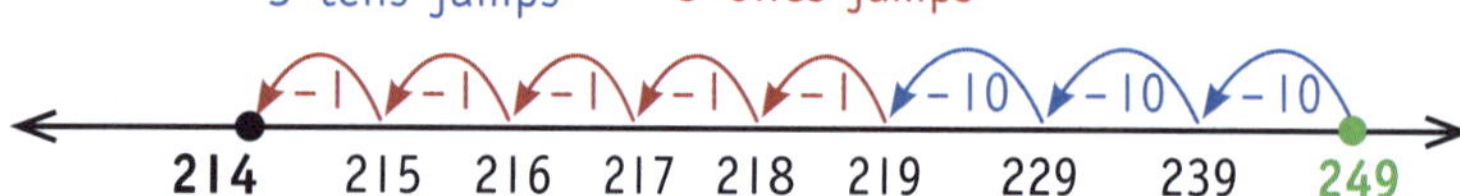

Example 3: 842 – 254 = 588

2 'hundreds' jumps, 5 'tens' jumps, 4 'ones' jumps

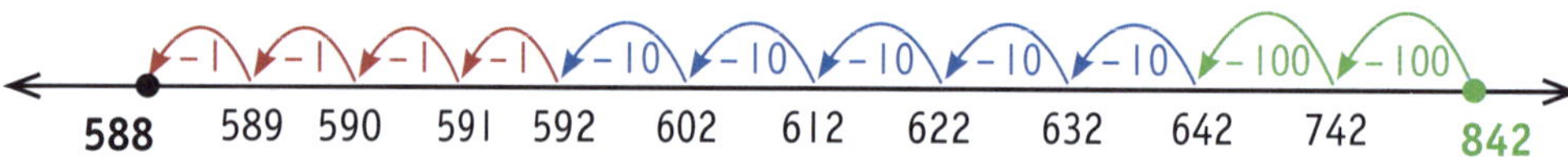

Example 4: 574 – 312 = ____

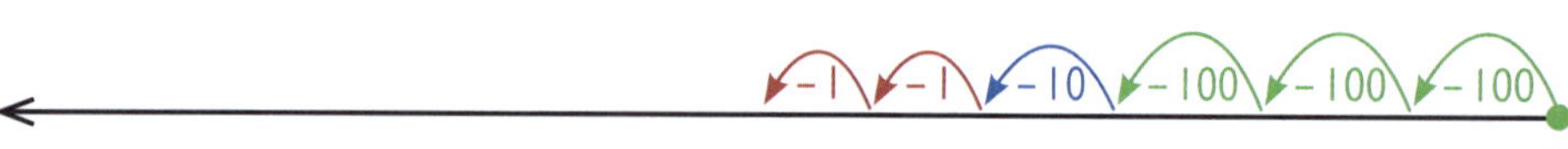

Solve using the number lines and the jump strategy.

- 747 – 143 = 604

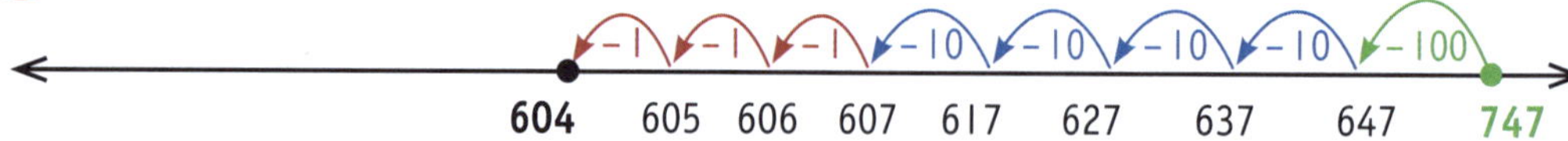

a 76 – 35 = ____

b 424 – 236 = ____

SELF CHECK Tick how you feel

Got it!	Need help...	I don't get it
 ☐	 ☐	☐

Check your answers
How many did you get correct?

CATCH UP MATHS YEAR 6 BOOK A © PASCAL PRESS ISBN: 9781925726183

PRACTICE

1 Solve using the jump strategy and the number lines.

653 − 107 = 546

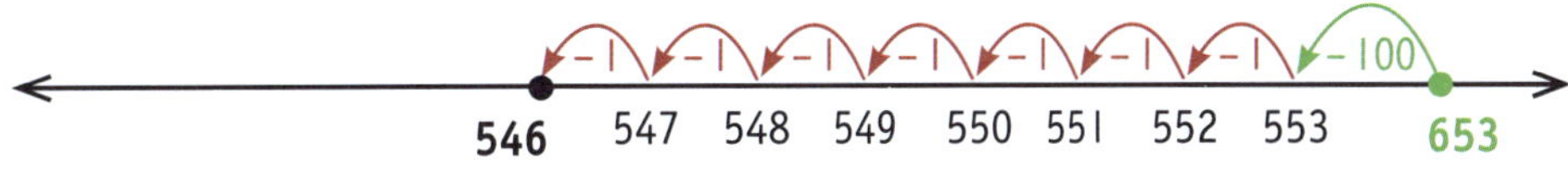

a 74 − 13 = ___

b 98 − 26 = ___

c 83 − 22 = ___

d 871 − 44 = _____

e 909 − 143 = _____

f 511 − 242 = _____

g 698 − 304 = _____

h 475 − 220 = _____

  ISBN: 9781925726183

USING THE JUMP STRATEGY
THREE-DIGIT AND FOUR-DIGIT NUMBERS

These subtraction problems have three-digit and four-digit numbers. You can solve them using the jump strategy.

Example 1:

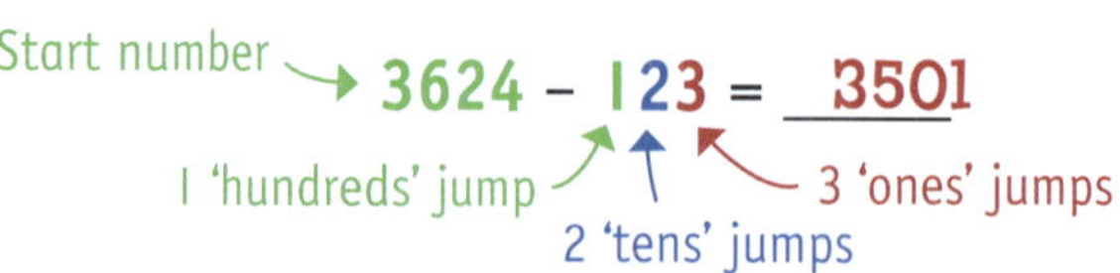

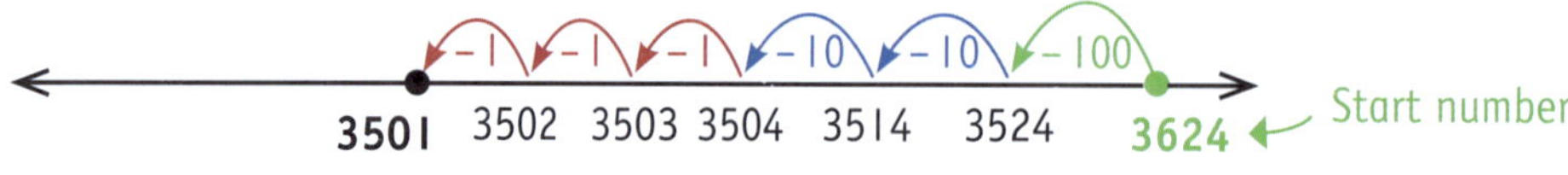

Example 2:

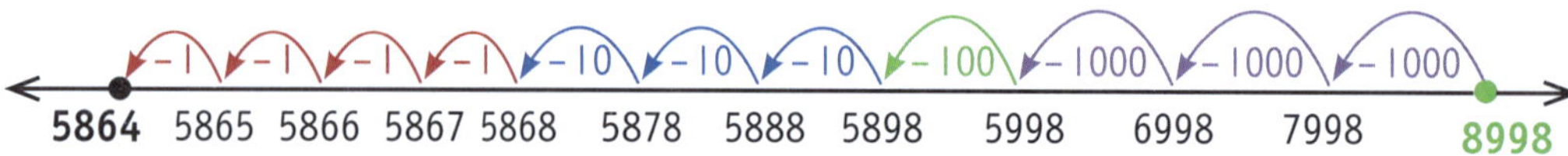

Example 3:

Start number → 4798 – 1232 = _____

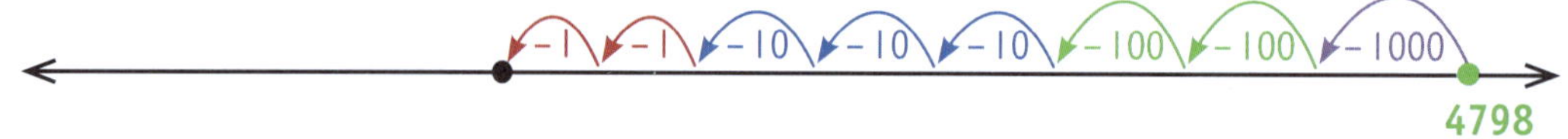

Solve using the number lines and the jump strategy.

● 5925 – 341 = 5584

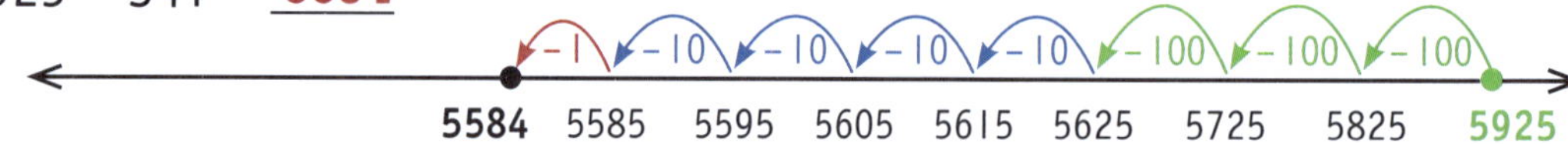

a 2998 – 1225 = _____

b 9774 – 530 = _____

Check your answers

How many did you get correct?

CATCH UP MATHS YEAR 6 BOOK A © PASCAL PRESS ISBN: 9781925726183

1 Solve using the jump strategy and the number lines.

8497 − 1532 = 6965

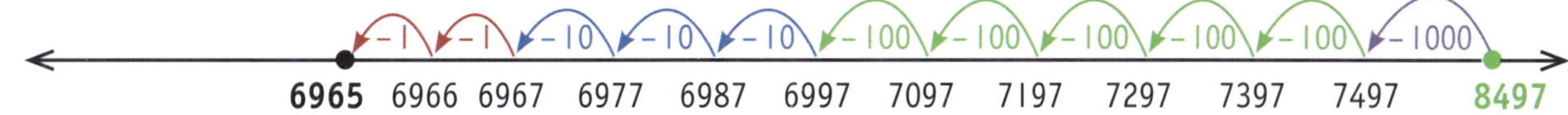

a 8844 − 243 = ______

b 2903 − 421 = ______

c 3196 − 304 = ______

d 7985 − 230 = ______

e 4258 − 1432 = ______

f 6329 − 1504 = ______

g 5935 − 2111 = ______

h 4979 − 3431 = ______

USING THE SPLIT STRATEGY
TWO-DIGIT AND THREE-DIGIT NUMBERS

When solving subtraction problems using the split strategy, we 'split' the numbers into their values.

Example 1:

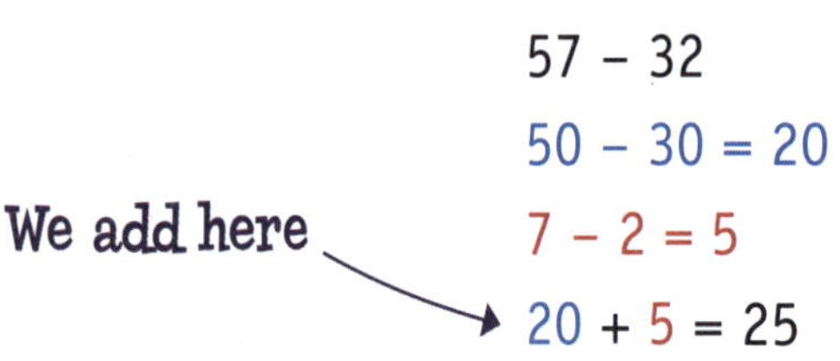

57 – 32
50 – 30 = 20
7 – 2 = 5
We add here → 20 + 5 = 25

Example 2:
Solve 247 – 35
200
40 – 30 = 10
7 – 5 = 2
200 + 10 + 2 = 212

Example 3:
Solve 64 – 21
60 – 20 = ____
4 – 1 = __
____ + __ = ____

Example 4:
Solve 789 – 175
700 – 100 = ____
80 – 70 = ____
9 – 5 = __
____ + ____ + __ = ____

Solve using the split strategy.

Check your answer on the video!

● 732 – 611
700 – 600 = 100
30 – 10 = 20
2 – 1 = 1
100 + 20 + 1 = 121

a 547 – 23
____ – ____ = ____
____ – ____ = ____
__ – __ = __
____ + ____ + __ = ____

b 562 – 51
____ – ____ = ____
____ – ____ = ____
__ – __ = __
____ + ____ + __ = ____

c 859 – 257
____ – ____ = ____
____ – ____ = ____
__ – __ = __
____ + ____ + __ = ____

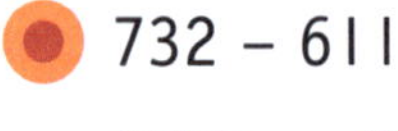

Check your answers
How many did you get correct?

CATCH UP MATHS YEAR 6 BOOK A © PASCAL PRESS ISBN: 9781925726183

PRACTICE

1 Use the split strategy to solve these subtraction problems.

549 − 28

500 − 0 = 500

40 − 20 = 20

9 − 8 = 1

500 + 20 + 1 = 521

a 83 − 21

___ − ___ = ___

___ − ___ = ___

___ + ___ = ___

b 96 − 85

___ − ___ = ___

___ − ___ = ___

___ + ___ = ___

c 76 − 44

___ − ___ = ___

___ − ___ = ___

___ + ___ = ___

d 68 − 47

___ − ___ = ___

___ − ___ = ___

___ + ___ = ___

e 793 − 82

___ − ___ = ___

___ − ___ = ___

___ − ___ = ___

___ + ___ + ___ = ___

f 497 − 73

___ − ___ = ___

___ − ___ = ___

___ − ___ = ___

___ + ___ + ___ = ___

g 542 − 22

___ − ___ = ___

___ − ___ = ___

___ − ___ = ___

___ + ___ + ___ = ___

h 689 − 74

___ − ___ = ___

___ − ___ = ___

___ − ___ = ___

___ + ___ + ___ = ___

i 856 − 245

___ − ___ = ___

___ − ___ = ___

___ − ___ = ___

___ + ___ + ___ = ___

j 763 − 452

___ − ___ = ___

___ − ___ = ___

___ − ___ = ___

___ + ___ + ___ = ___

USING THE SPLIT STRATEGY
THREE-DIGIT AND FOUR-DIGIT NUMBERS

Here are examples of how the split strategy can be used to solve subtraction problems with larger numbers.

SCAN to watch video

Example 1:

5876 – 564

5 has a value of 5000
8 has a value of 800
7 has a value of 70
6 has a value of 6
4 has a value of 4
6 has a value of 60
5 has a value of 500

Solve 5876 – 564

5000

800 – 500 = 300

70 – 60 = 10

6 – 4 = 2

5000 + 300 + 10 + 2 = 5312

Example 2: Solve 2943 – 1821

2000 – 1000 = 1000

900 – 800 = 100

40 – 20 = 20

3 – 1 = 2

1000 + 100 + 20 + 2 = 1122

Why do we add at the end?

Example 3: Solve 7385 – 273

7000

300 – 200 = ____

80 – 70 = ___

5 – 3 = __

_____ + _____ + ___ + __ = _____

Example 4: Solve 9873 – 4562

9000 – 4000 = _____

800 – 500 = _____

70 – 60 = ___

3 – 2 = __

_____ + _____ + ___ + __ = _____

Check your answer on the video!

Your turn

Solve using the split strategy.

● 6564 – 243

6000

500 – 200 = 300

60 – 40 = 20

4 – 3 = 1

6000 + 300 + 20 + 1 = 6321

a 8794 – 762

_____

_____ – _____ = _____

___ – ___ = ___

__ – __ = __

_____ + _____ + ___ + __ = _____

SELF CHECK Tick how you feel

Got it!	Need help...	I don't get it
☐	☐	☐

Check your answers

How many did you get correct? ☐

CATCH UP MATHS YEAR 6 BOOK A © PASCAL PRESS ISBN: 9781925726183

PRACTICE

1 Use the split strategy to solve these subtraction problems.

● 7894 – 3513

7000 – 3000 = 4000

800 – 500 = 300

90 – 10 = 80

4 – 3 = 1

4000 + 300 + 80 + 1 = 4381

a 8237 – 126

___ – ___ = ___

___ – ___ = ___

___ – ___ = ___

___ – ___ = ___

___ + ___ + ___ + ___ = ___

b 9983 – 742

___ – ___ = ___

___ – ___ = ___

___ – ___ = ___

___ – ___ = ___

___ + ___ + ___ + ___ = ___

c 6556 – 242

___ – ___ = ___

___ – ___ = ___

___ – ___ = ___

___ – ___ = ___

___ + ___ + ___ + ___ = ___

d 7295 – 185

___ – ___ = ___

___ – ___ = ___

___ – ___ = ___

___ – ___ = ___

___ + ___ + ___ + ___ = ___

e 9873 – 5562

___ – ___ = ___

___ – ___ = ___

___ – ___ = ___

___ – ___ = ___

___ + ___ + ___ + ___ = ___

f 8839 – 6428

___ – ___ = ___

___ – ___ = ___

___ – ___ = ___

___ – ___ = ___

___ + ___ + ___ + ___ = ___

g 7864 – 3521

___ – ___ = ___

___ – ___ = ___

___ – ___ = ___

___ – ___ = ___

___ + ___ + ___ + ___ = ___

h 4987 – 2843

___ – ___ = ___

___ – ___ = ___

___ – ___ = ___

___ – ___ = ___

___ + ___ + ___ + ___ = ___

i 6593 – 4212

___ – ___ = ___

___ – ___ = ___

___ – ___ = ___

___ – ___ = ___

___ + ___ + ___ + ___ = ___

SUBTRACTION WITHOUT TRADING
TWO-DIGIT AND THREE-DIGIT NUMBERS

We can solve subtraction problems using only numbers and symbols.

Example 1:

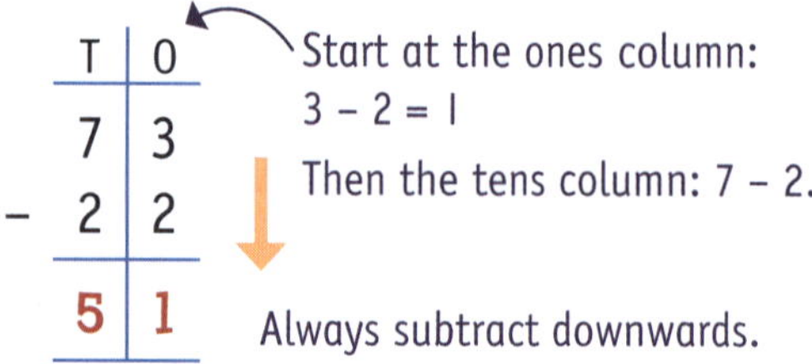

	T	O
	7	3
–	2	2
	5	1

Start at the ones column:
3 – 2 = 1
Then the tens column: 7 – 2.
Always subtract downwards.

First subtract the ones, then the tens, and then the hundreds.

Example 2:

	H	T	O
	8	6	5
–		3	4
	8	3	1

Start at the ones column:
5 – 4 = 1
Then the tens column: 6 – 3.
And then the hundreds column.
Always subtract downwards.

Example 3:

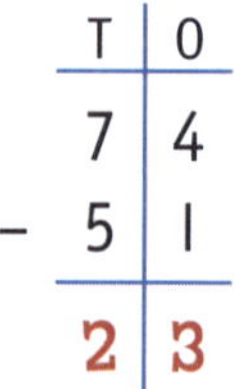

	T	O
	7	4
–	5	1
	2	3

Example 4:

	H	T	O
	6	9	8
–	4	1	4
	2	8	4

Example 5:

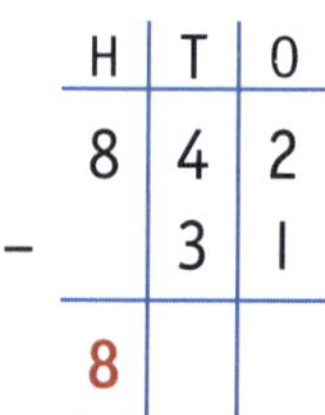

	H	T	O
	8	4	2
–		3	1
	8		

Example 6:

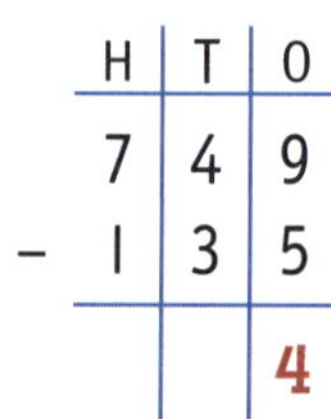

	H	T	O
	7	4	9
–	1	3	5
			4

Check your answer on the video!

Solve these subtraction algorithms.

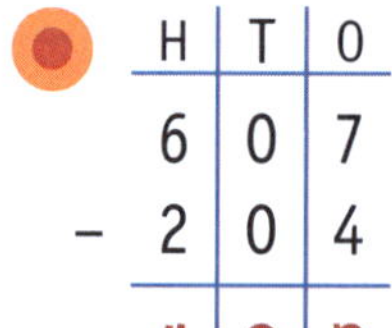

	H	T	O
	6	0	7
–	2	0	4
	4	0	3

a

	H	T	O
	8	3	6
–	4	1	4

b

	H	T	O
	9	7	3
–		2	2

c

	T	O
	4	9
–	2	5

d

	H	T	O
	5	3	6
–	1	2	3

e

	T	O
	7	8
–	4	0

SELF CHECK Tick how you feel

Got it!	Need help...	I don't get it
☐	☐	☐

Check your answers
How many did you get correct? ☐

CATCH UP MATHS YEAR 6 BOOK A © PASCAL PRESS ISBN: 9781925726183

Solve these subtraction algorithms.

Example:

	H	T	O
	2	7	0
−	1	2	0
	1	**5**	**0**

a

	T	O
	5	7
−	2	1

b

	T	O
	6	9
−	4	9

c

	H	T	O
	4	9	3
−		5	1

d

	H	T	O
	6	9	5
−		4	4

e

	H	T	O
	5	8	9
−	2	0	9

f

	H	T	O
	8	5	7
−		2	4

g

	H	T	O
	7	6	3
−	4	6	1

2 Now solve these subtraction algorithms.

Example: 320 − 120 = **200**

a 649 − 49 = ____

b 87 − 24 = ____

c 193 − 52 = ____

d 590 − 250 = ____

e 349 − 47 = ____

f 437 − 123 = ____

g 65 − 42 = ____

h 793 − 242 = ____

i 805 − 705 = ____

j 893 − 82 = ____

k 749 − 36 = ____

Fill in the missing digits.

Example: 53**4** − **2**3 = **5**11

a _ 8 − 2 _ = 54

b 4 _ 9 − 22 _ = _ 65

c 994 − 4 _ 1 = _ 2 _

d 8 _ 2 − _ 30 = 32 _

e 9 _ _ − _ 7 _ = _ 82

f 5 _ 4 − 2 _ _ = 413

g 65 _ − _ _ 1 = 311

SUBTRACTION WITHOUT TRADING

THREE-DIGIT AND FOUR-DIGIT NUMBERS

These subtraction algorithms with three-digit and four-digit numbers use only numbers and the subtraction symbol.

Example 1:

	Th	H	T	O
	5	4	9	3
–		3	7	2
	5	**1**	**2**	**1**

Always subtract downwards.

Start at the ones column.
Then move to the tens column.
Next, the hundreds.
Finally, the thousands.

Example 2:

	Th	H	T	O
	9	7	6	5
–		4	5	1
	9	**3**	**1**	**4**

Example 3:

	Th	H	T	O
	8	9	4	3
–	4	3	2	1
	4	**6**	**2**	**2**

Example 4:

	Th	H	T	O
	5	8	2	4
–		3	0	3
		5		**1**

Example 5:

	Th	H	T	O
	7	9	6	5
–	4	3	5	1
	3		**1**	

Solve these subtraction algorithms.

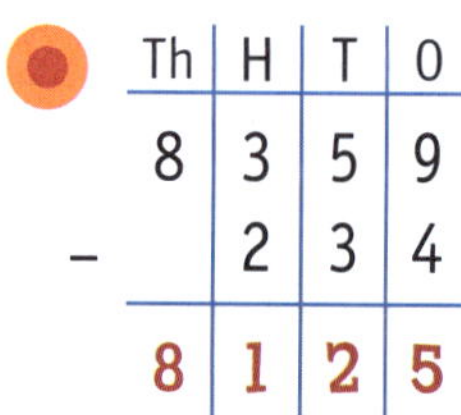

	Th	H	T	O
	8	3	5	9
–		2	3	4
	8	**1**	**2**	**5**

a

	Th	H	T	O
	9	7	6	2
–		4	5	0

b

	Th	H	T	O
	5	9	4	5
–	1	5	3	2

c

	Th	H	T	O
	6	9	7	4
–	3	8	2	3

d

	Th	H	T	O
	7	4	6	7
–		3	5	0

e

	Th	H	T	O
	2	0	8	6
–	1	0	3	6

SELF CHECK Tick how you feel

Got it! ☐ Need help... ☐ I don't get it ☐

Check your answers

How many did you get correct? ☐

CATCH UP MATHS YEAR 6 BOOK A © PASCAL PRESS ISBN: 9781925726183

PRACTICE

1 Solve these subtraction algorithms.

Example:
```
  7 2 5 0
−   1 4 0
---------
  7 1 1 0
```

a
```
  5 9 6 5
−   2 5 3
---------
```

b
```
  6 9 5 4
−   7 2 4
---------
```

c
```
  8 2 5 8
−   1 4 3
---------
```

d
```
  5 8 2 8
− 4 8 0 3
---------
```

e
```
  2 9 7 4
− 1 5 2 3
---------
```

f
```
  9 9 8 3
− 8 5 4 2
---------
```

g
```
  7 6 6 4
− 7 1 5 2
---------
```

2 Solve, then use addition to check your answer.
Hint: If the added total equals the top number you subtracted from, you are correct!

Example:
```
  8 3 9 5        6 2 4 2
− 2 1 5 3      + 2 1 5 3
---------      ---------
  6 2 4 2        8 3 9 5
```

a
```
  7 6 5 4
− 3 4 1 4      + 3 4 1 4
---------      ---------
```

b
```
  8 9 9 4
−   6 3 2      +   6 3 2
---------      ---------
```

c
```
  9 5 2 8
− 4 1 1 2      + 4 1 1 2
---------      ---------
```

d
```
  9 8 9 5
−   7 3 3      +   7 3 3
---------      ---------
```

e
```
  6 8 4 8
−   7 2 3      +   7 2 3
---------      ---------
```

f
```
  5 9 6 9
−   6 5 4      +   6 5 4
---------      ---------
```

g
```
  5 6 3 4
− 5 6 2 0      + 5 6 2 0
---------      ---------
```

h
```
  3 4 6 3
− 1 2 5 0      + 1 2 5 0
---------      ---------
```

i
```
  9 7 6 3
−   4 2 2      +   4 2 2
---------      ---------
```

SUBTRACTION WITHOUT TRADING

FOUR-DIGIT AND FIVE-DIGIT NUMBERS

Here are examples of four-digit and five-digit subtraction algorithms.

Example 1:

	TT	Th	H	T	O
	8	4	5	3	7
–		2	1	1	5
	8	2	4	2	2

↓ Always subtract downwards.

- Start at the ones.
- Next work out the tens.
- Then work out the hundreds.
- Then work out the thousands.
- Then work out the ten thousands.

Example 2: Solve, then check using addition.

	TT	Th	H	T	O
	6	3	5	7	4
–		2	4	3	3
	6	1	1	4	1

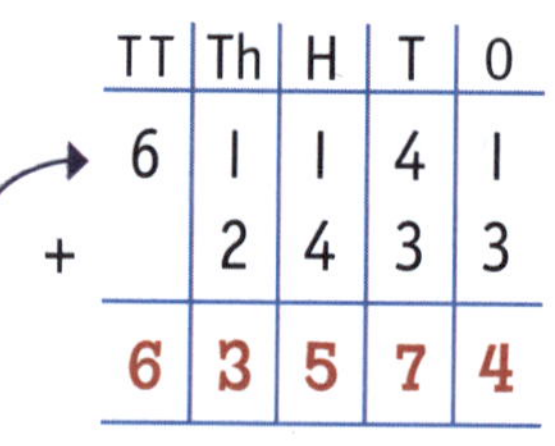

	TT	Th	H	T	O
	6	1	1	4	1
+		2	4	3	3
	6	3	5	7	4

✓

Example 3:
Solve, then check using addition.

	TT	Th	H	T	O
	8	6	5	7	3
–	2	1	3	5	1
		5	2		2

	TT	Th	H	T	O
		5	2		2
+	2	1	3	5	1

Example 4:
Solve, then check using addition.

	TT	Th	H	T	O
	9	9	4	7	5
–		3	4	3	2
	9	6			3

	TT	Th	H	T	O
	9	6			3
+		3	4	3	2

Solve, then check using addition.

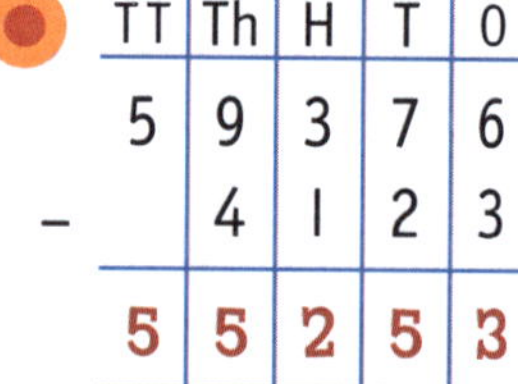

	TT	Th	H	T	O
	5	9	3	7	6
–		4	1	2	3
	5	5	2	5	3

	TT	Th	H	T	O
	5	5	2	5	3
+		4	1	2	3
	5	9	3	7	6

✓

a

	TT	Th	H	T	O
	9	8	4	7	4
–	4	3	1	3	1

	TT	Th	H	T	O
+	4	3	1	3	1

b

	TT	Th	H	T	O
	7	5	5	7	8
–		4	3	4	1

	TT	Th	H	T	O
+		4	3	4	1

SELF CHECK Tick how you feel

Got it!	Need help...	I don't get it
☐	☐	☐

Check your answers
How many did you get correct? ☐

CATCH UP MATHS YEAR 6 BOOK A © PASCAL PRESS ISBN: 9781925726183

PRACTICE

1 Solve these subtraction algorithms.

● 89435 − 4213 = **85222**

c 93586 − 2456 = ______

f 73958 − 42538 = ______

a 75467 − 4152 = ______

d 67569 − 51465 = ______

g 56584 − 46523 = ______

b 58245 − 4145 = ______

e 84932 − 53901 = ______

h 49874 − 7423 = ______

2 Solve, then use addition to check your answer.
Hint: If the added total equals the top number you subtracted from, you are correct!

● 95243 − 71131 = **24112**
24112 + 71131 = **95243**

b 73415 − 2403 = ______
______ + 2403 = ______

d 95013 − 4011 = ______
______ + 4011 = ______

a 69574 − 1463 = ______
______ + 1463 = ______

c 93974 − 43972 = ______
______ + 43972 = ______

e 56483 − 24371 = ______
______ + 24371 = ______

SUBTRACTION WITH TRADING
TWO-DIGIT AND THREE-DIGIT NUMBERS

Sometimes you need to use trading to solve subtraction problems.

Example 1:

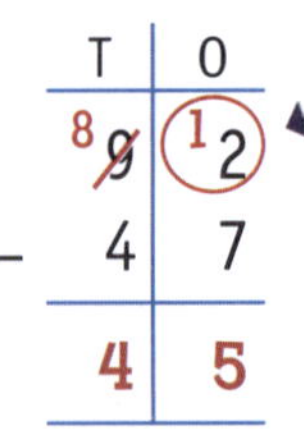

	T	O
	8 ~~9~~	(12)
–	4	7
	4	**5**

We cannot take 7 from 2, so trade 1 ten for 10 ones to make 12.

Now subtract 7 from 12.

Example 2:

	H	T	O
	5	1 ~~2~~	(15)
–	4	1	7
	1	**0**	**8**

Example 3:

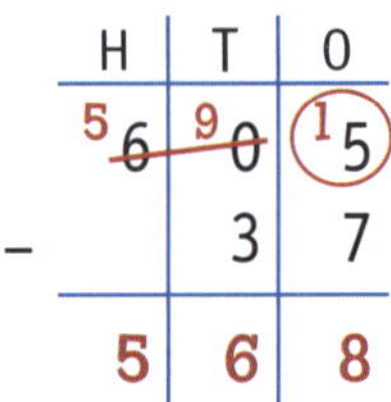

	H	T	O
	5 ~~6~~	9 ~~0~~	(15)
–		3	7
	5	**6**	**8**

Example 4:

	H	T	O
	4 ~~5~~	12 ~~3~~	(10)
–	2	5	3
	2	**7**	**7**

Example 5:

	H	T	O
	3 ~~4~~	10 ~~1~~	(13)
–	1	5	4
	2	**5**	**9**

Example 6:

	H	T	O
	6 ~~7~~	9 ~~0~~	(18)
–	1	7	9
			9

Example 7:

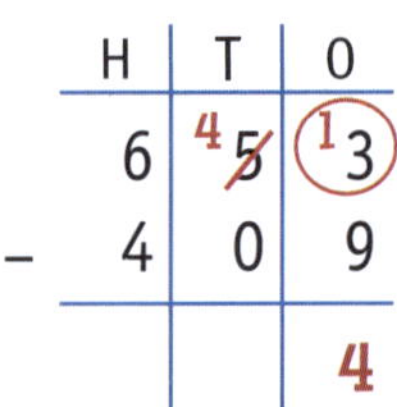

	H	T	O
	6	4 ~~5~~	(13)
–	4	0	9
			4

Check your answer on the video!

Solve these subtraction algorithms.

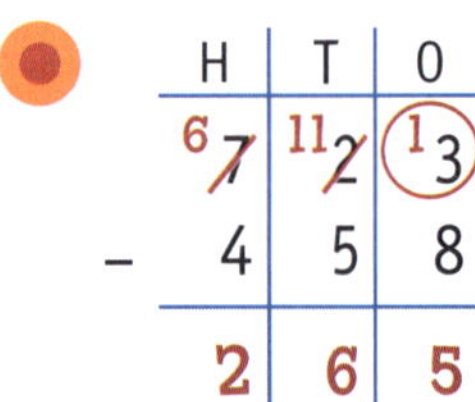

	H	T	O
	6 ~~7~~	11 ~~2~~	(13)
–	4	5	8
	2	**6**	**5**

a

	H	T	O
	8	4	3
–	7	5	4

b

	H	T	O
	9	0	5
–	2	6	4

c

	H	T	O
	8	7	9
–	6	8	9

d

	H	T	O
	3	2	6
–	1	5	6

e

	H	T	O
	7	3	0
–	2	9	3

SELF CHECK Tick how you feel

Got it!	Need help...	I don't get it
 ☐	 ☐	☐

Check your answers

How many did you get correct? ☐

CATCH UP MATHS YEAR 6 BOOK A © PASCAL PRESS ISBN: 9781925726183

PRACTICE

1 Solve these subtraction algorithms.

●	c	f	i
$^{7}\cancel{8}\ ^{1}4$	6 6	7 1 3	8 3 2
− 2 6	− 4 7	− 4 7	− 4 3
5 8			

a	d	g	j
9 4	4 8	9 0 2	3 5 6
− 3 6	− 2 9	− 3 6	− 1 7 8

b	e	h	k
7 2	5 0 6	9 1 3	6 0 6
− 4 3	− 3 8	− 2 7	− 4 3 7

2 Fill in the missing digits.

●	b	d	f
$^{4}\cancel{5}\ ^{11}\cancel{2}\ ^{1}4$	_ 0 3	7 1 4	9 _ _
− 2 4 7	− 3 _ _	− _ _ 6	− 5 2 6
2 7 7	5 4 1	2 2 _	3 _ 8

a	c	e	g
_ 3 6	6 2 3	5 _ 9	6 5 _
− 5 2 _	− 4 _ 7	− 4 6 _	− 1 _ 8
3 _ 7	_ 6 _	_ 2 2	_ 6 9

3 Use addition to check your answer to Question 2.
Hint: If the added total equals the top number you subtracted from, you are correct!

●	b	d	f
$^{1}2\ ^{1}4\ 7$	5 4 1	2 2 _	3 _ 8
+ 2 7 7	+ 3 _ _	+ _ _ 6	+ 5 2 6
5 2 4	_ 0 3	7 1 4	9 _ _

a	c	e	g
3 _ 7	_ 6 _	_ 2 2	_ 6 9
+ 5 2 _	+ 4 _ 7	+ 4 6 _	+ 1 _ 8
_ 3 6	6 2 3	5 _ 9	6 5 _

SUBTRACTION WITH TRADING
THREE-DIGIT AND FOUR-DIGIT NUMBERS

**Here is an example of four-digit subtraction.
We have to trade 1 ten for 10 ones and then 1 hundred for 10 tens.**

Example 1:

We cannot take 9 from 7, so trade 1 ten for 10 ones. Now subtract 9 from 17.

	Th	H	T	O
	7	~~9~~ 8	~~8~~ 17	14
–	2	3	9	9
	5	**5**	**8**	**5**

We cannot take 9 from 4, so trade 1 ten for 10 ones. Now we can subtract 9 from 14.

Example 2:

	Th	H	T	O
	9	8	~~7~~ 6	15
–	2	1	3	6
	7	**7**	**3**	**9**

Example 3:

	Th	H	T	O
	~~7~~ 6	~~2~~ 11	~~6~~ 15	11
–	5	3	7	5
	1	**8**	**8**	**6**

Example 4:

	Th	H	T	O
	6	6	~~3~~ 2	12
–		4	1	8
	6	**2**	**1**	**4**

Example 5:

	Th	H	T	O
	8	4	~~6~~ 5	13
–		3	5	7
	8	**1**	**0**	**6**

Example 6:

	Th	H	T	O
	5	2	~~3~~ 2	14
–		1	2	8
				6

Check your answer on the video!

Example 7:

	Th	H	T	O
	5	~~9~~ 8	10	3
–	2	5	7	2
				1

Solve these subtraction algorithms.

●

	Th	H	T	O
	8	3	~~6~~ 5	12
–		3	4	9
	8	**0**	**1**	**3**

a

	Th	H	T	O
	5	8	2	6
–	1	7	5	9

b

	Th	H	T	O
	6	0	4	8
–	3	2	5	9

c

	Th	H	T	O
	5	4	3	2
–		5	4	0

d

	Th	H	T	O
	5	7	7	5
–	5	2	9	6

e

	Th	H	T	O
	9	3	7	3
–		9	7	7

SELF CHECK Tick how you feel

Got it!	Need help...	I don't get it
☐	☐	☐

Check your answers
How many did you get correct? ☐

CATCH UP MATHS YEAR 6 BOOK A © PASCAL PRESS ISBN: 9781925726183

PRACTICE

1 Solve these subtraction algorithms.

●	2301 − 781 = 2220	c	4837 − 1989 = ____	f	3410 − 2249 = ____	i	8001 − 4930 = ____
a	9501 − 331 = ____	d	6341 − 4387 = ____	g	2873 − 1980 = ____	j	5003 − 1432 = ____
b	8734 − 549 = ____	e	5642 − 1652 = ____	h	7002 − 1341 = ____	k	6008 − 2050 = ____

2 Solve, then use addition to check your answer.

Hint: If the added total equals the top number you subtracted from, you are correct!

	Subtraction	Check		Subtraction	Check
●	2807 − 1439 = 1368	1368 + 1439 = 2807	d	8321 − 2689 = ____	____ + 2689 = ____
a	9602 − 5349 = ____	____ + 5349 = ____	e	4832 − 3999 = ____	____ + 3999 = ____
b	5243 − 178 = ____	____ + 178 = ____	f	6711 − 429 = ____	____ + 429 = ____
c	6020 − 4378 = ____	____ + 4378 = ____	g	8069 − 4309 = ____	____ + 4309 = ____

SUBTRACTION WITH TRADING
FIVE-DIGIT NUMBERS

This is a subtraction algorithm where we need to trade 1 ten for 10 ones and then 1 hundred for 10 tens.

SCAN to watch video

Example 1:

	TT	Th	H	T	O
	7	5	3 ~~4~~	12 ~~3~~	12
−	4	2	1	7	8
	3	3	2	5	4

We have traded twice in this algorithm.

Example 2:

	TT	Th	H	T	O
	5	6	8 ~~9~~	12 ~~3~~	14
−				4	7
	5	6	8	8	7

Example 4:

	TT	Th	H	T	O
	5 ~~6~~	13	7 ~~8~~	11 ~~2~~	11
−		4	3	8	9
	5	9	4	3	2

Example 6:

	TT	Th	H	T	O
	5	4	1 ~~2~~	11	5
−			3	9	0
				2	5

Check your answer on the video!

Example 3:

	TT	Th	H	T	O
	7	2	4 ~~5~~	12 ~~3~~	10
−			4	4	3
	7	2	0	8	7

Example 5:

	TT	Th	H	T	O
	7 ~~8~~	12 ~~3~~	11 ~~2~~	13 ~~4~~	12
−	4	8	9	7	3
	3	4	2	6	9

Example 7:

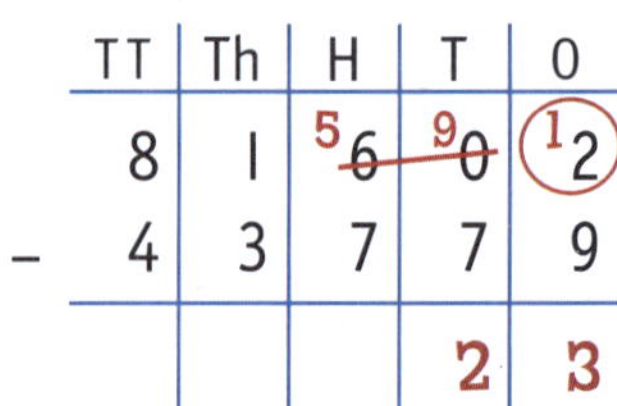

	TT	Th	H	T	O
	8	1	5 ~~6~~	9 ~~0~~	12
−	4	3	7	7	9
				2	3

Solve these subtraction algorithms.

	TT	Th	H	T	O
	7 ~~8~~	13 ~~4~~	12 ~~3~~	9 ~~0~~	11
−	2	6	4	9	3
	5	7	8	0	8

b

	TT	Th	H	T	O
	6	5	3	2	0
−				4	6

a

	TT	Th	H	T	O
	3	6	4	0	2
−	1	7	3	2	5

c

	TT	Th	H	T	O
	5	6	2	1	6
−		2	9	8	4

SELF CHECK Tick how you feel

Got it!	Need help...	I don't get it
☐	☐	☐

Check your answers

How many did you get correct? ☐

CATCH UP MATHS YEAR 6 BOOK A © PASCAL PRESS ISBN: 9781925726183

PRACTICE

1 Solve these subtraction algorithms.

Example: $^{1}\not{2}\,^{10}\not{1}\,^{11}\not{2}\,^{10}\not{1}\,^{1}1$ − 2 2 4 6 = 1 8 9 6 5

a 3 8 4 2 9 − 3 0 8 3 = ______

b 7 3 1 8 4 − 3 9 9 = ______

c 8 4 1 2 3 − 6 8 2 8 5 = ______

d 1 5 3 7 2 − 8 8 = ______

e 7 4 9 3 8 − 2 9 = ______

f 5 6 4 0 2 − 7 8 9 3 = ______

g 2 6 3 8 1 − 1 5 9 3 2 = ______

h 8 3 0 2 4 − 4 7 1 9 8 = ______

2 Use addition to check your answers for question 1.

Example: $^{1}1\ ^{1}8\ ^{1}9\ ^{1}6\ 5$ + 2 2 4 6 = 2 1 2 1 1

a ______ + 3 0 8 3 = ______

b ______ + 3 9 9 = ______

c ______ + 6 8 2 8 5 = ______

d ______ + 8 8 = ______

e ______ + 2 9 = ______

f ______ + 7 8 9 3 = ______

g ______ + 1 5 9 3 2 = ______

h ______ + 4 7 1 9 8 = ______

3 Fill in the missing numbers.

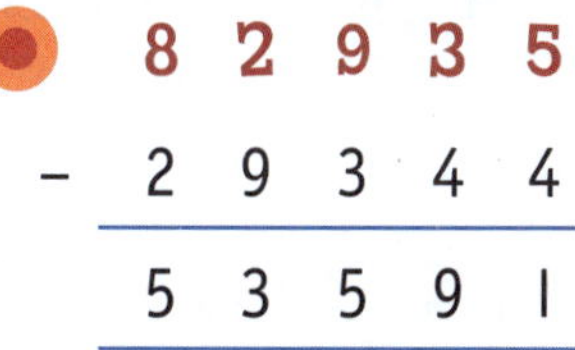
8 2 9 3 5 − 2 9 3 4 4 = 5 3 5 9 1

a 9 1 8 3 2 − ______ = 4 8 1 1 3

b ______ − 1 4 9 7 = 5 1 4 5 6

TRADING FROM HIGHER PLACE VALUES

When you are subtracting, sometimes you can't trade from the next place value.

Example 1:

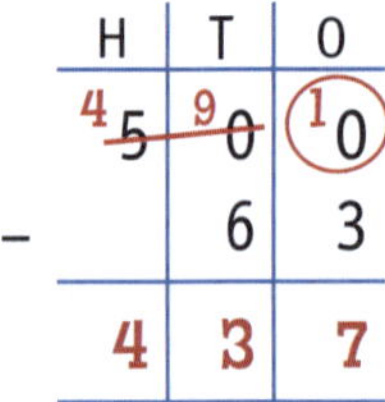

	H	T	O
	~~5~~ 4	~~0~~ 9	10
−		6	3
	4	3	7

Example 2:

	H	T	O
	~~8~~ 7	~~0~~ 9	10
−		5	1
	7	4	9

Example 3:

	Th	H	T	O
	~~9~~ 8	~~0~~ 9	~~0~~ 9	10
−		5	4	3
	8	4	5	7

Example 4:

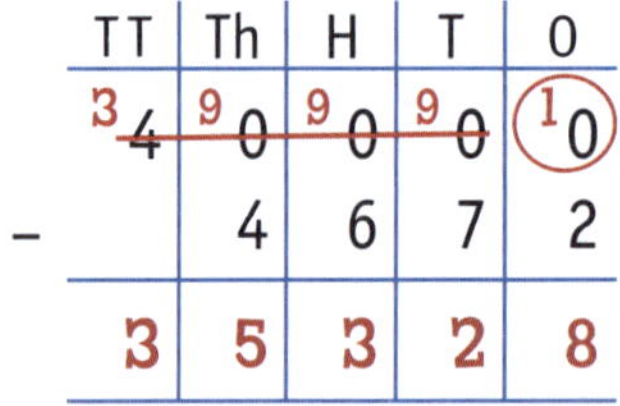

	TT	Th	H	T	O
	~~4~~ 3	~~0~~ 9	~~0~~ 9	~~0~~ 9	10
−		4	6	7	2
	3	5	3	2	8

Example 5:

	HT	TT	Th	H	T	O
	~~9~~ 8	~~0~~ 9	~~0~~ 9	~~0~~ 9	~~0~~ 9	10
−		4	7	5	2	4
	8	5	2	4	7	6

Example 6:

	Th	H	T	O
	~~5~~ 4	~~0~~ 9	~~0~~ 9	10
−		4	2	1

Check your answer on the video!

Example 7:

	HT	TT	Th	H	T	O
	~~2~~ 1	~~0~~ 9	~~0~~ 9	~~0~~ 9	~~0~~ 9	10
−	1	7	3	2	4	

Solve these subtraction algorithms.

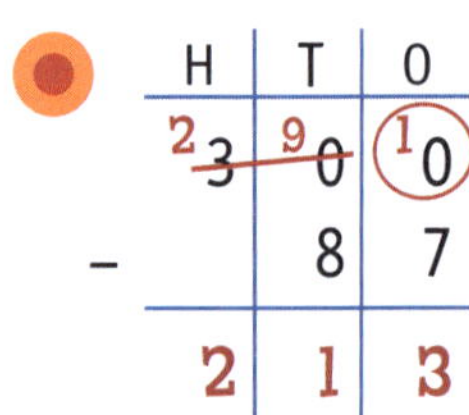

	H	T	O
	~~3~~ 2	~~0~~ 9	10
−		8	7
	2	1	3

a

	Th	H	T	O
	4	0	0	0
−	3	2	6	3

b

	TT	Th	H	T	O
	8	0	0	0	0
−	1	4	3	2	8

c

	HT	TT	Th	H	T	O
	1	0	0	0	0	0
−		4	3	6	4	5

Check your answers

How many did you get correct?

CATCH UP MATHS YEAR 6 BOOK A © PASCAL PRESS ISBN: 9781925726183

PRACTICE

1 Solve these subtraction algorithms.

Example:

```
  ⁸9 ⁹0 ⁹0 ⁹0 ⁹0 ¹0
-     4  3  8  2  6
   8  5  6  1  7  4
```

a
```
  1 0 0 0 0 0
-   3 9 8 7 3
```

b
```
  2 0 0 0 0
- 1 4 3 8 6
```

c
```
  3 0 0
-   6 3
```

d
```
  5 0 0 0
- 3 2 8 1
```

e
```
  6 4 0 0 0 0
- 3 2 5 6 0 4
```

f
```
  7 1 6 0 0 0
- 4 3 0 3 6 0
```

g
```
  4 0 0 0 0
- 3 0 2 0 4
```

h
```
  6 0 0 0 0
- 4 1 3 7 5
```

i
```
  2 9 0 0 0
- 1 4 3 8 3
```

j
```
  3 0 0 0 0
- 2 4 8 3 6
```

k
```
  7 0 0 0 0 0
- 4 3 4 5 6 3
```

2 Solve these money subtractions.

Example:

```
  $ 1 ⁹0 ⁹0 . ⁹0 ¹0
- $      5  8 .  4  3
         4  1 .  5  7
```

a
```
  $ 2 0 0 . 0 0
- $   5 7 . 4 5
```

b
```
  $ 1 5 0 . 0 0
- $   4 3 . 4 7
```

c
```
  $ 5 . 0 0
- $ 1 . 7 5
```

d
```
  $ 2 5 0 . 0 0
- $ 1 8 3 . 6 5
```

e
```
  $ 7 0 5 . 0 0
- $ 3 4 1 . 5 0
```

f
```
  $ 6 0 0 . 0 0
- $ 1 4 7 . 5 3
```

g
```
  $ 4 . 0 0
- $ 0 . 7 5
```

h
```
  $ 1 7 0 . 0 0
- $   4 3 . 5 0
```

  ISBN: 9781925726183

DIFFERENCE

The difference is the answer you get when you subtract one number from another. The difference is the amount one number is bigger or smaller than another number.

Example 1:

What is the difference between 65 and 14?

```
    6 5
  - 1 4
  -----
    5 1
  -----
```

The difference between 65 and 14 is 51.

Example 2:

What is the difference between 49 and 12?

```
    4 9
  - 1 2
  -----

  -----
```

The difference between 49 and 12 is ____.

Example 3:

What is the difference between 143 and 21?

```
    1 4 3
  -   2 1
  -------
    1 2 2
  -------
```

The difference between 143 and 21 is 122.

Example 4:

What is the difference between 634 and 63?

```
    6 3 4
  -   6 3
  -------

  -------
```

The difference between 634 and 63 is ____.

Example 5:

What is the difference between 1526 and 73?

```
    1 ⁴5̸ ¹2 6
  -       7 3
  -----------
    1  4  5 3
  -----------
```

The difference between 1526 and 73 is 1453.

Example 6:

What is the difference between 1495 and 143?

```
    1 4 9 5
  -   1 4 3
  ---------

  ---------
```

The difference between 1495 and 143 is ____.

Your turn Work out the difference.

● 7314 and 1262

```
    7 ²3̸ ¹1 4
  - 1  2  6 2
  -----------
    6  0  5 2
  -----------
```

Difference = 6052

a 6824 and 721

```
  - ________

    ________
```

Difference = ____

b 7136 and 2848

```
  - ________

    ________
```

Difference = ____

Check your answers
How many did you get correct?

CATCH UP MATHS YEAR 6 BOOK A © PASCAL PRESS ISBN: 9781925726183

PRACTICE

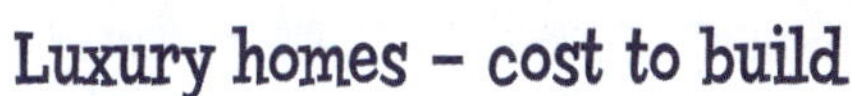

Luxury homes – cost to build

Strandford $727 986

Armany $843 289

Dion $947 384

Frendi $999 999

Belmoni $598 631

Tomayne $897 263

1 **Find the difference in cost.**

● Strandford and Dion

$	9	[3] ~~4~~	[16] ~~7~~	[12] ~~3~~	[17] ~~8~~	[1]4
– $	7	2	7	9	8	6
$	2	1	9	3	9	8

a Tomayne and Frendi

$

– $

b Belmoni and Armany

$

– $

c Strandford and Frendi

$

– $

2 **If you start with $1 000 000, how much change would you get?**

● Dion

$	~~1~~	[9] ~~0~~	[9] ~~0~~	[9] ~~0~~	[9] ~~0~~	[9] ~~0~~	[1]0
– $		9	4	7	3	8	4
$			5	2	6	1	6

a Armany

$ 1 0 0 0 0 0 0

– $

b Strandford

$ 1 0 0 0 0 0 0

– $

c Belmoni

$

– $

d Tomayne

$

– $

e Frendi

$

– $

  ISBN: 9781925726183

ROUNDING TO ESTIMATE ANSWERS

You can use rounding to get an approximate answer that is close to the actual answer.

Round to the nearest 10

Example 1:

531 – 173 is about 360

530 170

Example 2:

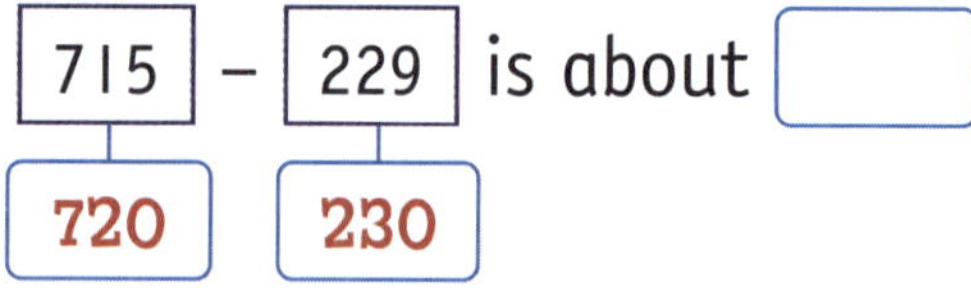

Example 3:

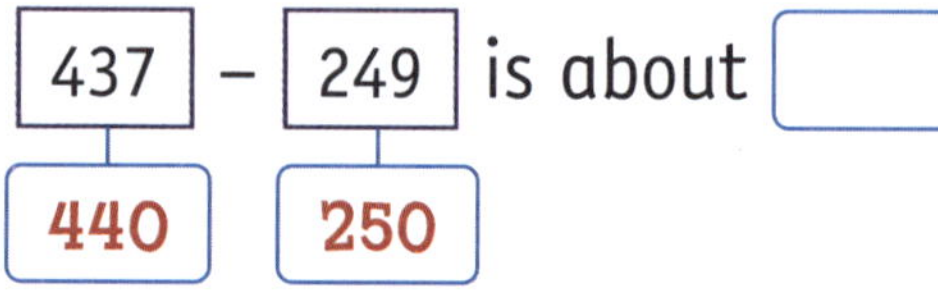

Round to the nearest 100

Example 4:

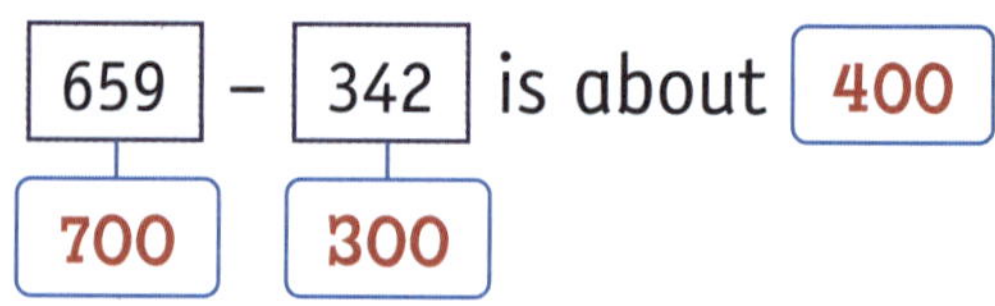

Example 5:

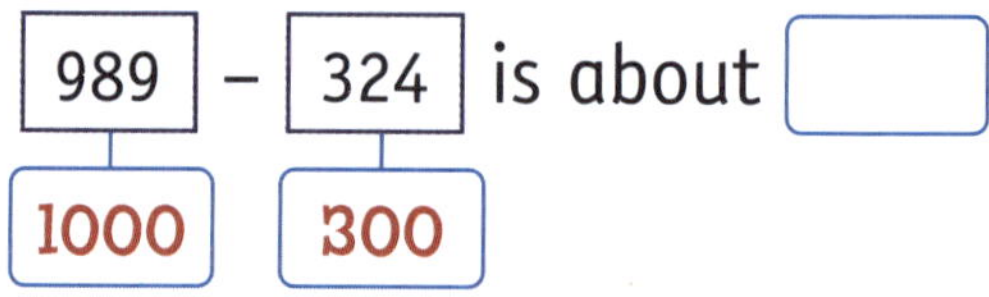

Example 6:

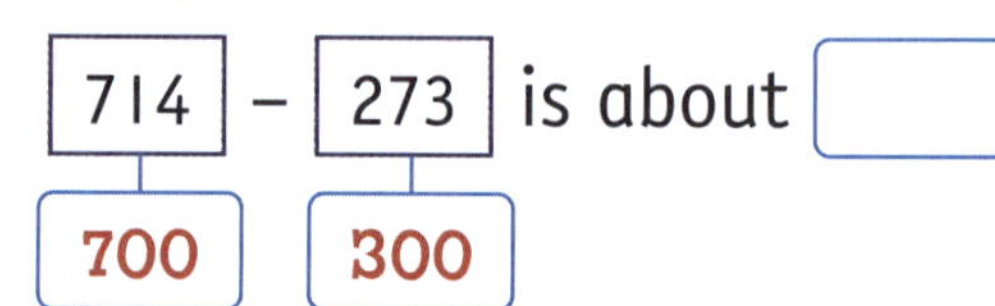

An estimate is near the answer, but it isn't perfectly accurate.

Your turn

1 Round to the nearest 10 and subtract.

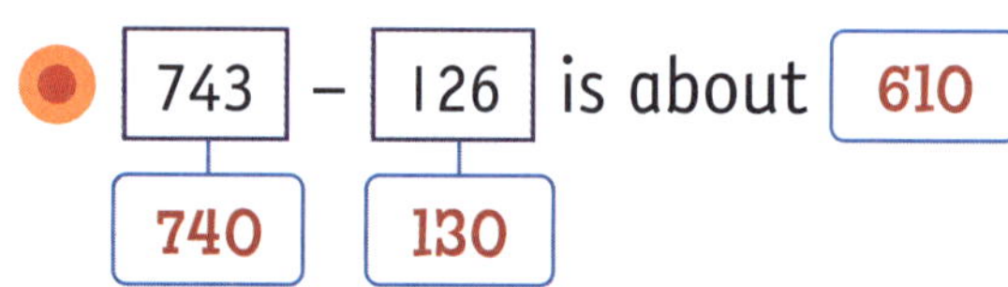

a

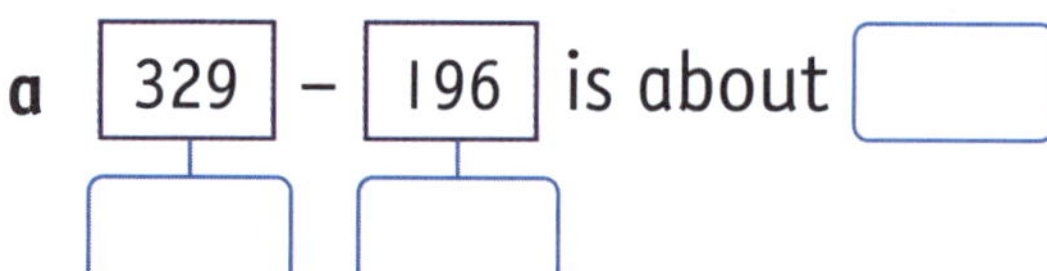

b 538 – 463 is about

2 Round to the nearest 100 and subtract.

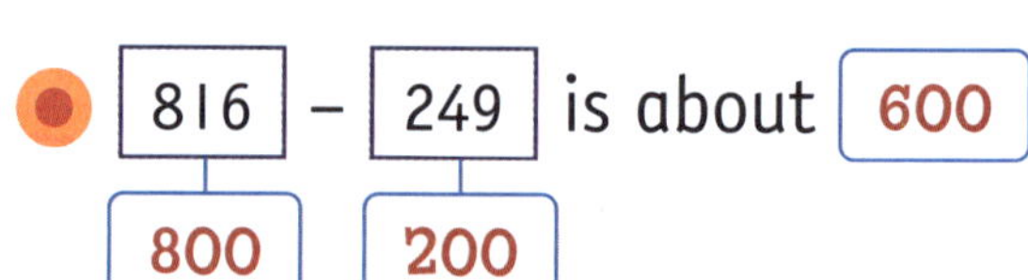

a 632 – 389 is about

b

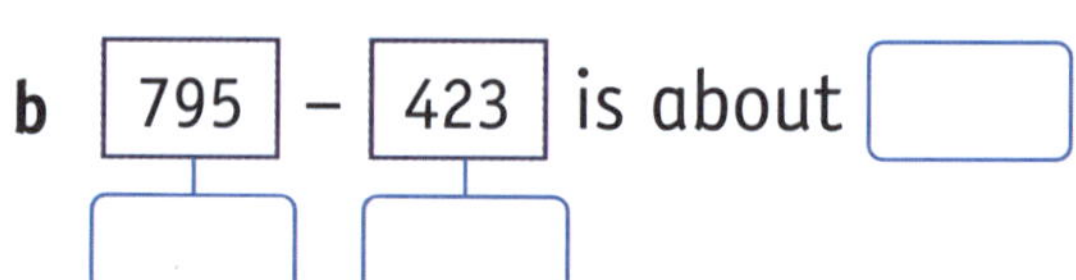

SELF CHECK Tick how you feel

Got it!	Need help...	I don't get it

Check your answers

How many did you get correct?

CATCH UP MATHS YEAR 6 BOOK A © PASCAL PRESS ISBN: 9781925726183

PRACTICE

1 Round each number to the nearest 10 and subtract to estimate the answer.

- Example: 4492 – 3157 is about 1330 (4490, 3160)
- **a** 6429 – 1362 is about ____ (____ , ____)
- **b** 7492 – 5635 is about ____ (____ , ____)
- **c** 143 – 125 is about ____ (____ , ____)
- **d** 341 – 202 is about ____ (____ , ____)
- **e** 5823 – 3217 is about ____ (____ , ____)

2 Round each number to the nearest 100 and subtract to estimate the answer.

- Example: 7463 – 5987 is about 1500 (7500, 6000)
- **a** 9289 – 3645 is about ____ (____ , ____)
- **b** 4943 – 2725 is about ____ (____ , ____)
- **c** 292 – 153 is about ____ (____ , ____)
- **d** 767 – 421 is about ____ (____ , ____)
- **e** 911 – 245 is about ____ (____ , ____)

3 Round each number to the nearest 100 and subtract to estimate the answer.

- **a** 15 432 – 10 987 is about ____ (____ , ____)
- **b** 84 179 – 38 894 is about ____ (____ , ____)
- **c** 73 284 – 44 215 is about ____ (____ , ____)
- **d** 19 246 – 11 426 is about ____ (____ , ____)

  ISBN: 9781925726183

WORD PROBLEMS USING SUBTRACTION

A word problem is when a maths problem is written in sentences instead of only numbers and symbols.

Follow these steps:

1 **Read the question.**

2 **Think about the question.**
- Do you need to add, subtract, multiply or divide?
- Is there more than one step or operation?

3 **Highlight important information.**
- Circle the numbers.
- Underline what you need to find out.
- Drawing a picture might help too.

4 **Solve it!**

When you read word problems, look for key words that help you work out how to solve the problem.

Here are some key words that tell you to subtract:
decrease, deduct, difference, fewer, less, left, minus, how much more, reduce, remain, subtract, take away

---

Example 1:
Annie had 37 apples, but 15 were rotten.
How many apples were not rotten?

Working out

$$\begin{array}{r} 37 \\ -\ 15 \\ \hline 22 \end{array}$$

Example 2:
Martha had 93 pencils and 72 were sharp.
How many pencils were blunt?

Working out

$$\begin{array}{r} 93 \\ -\ 72 \\ \hline 21 \end{array}$$

Example 3:
Peter caught 74 fish. His mate Blake knocked over the buckets of fish and 29 swam away.
How many fish did Peter have left?

Working out

$$\begin{array}{r} 74 \\ -\ 29 \\ \hline \end{array}$$

Example 4:
Talia swam 235 laps. Joanne swam 147 laps.
What is the difference between the laps they swam?

Working out

$$\begin{array}{r} \\ -\ \\ \hline \end{array}$$

CATCH UP MATHS YEAR 6 BOOK A © PASCAL PRESS ISBN: 9781925726183

Solve these word problems.

	Working out
● Rose the baker baked 753 chocolate chip cookies, but she burnt 122 of them. How many good cookies did Rose bake?	753 − 122 = 631
a Antonia kicked 37 goals in soccer and Elsie scored 21 goals. What is the difference in the goals they scored?	
b A farmer grew 2293 punnets of strawberries, but 129 went bad. How many punnets were not bad?	
c Xander bought a car for $59 624 and sold it for $39 515. What was the difference in price?	
d Ziva had $71 425 in the bank. She spent $21 325 on a motorbike. How much money does Ziva have left in the bank?	

SELF CHECK Tick how you feel

Got it!	Need help...	I don't get it
☐	☐	☐

Check your answers

How many did you get correct? ☐

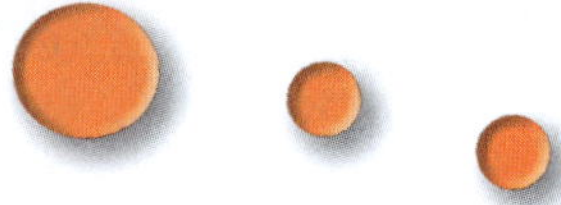

Solve these word problems.

	Working out
a Melbourne is 877 km from Sydney. Danny has driven 352 km. How many more kilometres does Danny have to drive to get to Melbourne?	
b Mark was driving from Darwin to Perth. He had driven 2812 km. How many more kilometres did Mark have to drive if Perth is 3848 km from Darwin?	
c Laila was flying from Sydney to Norway. She has flown 2943 km. How many more kilometres does Laila have to fly if Norway is 16 019 km from Sydney?	
d Nina won $754 218 in the lottery. She spent $314 020. How much money does Nina have left?	
e Xeni put her dog on a diet. His weight decreased from 73 kg to 54 kg. How much weight did the dog lose?	
f Makay ran 24 km less then Max. Max ran 37 km. How many kilometres did Makay run?	

CATCH UP MATHS YEAR 6 BOOK A © PASCAL PRESS ISBN: 9781925726183

	Working out
g Jake swam 112 km in May. Christian swam 236 km. How many more kilometres did Christian swim than Jake?	
h A car dealer bought a car for $68 249 and sold it for $73 268. How much profit did the car dealer make?	
i Shane has a pool. It holds 86 000 litres. It leaked, and now has 84 132 litres in it. How many litres leaked from Shane's pool?	
j Pia sells rugs. She sold one for $16 425 after buying it for $12 236. How much profit did Pia make?	
k Alyssa won $6 536 259 in the lottery. She spent $2 436 582. How much of her winnings does Alyssa have left?	
l Deborah hiked 13 413 metres up Mt Belmar. She had a break on her way back down after she had hiked 2599 m down the mountain. How many metres does Deborah have left to hike?	

SUBTRACTION REVIEW

1 Solve using the jump strategy and the number lines.

a 79 – 21 = ___

b 289 – 42 = _____

c 799 – 31 = _____

d 762 – 231 = _____

e 463 – 133 = _____

f 5213 – 213 = _______

g 7389 – 420 = _______

h 7459 – 2312 = _______

i 9113 – 3310 = _______

CATCH UP MATHS YEAR 6 BOOK A © PASCAL PRESS ISBN: 9781925726183

2 Use the split strategy to solve these subtractions.

a 84 – 21

____ – ____ = ____

__ – __ = __

____ + __ = _____

b 78 – 34

____ – ____ = ____

__ – __ = __

____ + __ = _____

c 499 – 38

_____ – _____ = _____

____ – ____ = ____

__ – __ = __

_____ + ____ + __ = _____

d 697 – 56

_____ – _____ = _____

____ – ____ = ____

__ – __ = __

_____ + ____ + __ = _____

e 784 – 153

_____ – _____ = _____

____ – ____ = ____

__ – __ = __

_____ + ____ + __ = _____

f 882 – 481

_____ – _____ = _____

____ – ____ = ____

__ – __ = __

_____ + ____ + __ = _____

g 7598 – 463

______ – ______ = ______

_____ – _____ = _____

____ – ____ = ____

__ – __ = __

______ + _____ + ____ + __ = ______

h 9897 – 544

______ – ______ = ______

_____ – _____ = _____

____ – ____ = ____

__ – __ = __

______ + _____ + ____ + __ = ______

i 6984 – 1273

______ – ______ = ______

_____ – _____ = _____

____ – ____ = ____

__ – __ = __

______ + _____ + ____ + __ = ______

j 8168 – 4034

______ – ______ = ______

_____ – _____ = _____

____ – ____ = ____

__ – __ = __

______ + _____ + ____ + __ = ______

k 9864 – 8731

______ – ______ = ______

_____ – _____ = _____

____ – ____ = ____

__ – __ = __

______ + _____ + ____ + __ = ______

REVIEW

Solve these subtractions without trading.

a $\begin{array}{r} 84 \\ -\ 42 \\ \hline \end{array}$

b $\begin{array}{r} 73 \\ -\ 21 \\ \hline \end{array}$

c $\begin{array}{r} 96 \\ -\ 23 \\ \hline \end{array}$

d $\begin{array}{r} 68 \\ -\ 33 \\ \hline \end{array}$

e $\begin{array}{r} 703 \\ -\ 201 \\ \hline \end{array}$

f $\begin{array}{r} 859 \\ -\ 257 \\ \hline \end{array}$

g $\begin{array}{r} 953 \\ -\ 12 \\ \hline \end{array}$

h $\begin{array}{r} 685 \\ -\ 43 \\ \hline \end{array}$

i $\begin{array}{r} 6873 \\ -\ 3142 \\ \hline \end{array}$

j $\begin{array}{r} 5754 \\ -\ 4324 \\ \hline \end{array}$

k $\begin{array}{r} 9372 \\ -\ 272 \\ \hline \end{array}$

l $\begin{array}{r} 8519 \\ -\ 7503 \\ \hline \end{array}$

4 Solve and use addition to check your answer.

a $\begin{array}{r} 85493 \\ -\ 2132 \\ \hline \end{array}$ $\begin{array}{r} \\ +\ 2132 \\ \hline \end{array}$

b $\begin{array}{r} 93724 \\ -\ 42523 \\ \hline \end{array}$ $\begin{array}{r} \\ +\ 42523 \\ \hline \end{array}$

c $\begin{array}{r} 71653 \\ -\ 20351 \\ \hline \end{array}$ $\begin{array}{r} \\ +\ 20351 \\ \hline \end{array}$

d $\begin{array}{r} 59458 \\ -\ 5344 \\ \hline \end{array}$ $\begin{array}{r} \\ +\ 5344 \\ \hline \end{array}$

e $\begin{array}{r} 64415 \\ -\ 34205 \\ \hline \end{array}$ $\begin{array}{r} \\ +\ 34205 \\ \hline \end{array}$

f $\begin{array}{r} 98347 \\ -\ 40235 \\ \hline \end{array}$ $\begin{array}{r} \\ +\ 40235 \\ \hline \end{array}$

CATCH UP MATHS YEAR 6 BOOK A © PASCAL PRESS ISBN: 9781925726183

5 Solve these subtractions with trading.

a 97 − 18 = ____

b 86 − 27 = ____

c 74 − 35 = ____

d 52 − 49 = ____

e 81 − 74 = ____

f 907 − 24 = ____

g 821 − 42 = ____

h 814 − 433 = ____

i 563 − 79 = ____

j 973 − 798 = ____

k 9342 − 2154 = ____

l 8361 − 453 = ____

m 5682 − 4593 = ____

n 7102 − 983 = ____

o 6033 − 4977 = ____

p 8456 − 4459 = ____

6 Solve these subtractions with trading.

a 84901 − 26948 = ____

b 73204 − 4568 = ____

c 99720 − 46853 = ____

7 Solve these subtractions with trading.

a 800 − 49 = ____

b 600 − 58 = ____

c 400 − 77 = ____

d 300 − 36 = ____

e 4000 − 328 = ____

f 9000 − 79 = ____

g 2000 − 1432 = ____

h 1000 − 471 = ____

  ISBN: 9781925726183

REVIEW

8 Solve these subtractions with trading.

a
```
  1 0 0 0 0
-   4 7 9 3
-----------
```

b
```
  3 0 0 0 0
-     3 2 4
-----------
```

c
```
  7 0 0 0 0
- 1 0 4 0 8
-----------
```

d
```
  4 0 0 0 0 0
-   4 3 4 5 0
-------------
```

e
```
  7 0 0 0 0 0
- 6 5 2 3 0 9
-------------
```

f
```
  3 0 0 0 0 0
-   4 2 5 1 5
-------------
```

9 What is the difference?

a 53 and 17 ____

b 47 and 33 ____

c 96 and 36 ____

d 731 and 45 ____

e 848 and 27 ____

f 653 and 49 ____

g 4573 and 1241 ____

h 5398 and 2593 ____

i 9809 and 2417 ____

10 Round each number to the nearest 10 and subtract to estimate the answer.

a 413 – 326 is about ____

b 907 – 483 is about ____

c 749 – 41 is about ____

d 531 – 79 is about ____

e
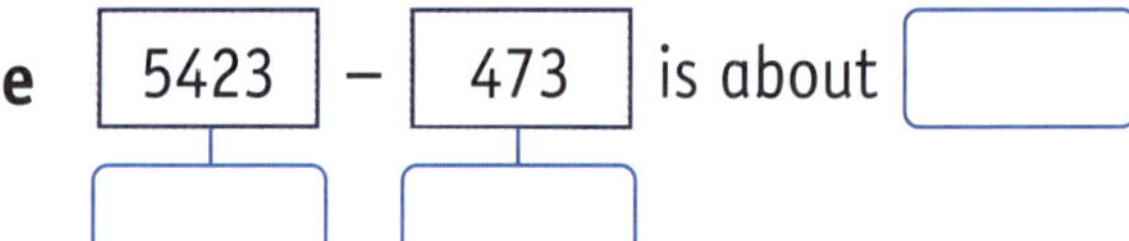

f
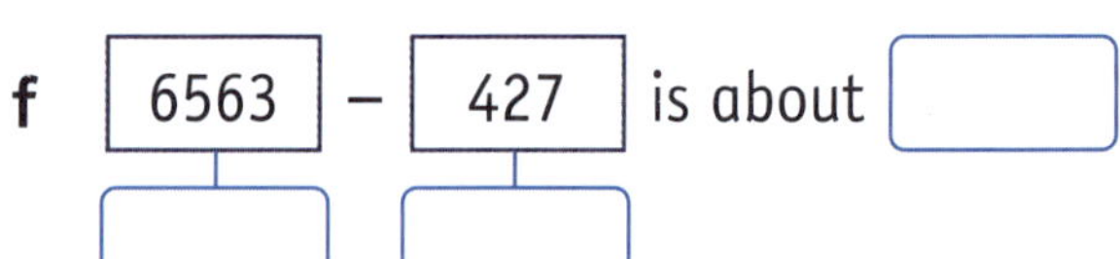

g
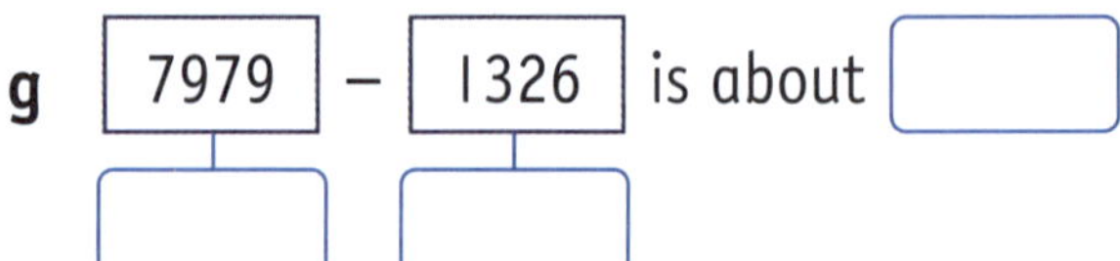

h
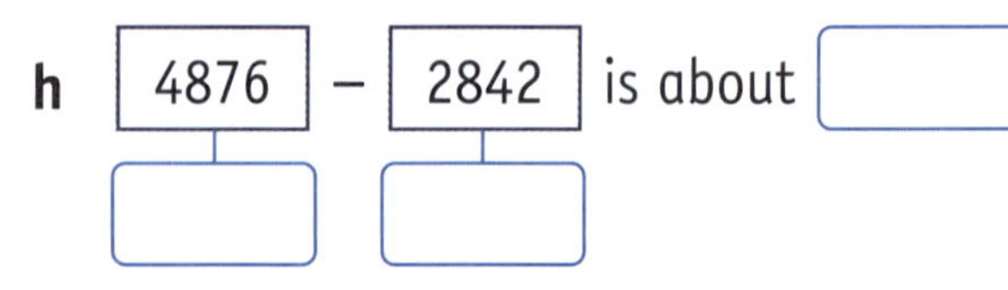

CATCH UP MATHS YEAR 6 BOOK A © PASCAL PRESS ISBN: 9781925726183

11 Round each number to the nearest 100 and subtract to estimate the answer.

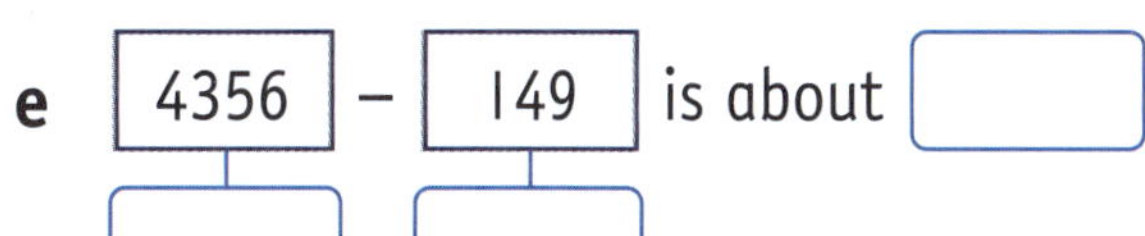

a 476 – 324 is about ______

b 814 – 193 is about ______

c 656 – 142 is about ______

d 711 – 423 is about ______

e 4356 – 149 is about ______

f 7267 – 389 is about ______

g 7142 – 534 is about ______

g 9871 – 3615 is about ______

12 Solve these word problems.

Working out

a Ava had 59 dolls and donated 24 to charity.
How many dolls does Ava have left?

______

b Ashton cycled 634 km in January and 816 km in February.
What is the difference between the distance cycled in these months?

______

c Syl had to drive 5638 km. She had driven 643 km.
How many more kilometres does Syl have to drive?

______

Working out

d Aria had $79 348 saved. She spent $25 291 on a new car.
How much does Aria have left in her savings?

______

e Mario won $895 268 in the lottery. He spent $624 254.
How much does Mario have left?

______

f George bought a house for $759 982 and sold it for $657 998.
What is the difference in her purchase and sale prices?

______

REVIEW

13 Fill in the missing digits.

a
```
  7 5
- 2
    1
```
e
```
  8
-   7
  4 2
```
i
```
  9 0 7
- 2   6
    5
```
m
```
  7 6   3
-     9 7
  7 1 2
```
b
```
  9
-   5
  4 7
```
f
```
  1 4 5
-   3
  1   3
```
j
```
    1 6
- 2
  3 2 2
```
n
```
  4 3   4
- 1   9 5
    5 1
```
c
```
  8 3
-   4
  2
```
g
```
  5 1 4
-   3
  4   8
```
k
```
  5 9   3
-     4
  5 7 7 6
```
o
```
  8 4 4
-   3   2
  4   9 6
```
d
```
  6
-   7
  2 7
```
h
```
  8   3
- 1 4
    9 8
```
l
```
  6   9 7
-   2   4
  1 8 8
```
p
```
  5 8   6
-     3 7
  3 2 8
```

14 Fill in the missing digits.

a
```
  6 4   7 3
-     1 3
  5 9 9   7
```
d
```
  8 2     3
-     7 3 6
  2 1 7 4
```
g
```
  4   2 3 0
-   5   2
  4 3 6   6
```
b
```
  7   9 2 3
-   4 6   4
  6 1   6
```
e
```
  4 5 3   9
-     5 4
  4 1   8 1
```
h
```
  9   4   6
- 4 7 3 2
    0   9 9
```
c
```
  9     1 1
- 7 3   2 7
    7 6   3
```
f
```
  2 5   6
-   1 3 7 4
  2   5   9
```
i
```
  7   4 9
-   4     8
  5 8 2 9 7
```


CATCH UP MATHS YEAR 6 BOOK A © PASCAL PRESS ISBN: 9781925726183

Here is Christian's maths test. Use addition to check his answers in the space below. Tick the correct answers and cross the wrong ones, and give him a mark out of 8.

Christian's maths test

1
$$\begin{array}{r} 723 \\ -\ 12 \\ \hline 611 \end{array}$$ ☐

2
$$\begin{array}{r} 897 \\ -\ 436 \\ \hline 463 \end{array}$$ ☐

3
$$\begin{array}{r} 6123 \\ -\ 759 \\ \hline 6636 \end{array}$$ ☐

4
$$\begin{array}{r} 7646 \\ -\ 4232 \\ \hline 3418 \end{array}$$ ☐

5
$$\begin{array}{r} 58250 \\ -\ 14623 \\ \hline 43637 \end{array}$$ ☐

6
$$\begin{array}{r} \$140.00 \\ -\ \$72.58 \\ \hline \$72.42 \end{array}$$ ☐

7
$$\begin{array}{r} \$250.00 \\ -\ \$65.43 \\ \hline \$195.67 \end{array}$$ ☐

8
$$\begin{array}{r} 725136 \\ -\ 147348 \\ \hline 687788 \end{array}$$ ☐

☐ out of 8

Add to check.

1
$$\begin{array}{r} \\ +\ 12 \\ \hline \end{array}$$

2
$$\begin{array}{r} \\ +\ 436 \\ \hline \end{array}$$

3
$$\begin{array}{r} \\ +\ 759 \\ \hline \end{array}$$

4
$$\begin{array}{r} \\ +\ 4232 \\ \hline \end{array}$$

5
$$\begin{array}{r} \\ +\ 14623 \\ \hline \end{array}$$

6
$$\begin{array}{r} \$\quad . \\ +\ \$72.58 \\ \hline \$\quad . \end{array}$$

7
$$\begin{array}{r} \$\quad . \\ +\ \$65.43 \\ \hline \$\quad . \end{array}$$

8
$$\begin{array}{r} \\ +\ 147348 \\ \hline \end{array}$$

PRODUCT, FACTORS & MULTIPLES

When you multiply numbers, the answer is called the **product**.
A **factor** is a number that can multiply with another to give a multiple.
A **multiple** is the answer you get when you multiply two numbers.

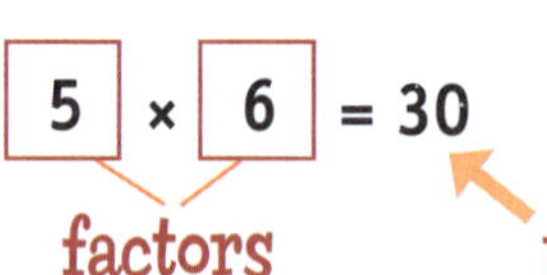

30 is also a multiple.
30 is a multiple of 5 because 5 × 6 = 30.
30 is a multiple of 6 because 6 × 5 = 30.

Example 1: What are the factors of 30?

1 × 30 = 30
2 × 15 = 30
3 × 10 = 30
5 × 6 = 30

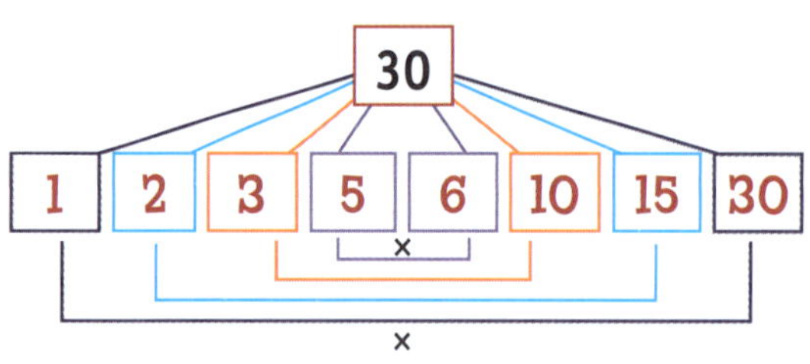

1, 2, 3, 5, 6, 10, 15 and 30 are the factors of 30.

30 is a multiple of 1, 2, 3, 5, 6, 10, 15 and 30.

Example 2:
What are the factors of 12?

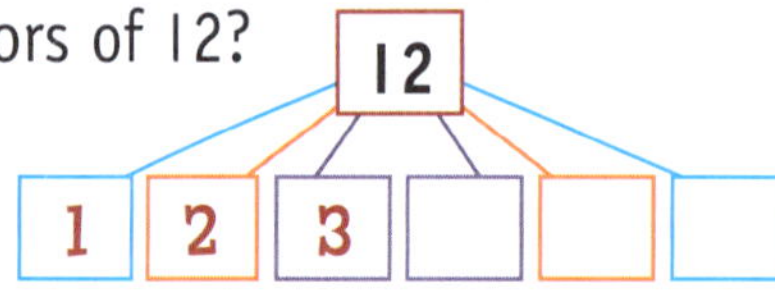

Example 3:
What are the factors of 9?

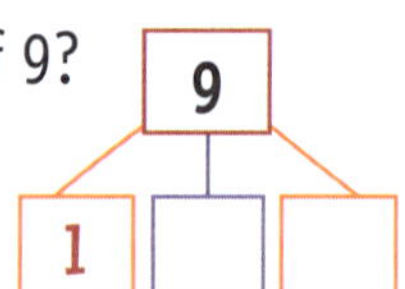

Example 4:
What are the factors of 16?

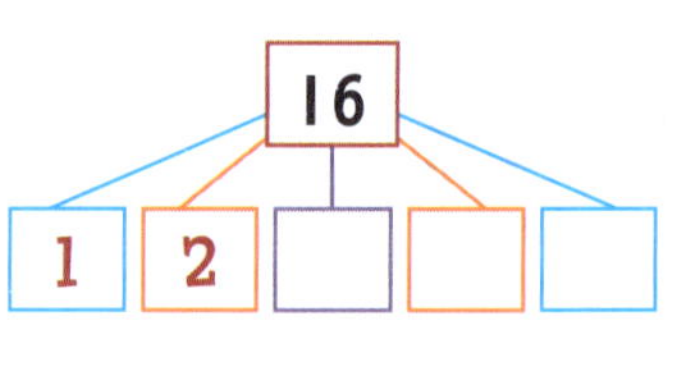

Your turn

1 Cross out the number that is not a factor of the number in the box.

● 12 1, 2, 3, 4, 6, 9, 12

b 9 1, 2, 3, 9

a 16 1, 2, 3, 4, 8, 16

2 Circle the number that is not a multiple of the number in the box.

● 2 1, 2, 4, 10, 12

b 10 10, 20, 30, 45, 50

a 5 5, 15, 21, 25, 35

c 6 6, 18, 24, 27, 30

Check your answers
How many did you get correct?

CATCH UP MATHS YEAR 6 BOOK A © PASCAL PRESS ISBN: 9781925726183

PRACTICE

1 Write the missing factors.

- 7 × _9_ = 63
- a 12 × ___ = 132
- b 8 × ___ = 48
- c 9 × ___ = 18
- d ___ × 4 = 32
- e ___ × 7 = 49
- f 8 × ___ = 72
- g 12 × ___ = 120
- h 6 × ___ = 36
- i 5 × ___ = 40
- j ___ × 9 = 63
- k 10 × ___ = 110
- l 11 × ___ = 110
- m ___ × 6 = 60
- n ___ × 7 = 42
- o 9 × ___ = 81
- p ___ × 7 = 28
- q 12 × ___ = 84
- r ___ × 12 = 12
- s ___ × 9 = 54
- t 4 × ___ = 24

2 Write the product.

- 7 and 3 _21_
- a 4 and 6 ___
- b 9 and 3 ___
- c 5 and 4 ___
- d 11 and 8 ___
- e 10 and 4 ___
- f 6 and 6 ___
- g 4 and 2 ___
- h 3 and 10 ___
- i 8 and 8 ___
- j 7 and 7 ___
- k 0 and 9 ___

3 Write the next five multiples.

- 4: 4, _8_, _12_, _16_, _20_, _24_
- a 6: 6, ___, ___, ___, ___, ___
- b 9: 9, ___, ___, ___, ___, ___
- c 10: 10, ___, ___, ___, ___, ___
- d 12: 12, ___, ___, ___, ___, ___
- e 7: 7, ___, ___, ___, ___, ___
- f 1: 1, ___, ___, ___, ___, ___
- g 8: 8, ___, ___, ___, ___, ___
- h 11: 11, ___, ___, ___, ___, ___
- i 5: 5, ___, ___, ___, ___, ___

4 Write the factors.

- 6: _1_, _2_, _3_, _6_
- a 14: ___, ___, ___, ___
- b 8: ___, ___, ___, ___
- c 10: ___, ___, ___, ___
- d 15: ___, ___, ___, ___
- e 24: ___, ___, ___, ___, ___, ___, ___, ___
- f 20: ___, ___, ___, ___, ___, ___
- g 36: ___, ___, ___, ___, ___, ___, ___, ___, ___
- h 50: ___, ___, ___, ___, ___, ___
- i 100: ___, ___, ___, ___, ___, ___, ___, ___, ___
- j 60: ___, ___, ___, ___, ___, ___, ___, ___, ___, ___, ___, ___

MULTIPLYING 2-DIGIT NUMBERS BY 1-DIGIT NUMBERS

Here are three different ways to multiply two-digit numbers by one-digit numbers.

Example 1: 23 × 6

Use known facts

23 × 6

20 × 6 = 120

120 + 6 + 6 + 6 (3 lots of 6)

= 138

Multiply by place value

23 × 6

= 2 tens × 6 + 6 threes

= 120 + 18

= 138

Use an area model

23 × 6

	20	3
6	120	18

120 + 18

= 138

Example 2: 46 × 5

Use known facts

46 × 5

40 × __ = ____

200 + __ + __ + __ + __ + __ + __ (6 lots of 5)

= ____

Multiply by place value

46 × 5

= 4 tens × __ + 5 ______

= ____ + ____

= ____

Use an area model

46 × 5

	40	6
5		

____ + ____

= ____

Check your answer on the video!

Your turn

Solve using the three different methods.

a 32 × 4

30 × 4 = ______

______

______

32 × 4

3 tens × 4 + ______

______

______

32 × 4

	30	2
4		

______

______

b 48 × 3

40 × 3 = ______

______

______

48 × 3

4 tens × 3 + ______

______

______

48 × 3

	40	8
3		

______

______

SELF CHECK Tick how you feel

Got it!	Need help...	I don't get it
 ☐	 ☐	☐

Check your answers

How many did you get correct? ☐

CATCH UP MATHS YEAR 6 BOOK A © PASCAL PRESS ISBN: 9781925726183

PRACTICE

1 Solve using the three different ways to multiply.

Use known facts	Multiply by place value	Use an area model
27 × 4 20 × 4 = 80 80 + 4 + 4 + 4 + 4 + 4 + 4 + 4 80 + 28 = 108	27 × 4 2 tens × 4 + 4 sevens = 80 + 28 = 108	27 × 4 20 7 4 | 80 | 28 80 + 28 = 108
a 64 × 2 ________ ________ ________	64 × 2 ________ ________ ________	64 × 2 ________
b 34 × 3 ________ ________ ________	34 × 3 ________ ________ ________	34 × 3 ________
c 71 × 6 ________ ________ ________	71 × 6 ________ ________ ________	71 × 6 ________
d 45 × 6 ________ ________ ________	45 × 6 ________ ________ ________	45 × 6 ________
e 52 × 3 ________ ________ ________	52 × 3 ________ ________ ________	52 × 3 ________

FORMAL ALGORITHMS

We can solve multiplication problems using an algorithm.

SCAN to watch video

Example 1:

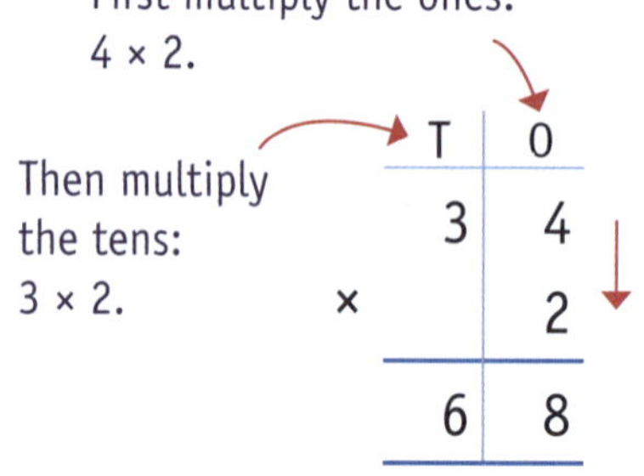

	T	O
	3	4
×		2
	6	8

Example 2:

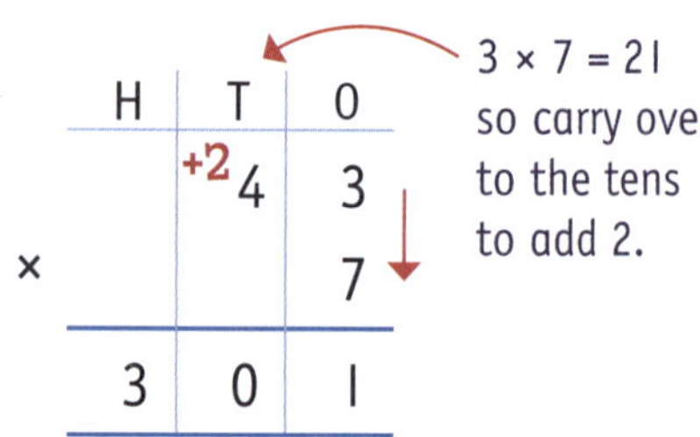

	H	T	O
		+2 4	3
×			7
	3	0	1

Example 3:

	H	T	O
		2	6
×			1
			6

Example 4:

	H	T	O
		6	3
×			2
			6

Example 5:

	H	T	O
		+2 4	4
×			5
			0

Example 6:

	H	T	O
		+2 6	7
×			4
			8

Solve these multiplications.

●

	H	T	O
		5	3
×			2
	1	0	6

a

	H	T	O
		7	1
×			4

b

	H	T	O
		3	7
×			5

c

	H	T	O
		4	9
×			3

d

	H	T	O
		8	5
×			5

e

	H	T	O
		1	6
×			7

Check your answers
How many did you get correct?

  ISBN: 9781925726183

PRACTICE

1 Find the product.

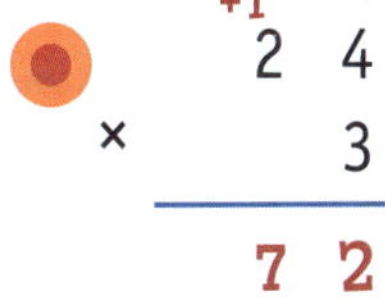

- Example: +1 over the 2; 24 × 3 = **72**
- a 32 × 3 = ____
- b 41 × 9 = ____
- c 73 × 3 = ____
- d 42 × 3 = ____
- e 91 × 5 = ____
- f 83 × 3 = ____
- g 52 × 3 = ____
- h 27 × 8 = ____
- i 46 × 4 = ____
- j 53 × 9 = ____
- k 37 × 6 = ____
- l 69 × 3 = ____
- m 74 × 6 = ____
- n 85 × 5 = ____
- o 93 × 4 = ____

2 Colour the correct answer.

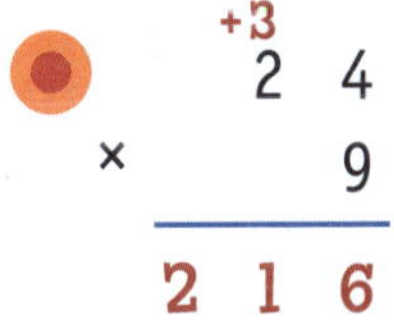

	Problem	Options	
Example	+3 over the 2; 24 × 9 = **216**	216 (coloured)	261
a	73 × 6 = ____	483	438
b	29 × 4 = ____	161	116
c	95 × 6 = ____	570	571
d	37 × 8 = ____	269	296
e	31 × 7 = ____	217	271
f	82 × 3 = ____	246	264
g	44 × 9 = ____	396	369
h	63 × 6 = ____	387	378

MULTIPLY 3-DIGIT & 4-DIGIT NUMBERS BY 1-DIGIT NUMBERS

Here are three ways to solve multiplication algorithms when you are multiplying three-digit and four-digit numbers by one-digit numbers.

Example 1:

Multiply by place value

572 × 3

= (500 × 3) + (70 × 3) + (2 × 3)

= 1500 + 210 + 6

= 1716

Formal Algorithm

572 × 3

$$\begin{array}{r} {}^{+2}5\ \ 7\ \ 2 \\ \times \qquad\ \ 3 \\ \hline 1\ \ 7\ \ 1\ \ 6 \\ \hline \end{array}$$

Area Model

572 × 3

	500	70	2
3	1500	210	6

1500 + 210 + 6

= 1716

Example 2:

Multiply by place value

2436 × 5

= (2000 × 5) + (400 × 5) + (30 × 5) + (6 × 5)

= 10 000 + 2000 + 150 + 30

= 12 180

Formal Algorithm

2436 × 5

$$\begin{array}{r} {}^{+2}2\ \ {}^{+1}4\ \ {}^{+3}3\ \ 6 \\ \times \qquad\qquad\ \ 5 \\ \hline 1\ \ 2\ \ 1\ \ 8\ \ 0 \\ \hline \end{array}$$

Area Model

2436 × 5

	2000	400	30	6
5	10 000	2000	150	30

10 000 + 2000 + 150 + 30

= 12 180

Example 3:

Multiply by place value

649 × 3

= (600 × __) + (40 × __) + (__ × 3)

= ______ + _____ + ____

= ______

Formal Algorithm

649 × 3

$$\begin{array}{r} {}^{+1}6\ \ {}^{+2}4\ \ 9 \\ \times \qquad\ \ 3 \\ \hline 7 \\ \hline \end{array}$$

Area Model

649 × 3

	600	40	9
3	1800		

1800 + _____ + ____

= ______

CATCH UP MATHS YEAR 6 BOOK A © PASCAL PRESS ISBN: 9781925726183

Solve.

351 × 4

= (300 × 4) + (50 × 4) + (1 × 4)

= 1200 + 200 + 4

= 1404

```
   +2
    3  5  1
×         4
-----------
 1  4  0  4
```

351 × 4

	300	50	1
4	1200	200	4

1200 + 200 + 4

= 1404

a 5932 × 4

= ______

= ______

= ______

```
    5  9  3  2
×            4
--------------

--------------
```

5932 × 4

	___	___	___	___
___				

______

= ______

b 3842 × 5

= ______

= ______

= ______

```
    3  8  4  2
×            5
--------------

--------------
```

3842 × 5

	___	___	___	___
___				

______

= ______

SELF CHECK Tick how you feel

Got it!	Need help...	I don't get it
 ☐	 ☐	☐

Check your answers
How many did you get correct? ☐

PRACTICE

1 Solve by multiplying by place value.

● 8247 × 2

= (8000 × 2) + (200 × 2) + (40 × 2) + (7 × 2)

= 16 000 + 400 + 80 + 14

= 16 484

a 7356 × 3

= ______________________

= ______________________

= __________

b 654 × 6

= ______________________

= ______________________

= __________

c 429 × 4

= ______________________

= ______________________

= __________

d 8945 × 8

= ______________________

= ______________________

= __________

2 Solve these formal algorithms.

●
```
     +1 +3
  3  1  2  5
×          7
-----------
2  1  8  7  5
```

b
```
  2  1  9  7
×          6
-----------

```

d
```
  7  2  0  5
×          5
-----------

```

a
```
     4  5  7
×          3
-----------

```

c
```
     8  9  3
×          4
-----------

```

e
```
     5  6  9
×          3
-----------

```


CATCH UP MATHS YEAR 6 BOOK A © PASCAL PRESS ISBN: 9781925726183

f
```
  6 9 1 0
×       8
_________

_________
```

g
```
    7 5 4
×       9
_________

_________
```

h
```
  4 6 0 7
×       2
_________

_________
```

Solve using area models.

● 3215 × 3

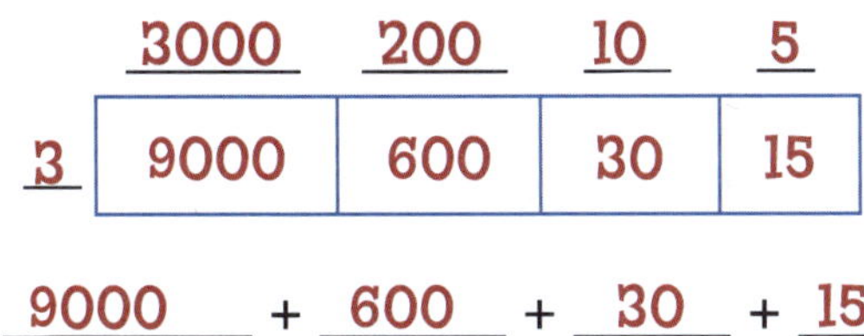

9000 + 600 + 30 + 15

= 9645

a 4974 × 5

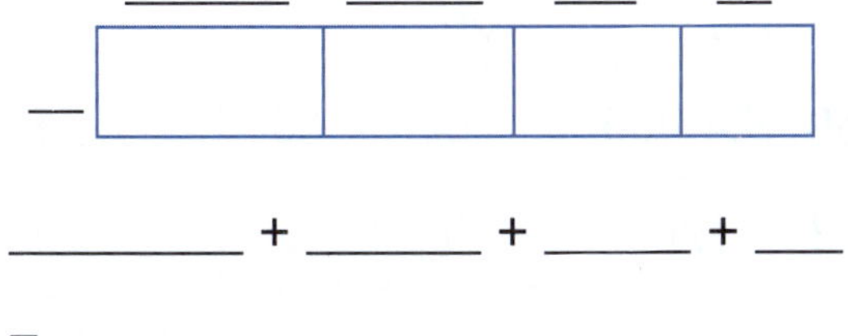

___ + ___ + ___ + ___

= ___

b 347 × 8

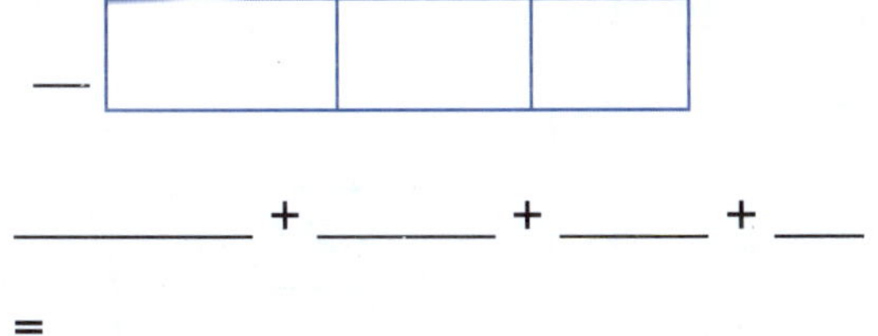

___ + ___ + ___ + ___

= ___

c 7963 × 8

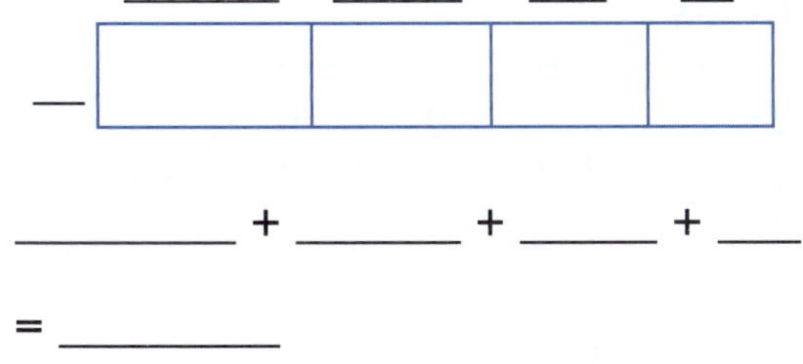

___ + ___ + ___ + ___

= ___

d 405 × 4

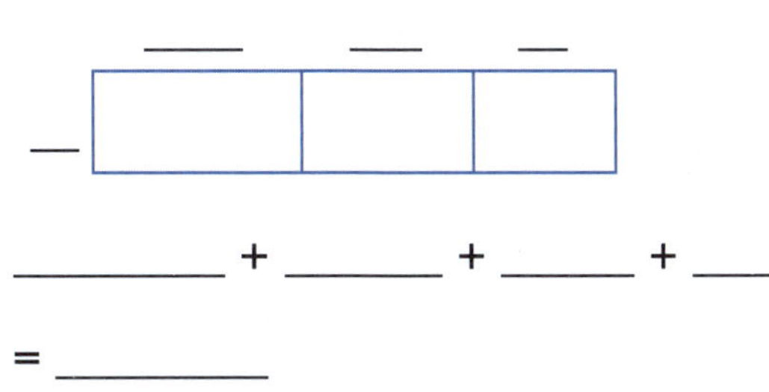

___ + ___ + ___ + ___

= ___

e 6490 × 9

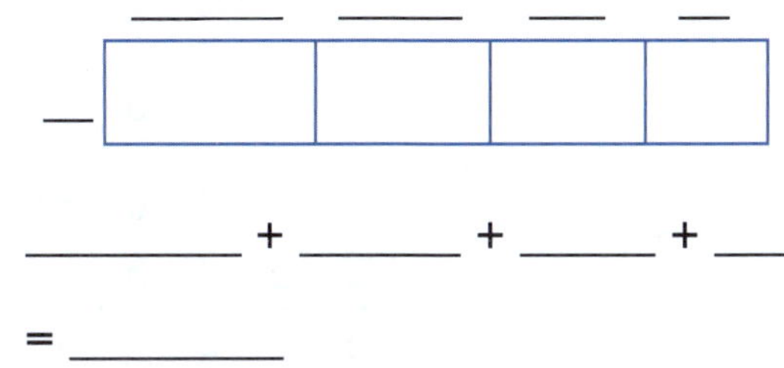

___ + ___ + ___ + ___

= ___

f 620 × 7

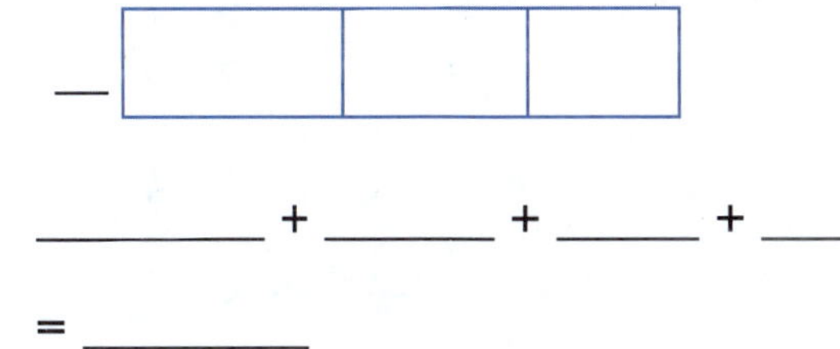

___ + ___ + ___ + ___

= ___

g 1599 × 6

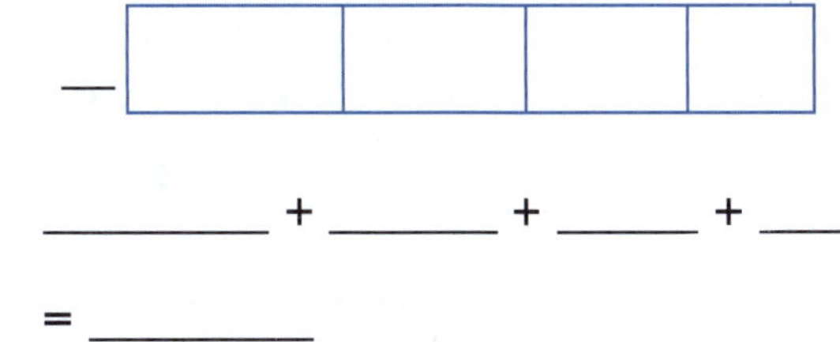

___ + ___ + ___ + ___

= ___

MULTIPLY 2-DIGIT AND 3-DIGIT NUMBERS BY 2-DIGIT NUMBERS

Here are two ways to solve multiplication algorithms when you are multiplying two-digit and three-digit numbers by two-digit numbers.

Long Multiplication | **Area Model**

Example 1:
46 × 25

Long Multiplication:

```
   +1
   +3
    4 6
×   2 5
-------
  2 3 0
+ 9 2 0
-------
1 1 5 0
```

Area Model:

	40	6	
20	800	120	920
5	200	30	230
			1150

920 + 230 = 1150

Example 2:
421 × 63

Long Multiplication:

```
     +1
      4 2 1
×       6 3
-----------
    1 ¹2 6 3
+ 2 5 2 6 0
-----------
  2 6 5 2 3
```

Area Model:

	400	20	1	
60	24 000	1200	60	25 ¹260
3	1200	60	3	1 263
				26 523

25 260 + 1263 = 26 523

Example 3:
57 × 32

Long Multiplication:

```
   +2
   +1
    5 7
×   3 2
-------
  1 1 4
+     0
-------

-------
```

Area Model:

	50	7	
30	1500		
2		14	______
			______

______ + ______ = ______

Example 4:
632 × 43

Long Multiplication:

```
    +1
      6 3 2
×       4 3
-----------
          6
+       8 0
-----------

-----------
```

Area Model:

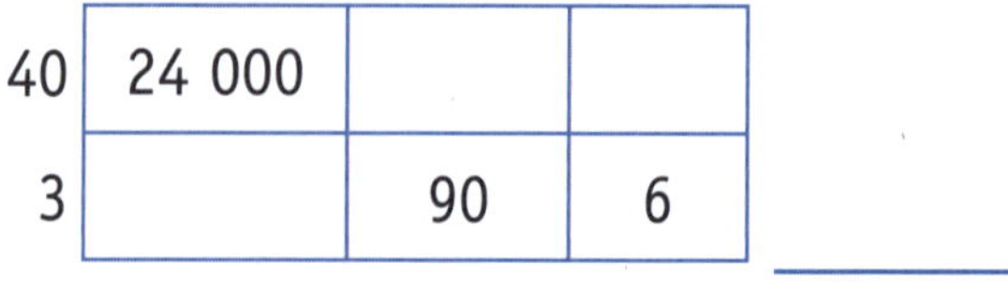

	600	30	2	
40	24 000			
3		90	6	______
				______

______ + ______ = ______

Check your answer on the video!

CATCH UP MATHS YEAR 6 BOOK A © PASCAL PRESS ISBN: 9781925726183

Solve using long multiplication and the area model.

594 × 76

```
     +6 +2
     +5 +2
      5  9  4
×        7  6
  ¹3 ¹5  6  4
+ 4  1  5  8  0
  4  5  1  4  4
```

	500	90	4
70	35 000	6300	280
6	3000	540	24

41 580
3 564
45 144

41 580 + 3564 = 45 144

a 64 × 38

```
      6  4
×     3  8
________
+ ________
  ________
```

	60	4
30		
8		

______
______

______ + ______ = ______

b 749 × 23

```
      7  4  9
×        2  3
________
+ ________
  ________
```

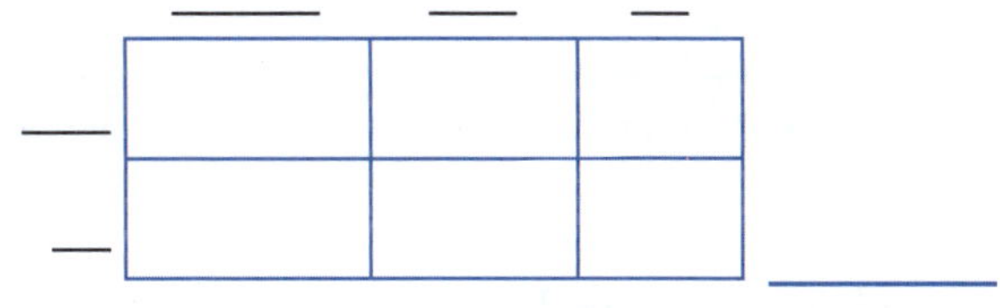

	___	___	___
___			
___			

______
______

______ + ______ = ______

c 93 × 46

```
      9  3
×     4  6
________
+ ________
  ________
```

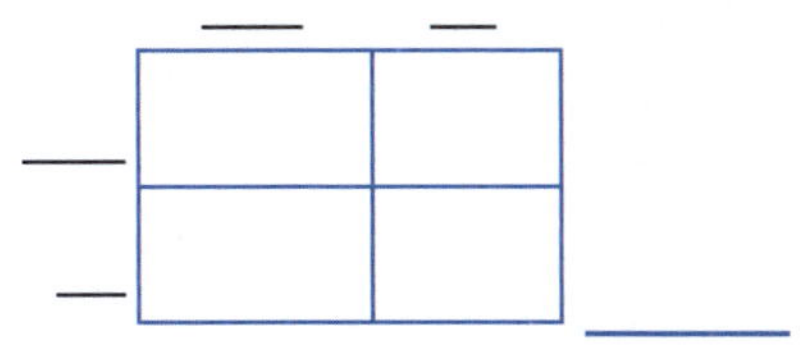

	___	___
___		
___		

______
______

______ + ______ = ______

SELF CHECK Tick how you feel

Got it!	Need help...	I don't get it
 ☐	 ☐	 ☐

Check your answers
How many did you get correct? ☐

PRACTICE

1 Solve using long multiplication.

Example:

```
      +1
   +2 +4
    4  2  6
×      3  8
-----------
  ¹3  4  0  8
+ 1  2  7  8  0
-----------
  1  6  1  8  8
```

a 271 × 43

b 593 × 97

c 42 × 99

d 81 × 37

e 59 × 66

f 846 × 59

g 709 × 42

h 32 × 48

2 Solve using area models.

Example: 374 × 42

	300	70	4	
40	12 000	2800	160	14 960
2	600	140	8	748
				15 708

14 960 + 748 = 15 708

a 38 × 43

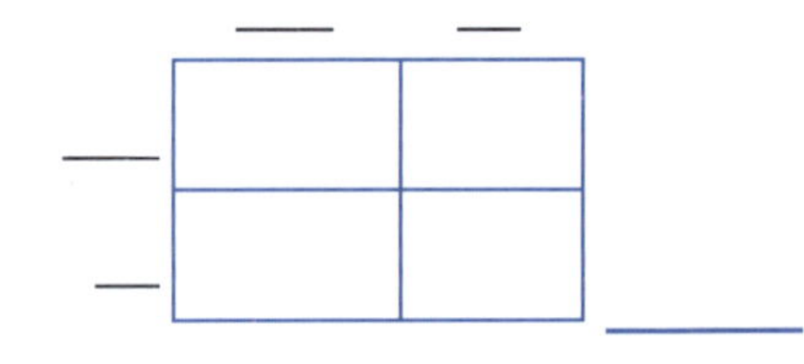

____ + ____ = ____

CATCH UP MATHS YEAR 6 BOOK A © PASCAL PRESS ISBN: 9781925726183

b 92 × 56

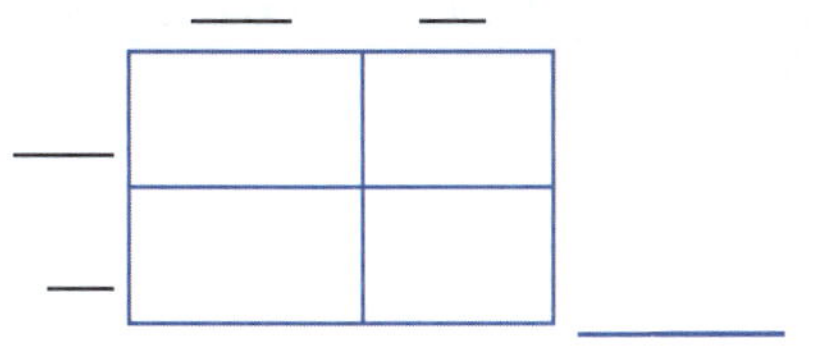

________ + ________ = ________

f 809 × 47

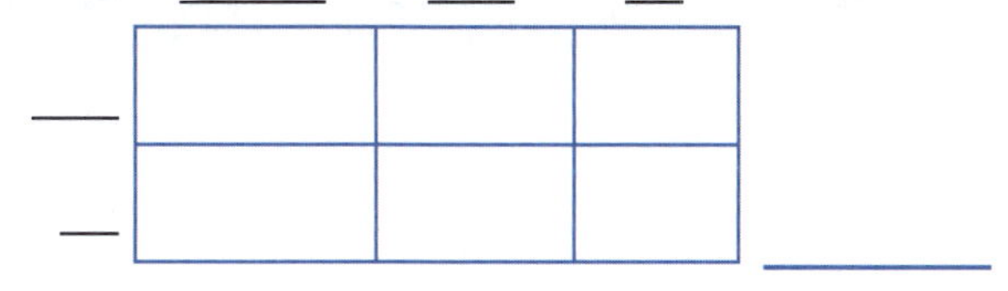

________ + ________ = ________

c 78 × 31

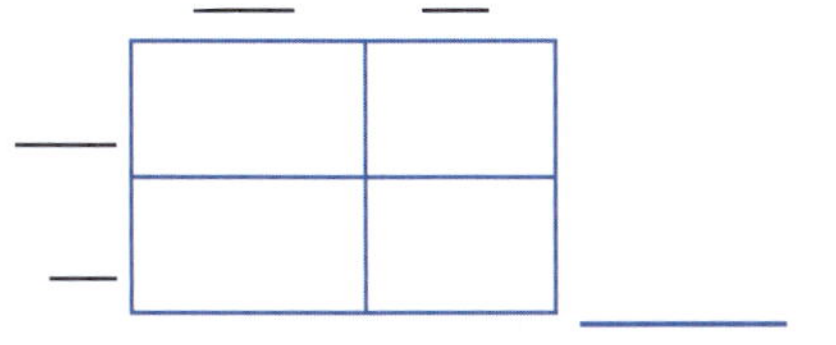

________ + ________ = ________

g 749 × 56

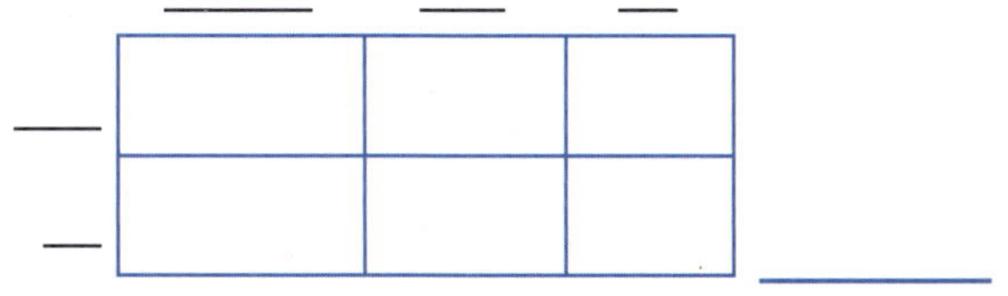

________ + ________ = ________

d 54 × 37

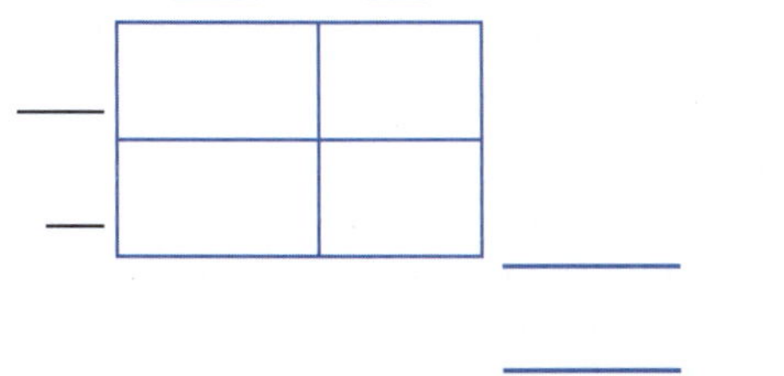

________ + ________ = ________

h 520 × 68

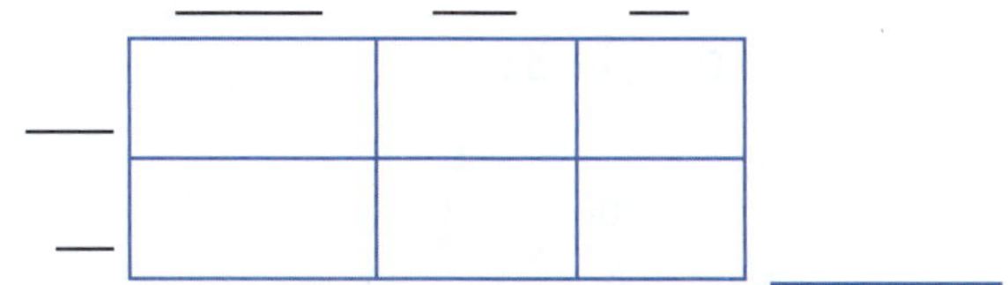

________ + ________ = ________

e 942 × 36

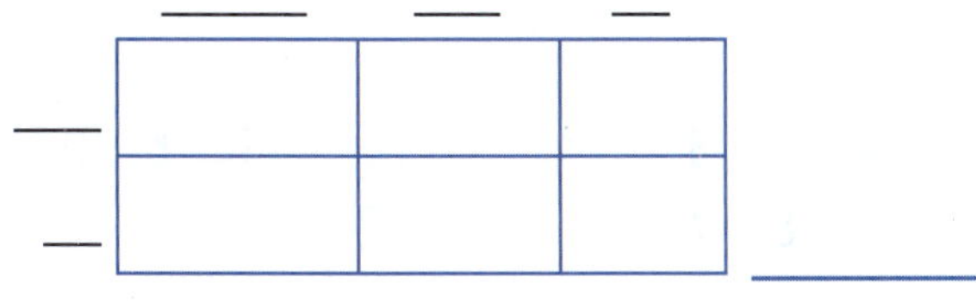

________ + ________ = ________

i 386 × 49

________ + ________ = ________

MULTIPLY 4-DIGIT NUMBERS BY 1-DIGIT AND 2-DIGIT NUMBERS

Here are examples of using formal algorithms to solve multiplication problems with four-digit numbers.

4-digit × 1-digit

Example 1:

```
    7 3 4 2
×         2
-----------
  1 4 6 8 4
```

Example 2:

```
   +3 +1 +3
    6 8 4 8
×         4
-----------
  2 7 3 9 2
```

Example 3:

```
       +1
    6 2 1 3
×         4
-----------
          2
```

4-digit × 2-digit

Example 4:

```
    +1    +1
    +2 +1 +3
     5  6  3  8
×           2  4
---------------
     2 ¹2 ¹5 5 2
+ 1  1  2  7 6 0
----------------
  1  3  5  3 1 2
```

Example 5:

```
     7  4  1  2
×           2  2
---------------
    ¹1 ¹4  8  2  4
+ 1  4  8  2  4  0
------------------
  1  6  3  0  6  4
```

Example 6:

```
   +1
    2 9 6 3
×       4 2
-----------
        2 6
+         0
-----------
          6
```

Example 7:

```
  +1
    4 8 2 0
×       5 2
-----------
          0
+         0
-----------
          0
```

Your turn Solve these multiplications.

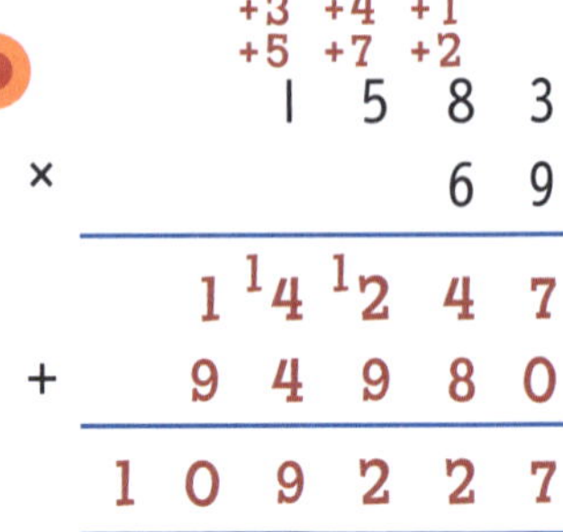

```
   +3 +4 +1
   +5 +7 +2
    1  5  8  3
×           6  9
----------------
   1 ¹4 ¹2  4  7
+  9  4  9  8  0
----------------
 1 0  9  2  2  7
```

a

```
    2 4 2 3
×       8 7
-----------
+
-----------

-----------
```

b

```
    5 6 0 5
×       4 9
-----------
+
-----------

-----------
```

SELF CHECK Tick how you feel

Got it!	Need help...	I don't get it
☐	☐	☐

Check your answers
How many did you get correct? ☐

CATCH UP MATHS YEAR 6 BOOK A © PASCAL PRESS ISBN: 9781925726183

PRACTICE

1 Solve.

Example: $2543 \times 8 = 20344$ (carries: +4 +3 +2)

c $6495 \times 3 =$ ______

f $2103 \times 2 =$ ______

a $3559 \times 6 =$ ______

d $5899 \times 7 =$ ______

g $6534 \times 4 =$ ______

b $2160 \times 8 =$ ______

e $1606 \times 5 =$ ______

h $9373 \times 9 =$ ______

2 Solve.

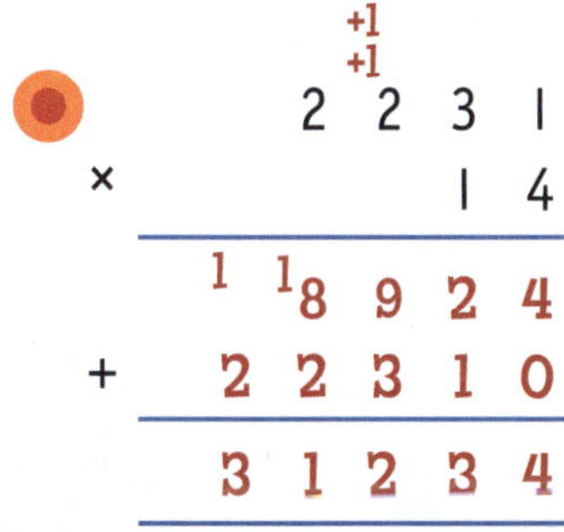

c 5787×33: ______ + ______ = ______

f 3099×80: ______ + ______ = ______

a 8120×32: ______ + ______ = ______

d 6876×54: ______ + ______ = ______

g 4650×17: ______ + ______ = ______

b 7908×20: ______ + ______ = ______

e 9342×91: ______ + ______ = ______

h 4631×85: ______ + ______ = ______

MULTIPLY BY 10, 100 AND 1000

When you multiply numbers by 10, add one zero
by 100, add two zeros
by 1000, add three zeros.

Example 1:

493 × 10 = 4930 ← Add one zero

493 × 100 = 49 300 ← Add two zeros

493 × 1000 = 493 000 ← Add three zeros

Example 2: 327

327 × 10 = 3270

327 × 100 = 32 700

327 × 1000 = 327 000

Example 3: 459

459 × 10 = 4590

459 × 100 = 45 900

459 × 1000 = 459 000

Example 4: 673

673 × 10 = 6730

673 × 100 = 67 300

673 × 1000 = 673 000

Example 5: 462

462 × 10 = 462_

462 × 100 = 46 2_ _

462 × 1000 = 462 000

Example 6: 840

840 × 10 = 840_

840 × 100 = 84 000

840 × 1000 = 840 _ _ _

Example 7: 731

731 × 10 = 7310

731 × 100 = 73 1_ _

731 × 1000 = 731 _ _ _

Check your answer on the video!

1 Multiply by 10.

● 75 — 750

a 49 ______

b 647 ______

c 182 ______

2 Multiply by 100.

● 69 — 6900

a 741 ______

b 176 ______

c 593 ______

3 Multiply by 1000.

● 747 — 747 000

a 623 ______

b 7 ______

c 42 ______

SELF CHECK Tick how you feel

Got it!	Need help...	I don't get it
☐	☐	☐

Check your answers

How many did you get correct? ☐

CATCH UP MATHS YEAR 6 BOOK A © PASCAL PRESS ISBN: 9781925726183

PRACTICE

1 Multiply by 10.

● 829	8290	d 5	______	h 506	______
a 63	______	e 711	______	i 2193	______
b 471	______	f 100	______	j 4571	______
c 109	______	g 16	______	k 8623	______

2 Multiply by 100.

● 735	73 500	d 40	______	h 100	______
a 2	______	e 119	______	i 2903	______
b 275	______	f 175	______	j 98	______
c 877	______	g 630	______	k 5215	______

3 Multiply by 1000.

● 497	497 000	d 103	______	h 208	______
a 7	______	e 17	______	i 5387	______
b 24	______	f 1582	______	j 574	______
c 520	______	g 2711	______	k 6258	______

4 Complete the table.

	Number	× 10	× 100	× 1000
●	209	2090	20 900	209 000
a	615			
b	26			
c	416			
d	8			
e	392			
f	815			
g	52			
h	607			

ORDER OF OPERATIONS

There are rules about the order for doing operations. Follow the rules when there is more than one operation in an algorithm.

Rules

1 Always do the work inside brackets ( ) first.
2 Orders, such as square numbers like 4^2
3 Multiplication × and Division ÷
4 Addition + and Subtraction −
5 Work out operations from left to right.

Example 1:

$(4 \times 5) + 3^2 + 6 - 9$
$= 20 + 9 + 6 - 9$
$= 26$

Example 2:

$3 \times 8 \div 2 - 4$
$= 24 \div 2 - 4$
$= 12 - 4$
$= 8$

Example 3:

$3 + 7 \times 9$
$= 3 + 63$
$= 66$

Example 4:

$39 + 7 \times (10 - 7)$
$= 39 + 7 \times$ (__)
$= 39 +$ ____
$=$ ____

Example 5:

$(13 + 8) \times (18 \div 2)$
= (____) × (____)
$= 189$

Example 6:

$86 - 4 \times (14 - 8)$
$= 86 - 4 \times$ (____)
$= 86 -$ ____
$=$ ____

Check your answer on the video!

Solve the following.

$(100 - 25) \times (48 + 10)$
= 75 × 480
= 36 000

a $(100 - 75) \div (6 - 1)$
= ____________
= ____________

b $5 \times 4 \times 5 - 26$
= ____________
= ____________

c $(25 + 42) \times 4 + 3$
= ____________
= ____________

Check your answers
How many did you get correct?

CATCH UP MATHS YEAR 6 BOOK A © PASCAL PRESS ISBN: 9781925726183

PRACTICE

Use the order of operations to work out the answers.

- $(8 \times 2) - 9$

  $= 16 - 9$

  $= 7$

a $2 \times (8 - 5)$

= ______

= ______

b $(18 \div 9) \times 6 + 7$

= ______

= ______

c $7 - 10 \div 2 + 3$

= ______

= ______

d $18 \div 9 \times 6 - 4$

= ______

= ______

e $2 \times (6 \div 2)$

= ______

= ______

f $24 \div 3 \times 6 - 2 + 4$

= ______

= ______

g $20 - 6 \times 3 + 2$

= ______

= ______

h $140 \div 7 + 4 \times 3 + 5$

= ______

= ______

i $160 \div 2 + 5 \times 2 - 6$

= ______

= ______

Write the missing numbers and symbols.

- $3 \times (6 \underline{+} 7) \div 3 = 13$

a $14 - 4 \div 2 \times \underline{\quad\quad} + 5 = 15$

b $(19 \underline{\quad} 4) \div 5 + 9 = 12$

c $4 \times \underline{\quad\quad} + 27 \div 3 = 33$

d $3 \times (2 \underline{\quad} 4) \div 9 = 2$

e $20 - 6 \div 2 \times 5 \underline{\quad} 3 = 8$

f $2 + 7 \underline{\quad} 7 = 51$

g $8 \times (5 \underline{\quad} 2) = 80$

h $48 \underline{\quad} 6 \times 4 = 24$

i $38 - 4 \underline{\quad} 3 = 26$

j $4 \times (8 \underline{\quad} 4) = 16$

k $8 + (10 \underline{\quad} 2) - 3 = 10$

l $39 + 6 \times (3 \underline{\quad} 8) = 183$

m $(100 \underline{\quad} 79) \times 4 = 84$

n $(8 \underline{\quad} 7) \times (12 - 2) = 150$

o $9 + (4 \underline{\quad} 10) - 7 + 2 = 44$

MULTIPLICATION REVIEW

1 Write the missing factors.

a ___ × 5 = 15
b 2 × ___ = 14
c 9 × ___ = 27
d 8 × ___ = 72
e 4 × ___ = 20
f ___ × 6 = 42
g ___ × 8 = 24
h ___ × 11 = 132
i 9 × ___ = 54
j 8 × ___ = 40
k 7 × ___ = 42
l ___ × 4 = 0

2 What is the product?

a 3 and 9 ___
b 4 and 1 ___
c 8 and 4 ___
d 6 and 3 ___
e 2 and 0 ___
f 5 and 6 ___
g 1 and 3 ___
h 8 and 8 ___
i 1 and 11 ___
j 6 and 8 ___
k 7 and 7 ___
l 5 and 12 ___

3 Write the next 5 multiples.

a 7: 7, ___, ___, ___, ___, ___
b 3: 3, ___, ___, ___, ___, ___
c 5: 5, ___, ___, ___, ___, ___
d 2: 2, ___, ___, ___, ___, ___
e 8: 8, ___, ___, ___, ___, ___
f 4: 4, ___, ___, ___, ___, ___
g 1: 1, ___, ___, ___, ___, ___
h 6: 6, ___, ___, ___, ___, ___

4 Write the factors.

a 30

___ × ___ = ___
___ × ___ = ___
___ × ___ = ___
___ × ___ = ___

b 10

___ × ___ = ___
___ × ___ = ___

c 32

___ × ___ = ___
___ × ___ = ___
___ × ___ = ___

d 16

___ × ___ = ___
___ × ___ = ___
___ × ___ = ___

e 20

___ × ___ = ___
___ × ___ = ___
___ × ___ = ___

f 28

___ × ___ = ___
___ × ___ = ___
___ × ___ = ___

  ISBN: 9781925726183

5 Solve using these three different ways to multiply.

Use known facts	Multiply by place value	Use an area model
a 24 × 5	24 × 5	24 × 5
b 63 × 8	63 × 8	63 × 8
c 54 × 7	54 × 7	54 × 7

6 Find the product.

a 84 × 2

b 21 × 3

c 42 × 2

d 71 × 8

e 92 × 4

f 17 × 8

g 93 × 6

h 49 × 7

i 36 × 8

j 29 × 9

k 62 × 7

l 89 × 9

REVIEW

7 Solve using the three different ways to multiply.

a 249 × 4 **Multiply by Place Value**

= (200 × 4) + (40 × 4) + (9 × 4)

= ______________

= ______________

249 × 4 **Formal Algorithm**

```
    2 4 9
×       4
---------

---------
```

249 × 4 **Area Model**

	200	40	9
4			

______ + ______ + ______

= ______

b 374 × 6 **Multiply by Place Value**

= ______________

= ______________

= ______________

374 × 6 **Formal Algorithm**

```
    3 7 4
×       6
---------

---------
```

374 × 6 **Area Model**

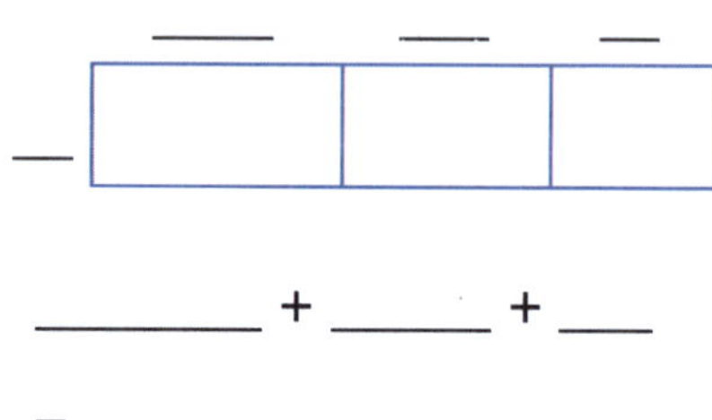

______ + ______ + ______

= ______

c 789 × 5 **Multiply by Place Value**

= ______________

= ______________

= ______________

789 × 5 **Formal Algorithm**

```
    7 8 9
×       5
---------

---------
```

789 × 5 **Area Model**

	___	___	___
___			

______ + ______ + ______

= ______

d 436 × 7 **Multiply by Place Value**

= ______________

= ______________

= ______________

436 × 7 **Formal Algorithm**

```
    4 3 6
×       7
---------

---------
```

436 × 7 **Area Model**

	___	___	___
___			

______ + ______ + ______

= ______

CATCH UP MATHS YEAR 6 BOOK A © PASCAL PRESS ISBN: 9781925726183

8 Solve these long multiplications.

a 57×45

d 46×50

g 270×89

b 42×73

e 71×43

h 503×27

c 69×96

f 437×33

i 652×46

9 Complete using area models.

a 47 × 35

	40	7
30		
5		

_____ + _____ = _____

b 62 × 33

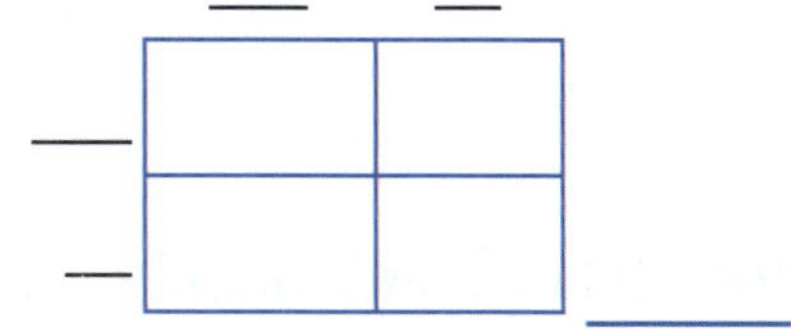

	__	__
__		
__		

_____ + _____ = _____

REVIEW

c 55 × 73

________ + ________ = ________

e 754 × 39

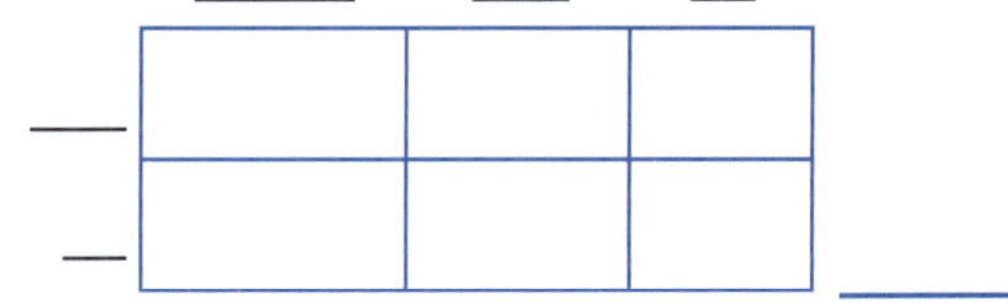

________ + ________ = ________

d 419 × 56

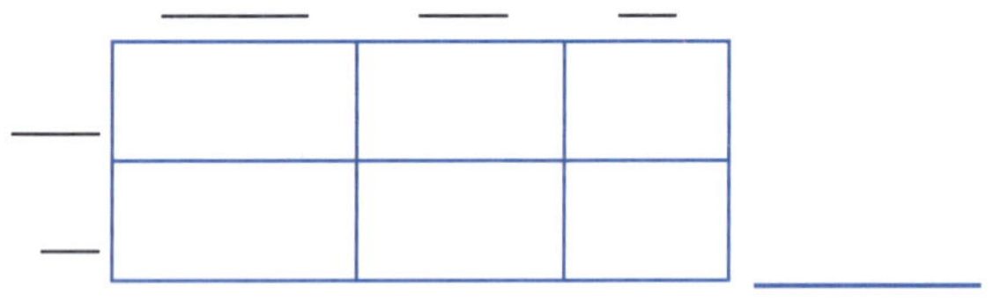

________ + ________ = ________

f 854 × 97

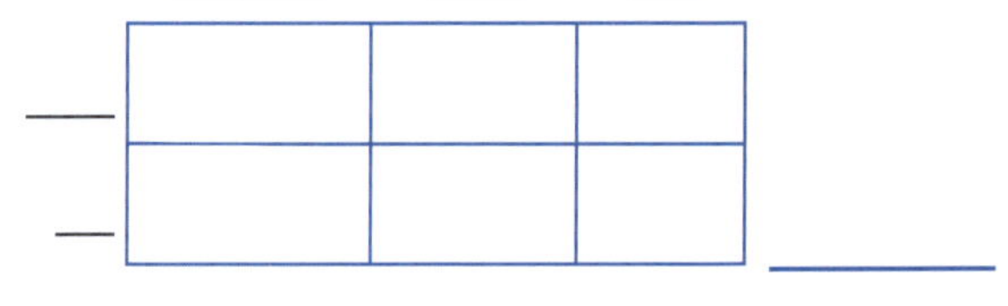

________ + ________ = ________

10 Solve these formal multiplications.

a 5962×2

c 8516×8

e 4711×7

b 4803×4

d 2462×9

f 2569×6

11 Solve these formal multiplications.

a 8938×42 +

b 6754×79 +

c 9004×26 +

CATCH UP MATHS YEAR 6 BOOK A © PASCAL PRESS ISBN: 9781925726183

d
```
      4 8 5 4
×         3 8
_____________

+ ___________
_____________
```

e
```
      6 2 5 9
×         7 4
_____________

+ ___________
_____________
```

f
```
      9 8 6 4
×         4 7
_____________

+ ___________
_____________
```

12 Fill in the table.

	Number	× 10	× 100	× 1000
a	6			
b	49			
c	317			
d	482			
e	5896			
f	4970			
g	689			
h	435			
i	28			
j	7			

13 Use the order of operations to work out the answers.

a (3 × 6) + 9 − 2 = ____

b 8 + (9 × 4) − 4 − 2 = ____

c 12 − 5 + 8 × 3 = ____

d 8 + 14 ÷ 2 − 5 = ____

e (20 + 8) ÷ 4 − 1 = ____

f 37 + 7 − (24 ÷ 6) = ____

g 50 − 15 ÷ (4 + 1) = ____

h 30 ÷ (5 + 5) × 9 + 5 = ____

i (25 + 14) − 8 + 2 = ____

j 100 − (4 × 6 + 3) = ____

k 25 + (10 − 5) × 2 + 1 = ____

l 40 + (5 + 5) × 4 = ____

m (8 + 7) × (30 ÷ 5) + 10 = ____

n (8 × 3 + 4) − 15 ÷ 3 = ____

o 8 + (3 × 7) − 4 − 5 = ____

p 40 + (7 + 3) × 3 = ____

q 50 ÷ (25 ÷ 5) + 5 × 4 = ____

r 16 − 15 ÷ 3 + 5 = ____

QUOTIENT, DIVISOR AND DIVIDEND

dividend ÷ divisor = quotient

The **dividend** is the number you are dividing.
The **divisor** is the number you are dividing by.
The **quotient** is the answer you get when you divide.

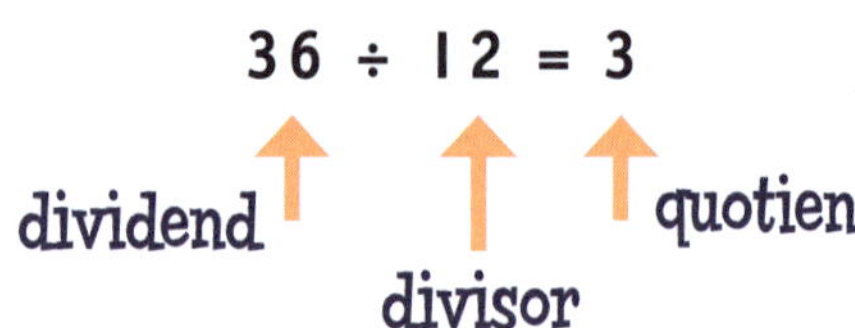

It's easier to talk about division and maths when you know what words to use.

Example 1:
The quotient is coloured blue.

25 ÷ 5 = **5**
30 ÷ 3 = **10**
144 ÷ 12 = **12**
100 ÷ 4 = **25**
90 ÷ 10 = **9**

Example 2:
Trace over the dividend in red.

24 ÷ 6 = 4
27 ÷ 3 = 9
81 ÷ 9 = 9
72 ÷ 9 = 8
60 ÷ 6 = 10

Example 3:
Use green to circle the divisor.

26 ÷ (13) = 2
144 ÷ 12 = 12
64 ÷ 8 = 8
33 ÷ 3 = 11
20 ÷ 4 = 5

Check your answer on the video!

1 Write the quotient.

- 56 ÷ 7 = 8
- **a** 24 ÷ 4 = ___
- **b** 30 ÷ 6 = ___
- **c** 32 ÷ 4 = ___
- **d** 100 ÷ 10 = ___
- **e** 16 ÷ 8 = ___

2 Write the divisor.

- 50 ÷ 5 = 10
- **a** 35 ÷ ___ = 5
- **b** 121 ÷ ___ = 11
- **c** 32 ÷ ___ = 8
- **d** 63 ÷ ___ = 7
- **e** 90 ÷ ___ = 10

3 Write the dividend.

- 108 ÷ 9 = 12
- **a** ____ ÷ 8 = 8
- **b** ____ ÷ 10 = 8
- **c** ____ ÷ 14 = 1
- **d** ____ ÷ 11 = 12
- **e** ____ ÷ 9 = 9

Check your answers
How many did you get correct?

CATCH UP MATHS YEAR 6 BOOK A © PASCAL PRESS ISBN: 9781925726183

PRACTICE

1 Circle the quotient.

- ● 6 ÷ 2 = 3
- **a** 16 ÷ 4 = 4
- **b** 70 ÷ 7 = 10
- **c** 9 ÷ 3 = 3
- **d** 21 ÷ 3 = 7
- **e** 8 ÷ 1 = 8
- **f** 4 ÷ 4 = 1
- **g** 18 ÷ 9 = 2
- **h** 40 ÷ 2 = 20
- **i** 100 ÷ 5 = 20
- **j** 32 ÷ 8 = 4
- **k** 60 ÷ 2 = 30

2 Tick the labels that have a quotient of 5.

● 25 ÷ 5 ✓ | 80 ÷ 5 | 30 ÷ 5 | 20 ÷ 4 | 45 ÷ 9 | 15 ÷ 3

40 ÷ 8 | 42 ÷ 6 | 5 ÷ 1 | 5 ÷ 5 | 30 ÷ 6 | 15 ÷ 5

3 Use red to circle the dividend and blue to circle the divisor. Then write the quotient.

- ● 8 ÷ 2 = 4
- **a** 12 ÷ 3 = ___
- **b** 81 ÷ 9 = ___
- **c** 90 ÷ 9 = ___
- **d** 30 ÷ 10 = ___
- **e** 42 ÷ 6 = ___
- **f** 70 ÷ 7 = ___
- **g** 42 ÷ 7 = ___
- **h** 120 ÷ 3 = ___
- **i** 49 ÷ 7 = ___
- **j** 21 ÷ 7 = ___
- **k** 30 ÷ 6 = ___

4 Complete the tables.

a ÷ b = quotient

	a	b	Quotient
●	12	2	6
a	4	1	
b	8	2	
c	6	2	
d	3	3	
e	7	1	

⬢ ÷ ▲ = quotient

	⬢	▲	Quotient
f	36	12	
g	49	7	
h	30	10	
i	15	5	
j	10	5	
k	8	8	

x ÷ y = quotient

	x	y	Quotient
l	42	6	
m	50	5	
n	16	4	
o	4	2	
p	30	5	
q	81	9	

5 Fill in the missing information.

	Dividend	Divisor	Quotient
●	25	5	5
a		7	4
b	63		9
c		10	10

	Dividend	Divisor	Quotient
d	169	13	
e		6	9
f	108	12	
g	72		6

FORMAL DIVISION

Formal division is where division problems are written using the $\overline{)\quad}$ symbol instead of ÷.

Example 1:

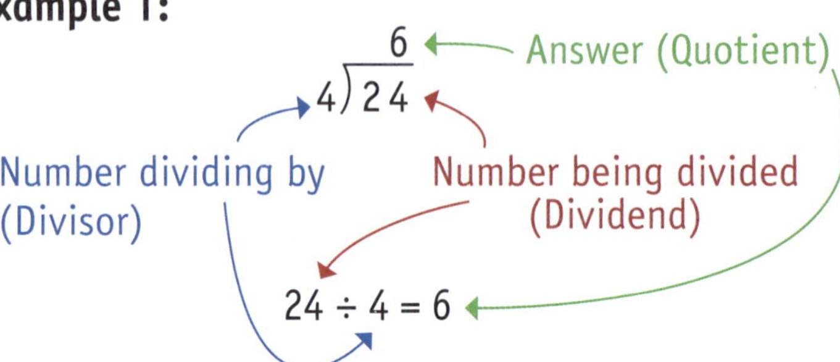

Example 2: Fill in the gaps.

$5\overline{)30}$ → 6

30 ÷ 5 = 6

6 × 5 = 30

Example 3: Fill in the gaps.

$6\overline{)42}$ → 7

42 ÷ 6 = 7

7 × 6 = 42

Example 4: Fill in the gaps.

$10\overline{)50}$

50 ÷ __ = __

5 × __ = 50

Example 5: Fill in the gaps.

$7\overline{)21}$

21 ÷ __ = __

__ × __ = 21

Write the answer.

● $7\overline{)70}$ → 10

a $8\overline{)80}$

b $9\overline{)27}$

c $4\overline{)40}$

d $6\overline{)60}$

e $9\overline{)72}$

f $10\overline{)20}$

g $9\overline{)81}$

Check your answers
How many did you get correct?

CATCH UP MATHS YEAR 6 BOOK A © PASCAL PRESS ISBN: 9781925726183

PRACTICE

1 Answer and then use multiplication to check.

	Division	Check
●	8)16 = 2	2 × 8 = 16
a	6)24	
b	8)56	
c	11)121	
d	12)108	
e	12)132	
f	7)63	

	Division	Check
g	2)8	
h	6)6	
i	10)90	
j	7)77	
k	5)50	
l	7)49	
m	3)21	

2 Fill in the missing numbers.

● 5)40 = 8

b)32 = 8

d 10)120

f 10) = 5

h)63 = 9

a 6)66

c 9) = 8

e)66 = 11

g 8) = 7

i 2) = 5

3 Use the numbers in the multiplication to complete the division algorithms.

● 2 × 8 = 16
16 ÷ 8 = 2 8)16 = 2

a 9 × 3 = 27
___ ÷ ___ = ___)

b 3 × 12 = 36
___ ÷ ___ = ___)

c 5 × 2 = 10
___ ÷ ___ = ___)

d 7 × 8 = 56
___ ÷ ___ = ___)

e 10 × 6 = 60
___ ÷ ___ = ___)

DIFFERENT WAYS TO WRITE DIVISION

Here are three different ways of writing division.

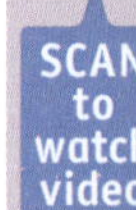

Example 1:
Write 32 divided by 4 in three different ways.

$32 \div 4$ $\quad 4\overline{)32}$ $\quad \frac{32}{4}$

These all mean the same thing: 32 shared among 4.

Example 2:
Write 36 divided by 5 in three different ways.

$36 \div 5$ $\quad 5\overline{)36}$ $\quad \frac{36}{5}$

Example 3:
Write 22 divided by 7 in three different ways.

$22 \div 7$ $\quad 7\overline{)22}$ $\quad \frac{22}{7}$

Example 4:
Write 42 divided by 6 in three different ways.

$42 \div 6$ $\quad 6\overline{)}$ $\quad \frac{42}{}$

Example 5:
Write 30 divided by 3 in three different ways.

$\_\_\_ \div 3$ $\quad 3\overline{)}$ $\quad \frac{}{3}$

Match the divisions.

● $49 \div 7$	$3\overline{)92}$
a $21 \div 4$	$\frac{20}{5}$
b $32 \div 8$	$8\overline{)32}$
c $64 \div 8$	$7\overline{)49}$
d $20 \div 5$	$\frac{47}{3}$
e $47 \div 3$	$8\overline{)64}$
f $92 \div 3$	$\frac{21}{4}$

Check your answers
How many did you get correct?

CATCH UP MATHS YEAR 6 BOOK A © PASCAL PRESS ISBN: 9781925726183

PRACTICE

1 Write the divisions as fractions.

- ● $3\overline{)42} = \frac{42}{3}$
- a $62 \div 4 =$ ___
- b $90 \div 9 =$ ___
- c $4\overline{)84} =$ ___
- d $6\overline{)36} =$ ___
- e $4\overline{)81} =$ ___
- f $21 \div 3 =$ ___
- g $2\overline{)26} =$ ___
- h $5\overline{)54} =$ ___
- i $29 \div 5 =$ ___
- j $101 \div 10 =$ ___
- k $9\overline{)72} =$ ___

2 Record the divisions using the $\overline{)}$ symbol.

- ● $37 \div 2 = 2\overline{)37}$
- a $\frac{21}{4} =$ ___
- b $83 \div 9 =$ ___
- c $\frac{70}{4} =$ ___
- d $29 \div 4 =$ ___
- e $\frac{33}{8} =$ ___
- f $121 \div 11 =$ ___
- g $91 \div 10 =$ ___
- h $63 \div 10 =$ ___
- i $87 \div 3 =$ ___
- j $\frac{32}{6} =$ ___
- k $27 \div 3 =$ ___

3 Write the divisions using the ÷ symbol.

- ● $3\overline{)29} = 29 \div 3$
- a $6\overline{)37} =$ ___
- b $\frac{43}{7} =$ ___
- c $\frac{46}{5} =$ ___
- d $4\overline{)53} =$ ___
- e $\frac{84}{5} =$ ___
- f $\frac{24}{4} =$ ___
- g $8\overline{)27} =$ ___
- h $\frac{75}{5} =$ ___
- i $6\overline{)89} =$ ___
- j $\frac{39}{3} =$ ___
- k $6\overline{)42} =$ ___

4 Fill in the tables.

	Fraction	$\overline{)}$	÷
●	$\frac{54}{5}$	$5\overline{)54}$	$54 \div 5$
a		$3\overline{)33}$	
b			$27 \div 9$
c	$\frac{63}{10}$		
d		$4\overline{)36}$	
e			$81 \div 9$

	Fraction	$\overline{)}$	÷
f			$72 \div 8$
g		$5\overline{)32}$	
h	$\frac{50}{7}$		
i		$7\overline{)56}$	
j	$\frac{44}{11}$		
k			$70 \div 10$

DIVISION WITH REMAINDERS

When a number cannot be divided exactly, the leftover is called the remainder.

Example 1:

22 ÷ 3 = 7 remainder 1

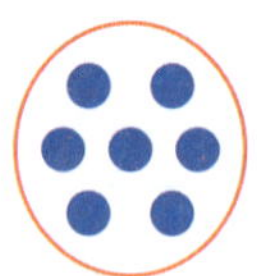
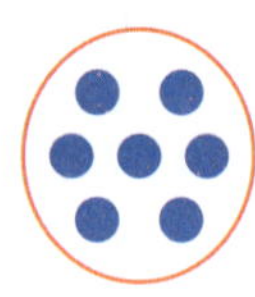
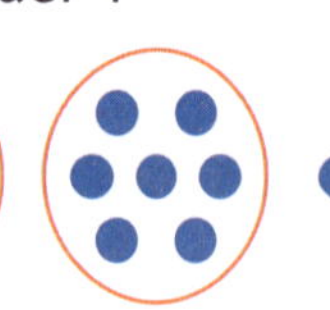

You can use multiplication to check your answer, then add the remainder at the end: 7 × 3 = 21

Add the remainder, 1: 21 + 1 = 22

Example 2: Share 32 jellybeans equally among 5 people.

32 ÷ 5 = 6 remainder 2

Check: 6 × 5 = 30

30 + 2 = 32

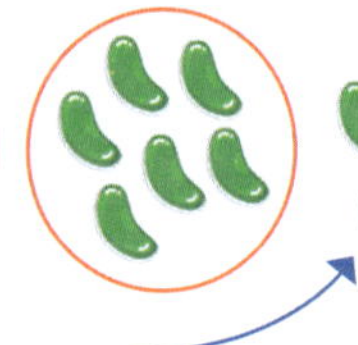

Example 3: Share 13 flowers equally among 2 people.

13 ÷ 2 = __ remainder __

Check: __ × 2 = ___

___ + ___ = 13

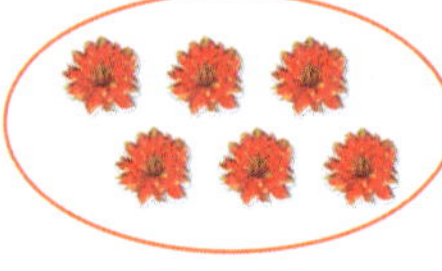

Example 4: Share 24 marbles equally among 5 people.

24 ÷ 5 = __ remainder __

Check: __ × 5 = ___

___ + ___ = 24

Solve.

28 ÷ 3

= 9 remainder 1

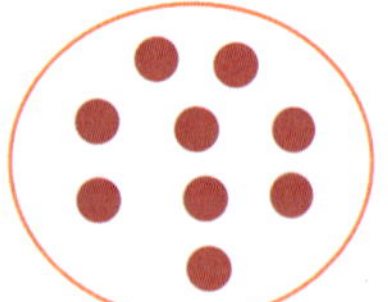
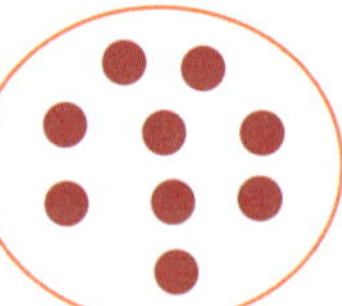
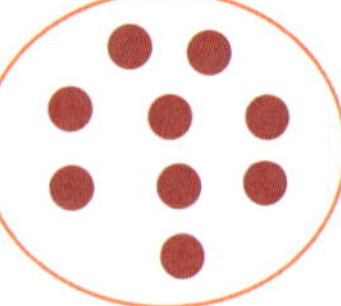

a 17 ÷ 2

= ___ remainder __

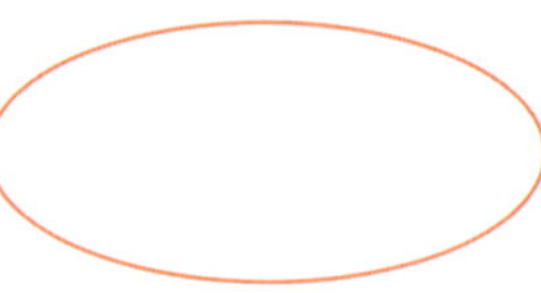
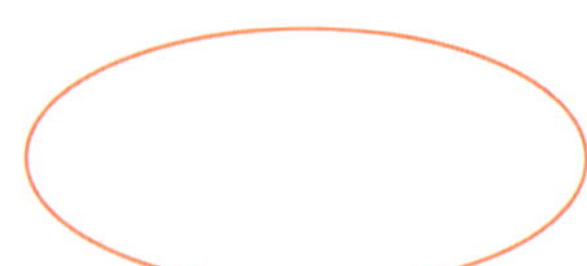

SELF CHECK Tick how you feel

Got it!	Need help...	I don't get it
 ☐	 ☐	☐

Check your answers — How many did you get correct? ☐

CATCH UP MATHS YEAR 6 BOOK A © PASCAL PRESS ISBN: 9781925726183

1 Fill in the gaps.

- 34 ÷ 10 = 3 remainder 4
- a 28 ÷ 8 = __ remainder __
- b 73 ÷ 9 = __ remainder __
- c 63 ÷ 10 = __ remainder __
- d 24 ÷ 7 = __ remainder __
- e 42 ÷ 4 = __ remainder __
- f 71 ÷ 10 = __ remainder __
- g 89 ÷ 10 = __ remainder __

2 Solve these divisions with remainders.

- 3)38 = 12 r 2
- a 7)46 = ___ r
- b 8)83 = ___ r
- c 6)27 = ___ r
- d 9)60 = ___ r
- e 4)23 = ___ r
- f 4)35 = ___ r
- g 6)74 = ___ r
- h 5)31 = ___ r
- i 4)29 = ___ r

3 Use the first number sentence to fill in the missing numbers.

- (8 × 3) + 2 = 26
  26 ÷ 3 = 8 remainder 2
- a (4 × 10) + 6 = 46
  46 ÷ 10 = __ remainder __
- b (6 × 8) + 4 = 52
  52 ÷ 8 = __ remainder __
- c (5 × 7) + 3 = 38
  38 ÷ 7 = __ remainder __

4 Write the number sentence and the answer.

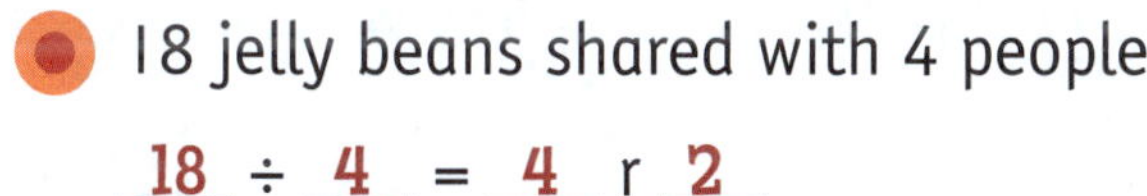

- 18 jelly beans shared with 4 people
  18 ÷ 4 = 4 r 2
- a 23 balls shared with 3 people
  ___ ÷ ___ = ___ r ___
- b 37 lollies shared with 5 people
  ___ ÷ ___ = ___ r ___
- c 26 balls shared among 7 people
  ___ ÷ ___ = ___ r ___
- d 29 apples shared among 8 people
  ___ ÷ ___ = ___ r ___
- e 74 oranges shared among 9 people
  ___ ÷ ___ = ___ r ___
- f 69 counters shared among 10 people
  ___ ÷ ___ = ___ r ___
- g 87 markers shared among 7 people
  ___ ÷ ___ = ___ r ___
- h 153 marbles shared among 12 people
  ___ ÷ ___ = ___ r ___
- i 189 pencils shared among 13 people
  ___ ÷ ___ = ___ r ___

DIVISION OF 2-DIGIT NUMBERS

Here are examples of dividing two-digit numbers by one-digit numbers.

Example 1:

Share 37 marbles among 3 people.	Give 10 marbles to each person.	Now share 7 marbles among 3 people.
$3\overline{)37}$	1 $3\overline{)37}$	1 2 r 1 $3\overline{)37}$

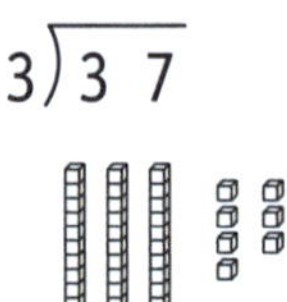

Example 2:

Share 73 balls among 3 people.	Give 20 balls to each person.	There are 13 balls left to share.
$3\overline{)73}$	2 tens → 2 $3\overline{)7\,{}^{1}3}$ ← Trade 1 ten for 10 ones	2 4 r 1 $3\overline{)7\,{}^{1}3}$

Example 3:	Example 4:	Example 5:	Example 6:
1 8 r 1 $4\overline{)7\,{}^{3}3}$	1 2 r 4 $5\overline{)6\,{}^{1}4}$	1 r $7\overline{)9\,{}^{2}1}$	1 r $5\overline{)54}$

Check your answer on the video!

Solve.

● 1 3 r 2
 $4\overline{)5\,{}^{1}4}$

	c $6\overline{)87}$	**f** $9\overline{)29}$
a $2\overline{)49}$	**d** $5\overline{)73}$	**g** $5\overline{)98}$
b $3\overline{)57}$	**e** $6\overline{)84}$	**h** $7\overline{)82}$

Check your answers
How many did you get correct?

CATCH UP MATHS YEAR 6 BOOK A © PASCAL PRESS ISBN: 9781925726183

PRACTICE

Solve these algorithms that have no remainders.

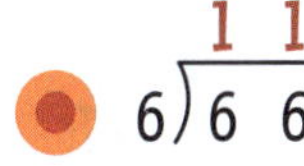

6)66 (answer 11)	b 3)39	d 6)42	f 2)74
a 7)77	c 4)48	e 7)49	g 6)54

Answer these division algorithms. You will need to trade.

3)5²4 (answer 18)	d 5)63	h 9)93	l 8)94
a 4)51	e 7)84	i 6)74	m 4)67
b 7)92	f 9)86	j 4)63	n 2)94
c 6)95	g 3)42	k 2)57	o 5)67

Check your answers to Question 2 using multiplication.
Don't forget to add the remainder.

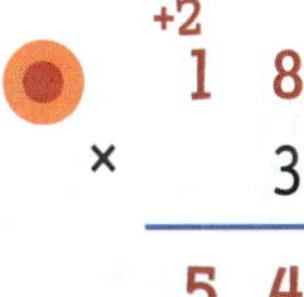

18 × 3 = 54 (+2)	d × 5	h × 9	l × 8
a × 4	e × 7	i × 6	m × 4
b × 7	f × 9	j × 4	n × 2
c × 6	g × 3	k × 2	o × 5

DIVISION OF 3-DIGIT NUMBERS

Here are some examples of dividing three-digit numbers by single-digit numbers.

Example 1: If I arrange 658 fish equally into 3 fish tanks, how many fish will be in each tank?

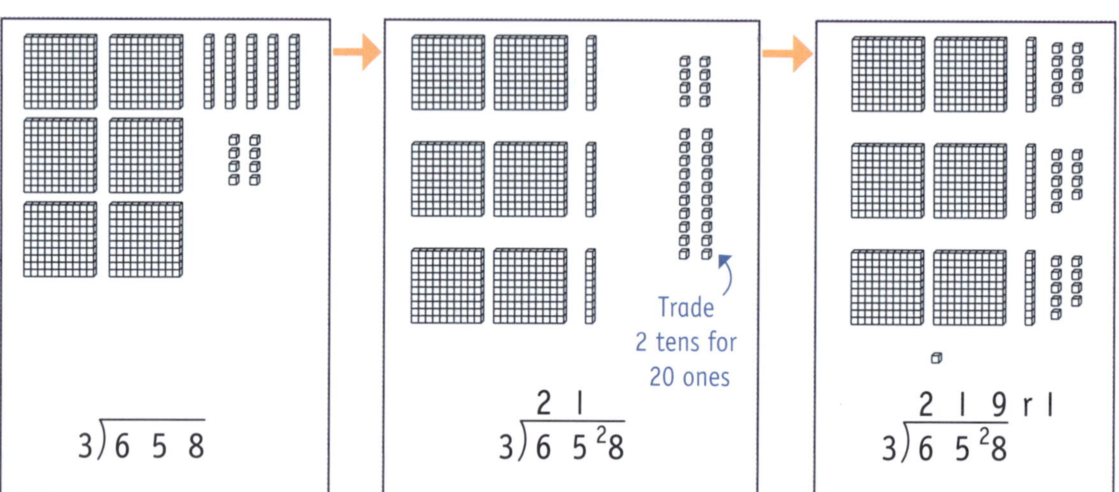

There will be 219 fish in each tank with 1 fish left over.

$658 \div 3 = 219$ remainder 1

Example 2:

$\begin{array}{r} 1\ 4\ 4 \text{ r } 3 \\ 4\overline{)5\ {}^{1}7\ {}^{1}9} \end{array}$

Example 3:

$\begin{array}{r} 5\ 2 \text{ r } 7 \\ 8\overline{)4\ 2\ {}^{2}3} \end{array}$

Example 4:

$\begin{array}{r} 1\ \ \text{ r } \\ 5\overline{)6\ {}^{1}4\ {}^{4}2} \end{array}$

Example 5:

$\begin{array}{r} 1\ \ \text{ r } \\ 6\overline{)7\ {}^{1}3\ {}^{1}7} \end{array}$

Solve.

● $\begin{array}{r} 1\ 2\ 6 \text{ r } 2 \\ 4\overline{)5\ {}^{1}0\ {}^{2}6} \end{array}$

a $2\overline{)3\ 5\ 9}$

b $3\overline{)4\ 2\ 5}$

c $2\overline{)7\ 3\ 3}$

d $5\overline{)5\ 3\ 7}$

e $6\overline{)4\ 2\ 3}$

f $7\overline{)4\ 9\ 5}$

g $5\overline{)5\ 0\ 9}$

h $6\overline{)7\ 4\ 3}$

Check your answers
How many did you get correct?

CATCH UP MATHS YEAR 6 BOOK A © PASCAL PRESS ISBN: 9781925726183

PRACTICE

1 Solve.

- ● $3\overline{)2\,0\,{}^{2}7}$ = 69
- **a** $7\overline{)6\,4\,3}$
- **b** $6\overline{)4\,9\,1}$
- **c** $5\overline{)3\,4\,8}$
- **d** $8\overline{)7\,4\,5}$
- **e** $4\overline{)2\,5\,7}$
- **f** $2\overline{)1\,4\,6}$
- **g** $3\overline{)2\,8\,5}$
- **h** $5\overline{)6\,9\,2}$
- **i** $4\overline{)5\,9\,9}$
- **j** $2\overline{)7\,4\,3}$
- **k** $6\overline{)8\,3\,7}$
- **l** $9\overline{)9\,8\,4}$
- **m** $3\overline{)8\,5\,4}$
- **n** $5\overline{)6\,5\,3}$

2 Check the answers to Question 1 using multiplication.

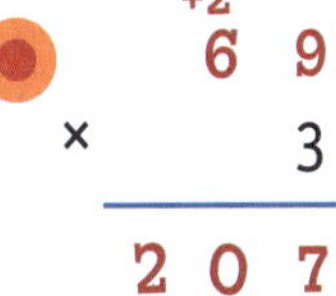

● 69 (+2) × 3 = 207

- **a** × 7 = ____
- **b** × 6 = ____
- **c** × 5 = ____
- **d** × 8 = ____
- **e** × 4 = ____
- **f** × 2 = ____
- **g** × 3 = ____
- **h** × 5 = ____
- **i** × 4 = ____
- **j** × 2 = ____
- **k** × 6 = ____
- **l** × 9 = ____
- **m** × 3 = ____
- **n** × 5 = ____

3 Complete.

● ★ = ■ ÷ 3

■	300	270	240	180	120	150	90
★	100	90	80	60	40	50	30

a ▲ = ● ÷ 2

●	240	120	600	800	100	160	300
▲							

b ■ = ★ ÷ 6

★	120	180	240	480	600	420	300
■							

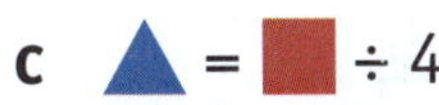

c ▲ = ■ ÷ 4

■	400	480	160	200	240	280	440
▲							

  ISBN: 9781925726183

RECORDING REMAINDERS AS FRACTIONS AND DECIMALS

You can write the remainder from a division as a fraction or a decimal.

Remainder as a fraction

$4\overline{)2\,5\,{}^{1}7}$ = 64 r 1 $= 64\frac{1}{4}$

To write a remainder as a fraction, put the remainder over the divisor. Here, the remainder, 1, is over the divisor, 4.

Remainder as a decimal

$4\overline{)2\,5\,{}^{1}7}$ = 64 r 1 $= 64.25$

To write a remainder as a decimal, divide the numerator (remainder) by the denominator (divisor). $1 \div 4 = 0.25$

Example 1:

$3\overline{)4\,{}^{1}9\,{}^{1}1}$ = 163 r 2

$= 163\frac{2}{3}$ or 163.66

Example 2:

$8\overline{)6\,4\,7}$ = 80 r 7

$= 80\frac{7}{8}$ or 80.875

Example 3:

$5\overline{)4\,2\,{}^{2}7}$ = 85 r 2

= 85— or 85.____

Example 4:

$2\overline{)7\,{}^{1}5\,{}^{1}7}$ = 378 r 1

= 378— or 378.____

Check your answer on the video!

Some fraction and decimal equivalents

$\frac{1}{8} = 0.125$

$\frac{1}{5} = 0.2$

$\frac{1}{4} = 0.25$

$\frac{1}{3} = 0.33$

$\frac{2}{5} = 0.4$

$\frac{1}{2} = 0.5$

$\frac{2}{3} = 0.66$

$\frac{3}{4} = 0.75$

1 Solve, then write the remainder as a fraction.

- $3\overline{)4\,{}^{1}9\,{}^{1}4}$ = 164 r 2 $= 164\frac{2}{3}$

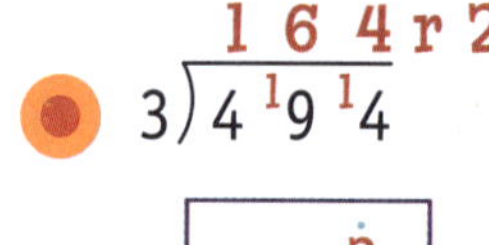

a $5\overline{)4\,9\,3}$ = ____

b $2\overline{)5\,2\,1}$ = ____

2 Solve, then write the remainder as a decimal.

- $8\overline{)4\,8\,3}$ = 60 r 3 = 60.375

a $4\overline{)5\,1\,3}$ = ____

b $2\overline{)8\,4\,1}$ = ____

Check your answers
How many did you get correct?

CATCH UP MATHS YEAR 6 BOOK A © PASCAL PRESS ISBN: 9781925726183

1 Fill in the table.

	Dividend	Divisor	Quotient	Remainder	Quotient and Remainder as a Fraction	Quotient and Remainder as a Decimal
●	26	4	6	2	$6\frac{2}{4}$	6.50
a	17	2				
b	33	5				
c	57	4				
d	37	3				

2 Solve and record the remainder as a decimal.

● $4\overline{)3\,4^{2}9}$ = 8 7 r 1

= 87 remainder 1 = 87.25

a $4\overline{)1\,4\,7}$ r

= ____ remainder __ = ______

b $5\overline{)4\,4\,9}$ r

= ____ remainder __ = ______

c $8\overline{)8\,3\,6}$ r

= ____ remainder __ = ______

d $2\overline{)8\,3\,7}$ r

= ____ remainder __ = ______

e $4\overline{)9\,8\,1}$ r

= ____ remainder __ = ______

3 Solve and record the remainder as a fraction.

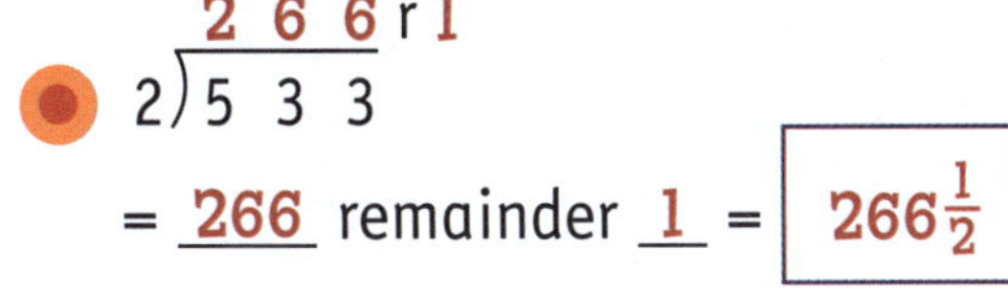

● $2\overline{)5\,3\,3}$ = 2 6 6 r 1

= 266 remainder 1 = $266\frac{1}{2}$

a $4\overline{)7\,4\,1}$ r

= ____ remainder __ = ☐

b $8\overline{)6\,7\,3}$ r

= ____ remainder __ = ☐

c $5\overline{)4\,9\,6}$ r

= ____ remainder __ = ☐

d $3\overline{)5\,3\,2}$ r

= ____ remainder __ = ☐

e $2\overline{)1\,9\,1}$ r

= ____ remainder __ = ☐

AVERAGES

The average is a number that you work out by adding a group of numbers and then dividing by how many numbers are in the group.

SCAN to watch video

Example 1:

What is the average of 2, 7 and 9?

2 + 7 + 9 = 18 ← Add the numbers.

18 ÷ 3 = 6 ← Divide by 3 because there are 3 numbers in the group.

The average of 2, 7 and 9 is 6.

Example 2:

What is the average of 10, 12 and 17?

10 + 12 + 17 = 39

39 ÷ 3 = 13

Example 3:

Find the average: 6, 4, 5, 7, 13.

6 + 4 + 5 + 7 + 13 = 35

35 ÷ 5 = 7

Example 4:

What is the average of 5, 7, 3, 9, 6 and 6?

5 + 7 + 3 + 9 + 6 + 6 = 36

36 ÷ 6 = 6

Example 5:

What is the average of 4, 2, 5 and 1?

4 + 2 + 5 + 1 = ___

___ ÷ 4 = ___

Example 6:

Find the average: 7, 3, 5, 4, 1.

7 + 3 + 5 + 4 + 1 = ___

___ ÷ 5 = ___

Example 7:

What is the average of 6, 7, 5, 4, 2 and 6?

6 + 7 + 5 + 4 + 2 + 6 = ___

___ ÷ 6 = ___

Work out the average of each set of numbers.

● 2, 4, 6, 8, 10

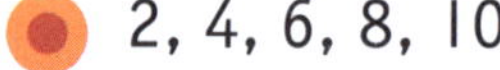

30 ÷ 5 = 6

a 3, 5, 7, 9

____

____

b 4, 8, 6, 3, 4

____

____

c 9, 11, 15, 6, 4

____

____

Check your answers

How many did you get correct?

CATCH UP MATHS YEAR 6 BOOK A © PASCAL PRESS ISBN: 9781925726183

1 Find the average of each set of numbers.

- 10, 15, 12, 4, 9, 4

  10 + 15 + 12 + 4 + 9 + 4

  54 ÷ 6 = 9

a 1, 8, 12, 9, 5

b 7, 6, 10, 9

c 8, 4, 12, 6, 4, 2

d 0, 1, 1, 3, 0, 3, 0, 0

e 14, 13, 12, 11, 10

f 9, 11, 6, 10, 14

g 7, 9, 6, 10

h 12, 15, 10, 9, 4

i 8, 1, 12, 10, 4

j 11, 7, 8, 10, 4

k 11, 7, 5, 4, 3

l 49, 45, 48, 46

m 7, 10, 8, 9, 6, 2

n 16, 17, 19, 15, 15, 18, 19

o 32, 34, 38, 33, 35, 36, 37, 35

DIVIDING BY 10, 100 AND 1000

To divide by 10, move the decimal point one place to the left.

732 ÷ 10

73.2

732 ÷ 10 = 73.2

To divide by 100, move the decimal point two places to the left.

487 ÷ 100

4.87

487 ÷ 100 = 4.87

To divide by 1000, move the decimal point three places to the left.

384 ÷ 1000

.384

384 ÷ 1000 = 0.384

Example 1:

43 ÷ 10

4.3

43 ÷ 10 = 4.3

43 ÷ 100

.43

43 ÷ 100 = 0.43

43 ÷ 1000

.043

43 ÷ 1000 = 0.043

Example 2:

848 ÷ 10

84.8

848 ÷ 10 = ______

848 ÷ 100

8.48

848 ÷ 100 = ______

848 ÷ 1000

.848

848 ÷ 1000 = ______

Example 3:

Check your answer on the video!

652 ÷ 10

65.2

652 ÷ 10 = ______

652 ÷ 100

6.52

652 ÷ 100 = ______

652 ÷ 1000

.652

652 ÷ 1000 = ______

Complete the table.

	Number	÷ 10	÷ 100	÷ 1000
●	23	2.3	0.23	0.023
a	45			
b	124			
c	865			

SELF CHECK Tick how you feel

Got it!	Need help...	I don't get it
☐	☐	☐

Check your answers
How many did you get correct? ☐

CATCH UP MATHS YEAR 6 BOOK A © PASCAL PRESS ISBN: 9781925726183

PRACTICE

1 Complete the table.

	Number	÷ 10	÷ 100	÷ 1000
●	2	0.2	0.02	0.002
a	96			
b	8			
c	154			
d	27			
e	357			
f	4215			
g	6387			
h	9			
i	14 359			

2 Divide by 10.

● 6 0.6

a 72 ______

b 159 ______

c 2687 ______

d 53 ______

e 475 ______

f 6824 ______

g 74 372 ______

h 5945 ______

i 1638 ______

j 440 ______

k 82 437 ______

3 Divide by 100.

● 32 574 325.74

a 4 ______

b 249 ______

c 1943 ______

d 44 ______

e 74 273 ______

f 81 510 ______

g 17 ______

h 72 593 ______

i 560 ______

j 69 420 ______

k 826 ______

4 Divide by 1000.

● 52 437 52.437

a 6 ______

b 182 ______

c 7111 ______

d 75 264 ______

e 950 ______

f 8256 ______

g 19 ______

h 28 ______

i 6359 ______

j 42 003 ______

k 778 ______

DIVISION REVIEW

1 Write the quotient.

a $24 \div 3 =$ ___ c $45 \div 5 =$ ___ e $77 \div 11 =$ ___ g $84 \div 12 =$ ___

b $60 \div 10 =$ ___ d $72 \div 8 =$ ___ f $90 \div 9 =$ ___ h $44 \div 11 =$ ___

2 Tick the labels that have a quotient of 4.

$8 \div 2$ | $17 \div 3$ | $40 \div 10$ | $48 \div 12$ | $20 \div 4$ | $16 \div 4$ | $48 \div 4$

$20 \div 5$ | $28 \div 4$ | $18 \div 4$ | $24 \div 6$ | $36 \div 9$ | $32 \div 8$ | $48 \div 11$

3 Fill in the tables.

	Fraction	$\overline{)}$	÷
a		$2\overline{)63}$	
b	$\frac{47}{5}$		
c			$23 \div 8$

	Fraction	$\overline{)}$	÷
d		$8\overline{)41}$	
e	$\frac{22}{4}$		
f			$63 \div 7$

4 Fill in the missing numbers.

a $\begin{array}{r} 12 \\ 6\overline{)} \end{array}$ b $\begin{array}{r} 12 \\ \overline{)60} \end{array}$ c $4\overline{)44}$ d $\begin{array}{r} 8 \\ 7\overline{)} \end{array}$ e $\begin{array}{r} 12 \\ 9\overline{)} \end{array}$

5 Fill in the missing numbers.

a $5 \times 3 =$ ___
___ ÷ ___ = ___ $\begin{array}{r} 5 \\ 3\overline{)} \end{array}$

b $6 \times 2 =$ ___
___ ÷ ___ = ___ $\overline{)}$

c $7 \times 8 =$ ___
___ ÷ ___ = ___ $\overline{)}$

d $11 \times 10 =$ ___
___ ÷ ___ = ___ $\overline{)}$

e $7 \times 12 =$ ___
___ ÷ ___ = ___ $\overline{)}$

f $2 \times 9 =$ ___
___ ÷ ___ = ___ $\overline{)}$

6 Write the missing numbers.

a $37 \div 3 =$ ___ remainder ___ d $43 \div 7 =$ ___ remainder ___

b $62 \div 10 =$ ___ remainder ___ e $35 \div 10 =$ ___ remainder ___

c $74 \div 9 =$ ___ remainder ___ f $74 \div 10 =$ ___ remainder ___

7 Use the first number sentence to fill in the missing numbers.

a (9 × 3) + 2 = 29

29 ÷ 3 = ___ remainder ___

b (2 × 7) + 6 = 20

20 ÷ 7 = ___ remainder ___

c (4 × 9) + 8 = 44

44 ÷ 9 = ___ remainder ___

d (8 × 7) + 5 = 61

61 ÷ 7 = ___ remainder ___

8 Solve.

a $4\overline{)6\ 4}$

b $2\overline{)2\ 6}$

c $3\overline{)7\ 8}$

d $7\overline{)2\ 8}$

e $4\overline{)8\ 7}$

f $8\overline{)2\ 9}$

g $5\overline{)6\ 3}$

h $6\overline{)5\ 7}$

9 Check your answers to Question 8 using multiplication. Add any remainders.

a × 4 ______

b × 2 ______

c × 3 ______

d × 7 ______

e × 4 ______

f × 8 ______

g × 5 ______

h × 6 ______

10 Solve.

a $4\overline{)4\ 2\ 4}$

b $5\overline{)5\ 8\ 5}$

c $6\overline{)7\ 5\ 6}$

d $9\overline{)2\ 5\ 2}$

e $5\overline{)6\ 4\ 6}$

f $7\overline{)2\ 7\ 7}$

g $4\overline{)6\ 2\ 5}$

h $8\overline{)6\ 7\ 3}$

11 Check your answers to Question 10 using multiplication. Add any remainders.

a × 4 ______

b × 5 ______

c × 6 ______

d × 9 ______

e × 5 ______

f × 7 ______

g × 4 ______

h × 8 ______

REVIEW

12 Complete.

a ★ = ■ ÷ 5

■	250	500	350	300	100	600	550
★							

b ⬢ = ● ÷ 6

●	240	120	300	600	720	18	360
⬢							

13 Solve, then write the remainder as a fraction.

a $4\overline{)726}$ r = ☐

b $3\overline{)511}$ r = ☐

c $2\overline{)765}$ r = ☐

d $3\overline{)842}$ r = ☐

e $5\overline{)873}$ r = ☐

f $3\overline{)428}$ r = ☐

14 Solve, then write the remainder as a decimal.

a $3\overline{)742}$ r = _____ . _____

b $4\overline{)849}$ r = _____ . _____

c $5\overline{)756}$ r = _____ . _____

d $2\overline{)845}$ r = _____ . _____

e $5\overline{)122}$ r = _____ . _____

f $8\overline{)670}$ r = _____ . _____

15 Fill in the table.

	Dividend	Divisor	Quotient	Remainder	Quotient and Remainder as a Fraction	Quotient and Remainder as a Decimal
a	29	8				
b	37	4				
c	85	3				
d	79	2				
e	48	5				

CATCH UP MATHS YEAR 6 BOOK A © PASCAL PRESS ISBN: 9781925726183

16 Find the average.

a 6 and 2

d 5, 2, 3, 6

b 4, 9, 2

e 7, 4, 3, 2

c 35, 36, 33, 39, 37

f 21, 22, 22, 20, 21, 22, 21, 20, 20

17 Divide by 10.

a 7 ________

b 42 ________

c 739 ________

d 1493 ________

e 72 673 ________

f 3781 ________

g 3313 ________

h 92 578 ________

i 6 ________

18 Divide by 100.

a 5 ________

b 63 ________

c 973 ________

d 9441 ________

e 81 810 ________

f 5243 ________

g 2624 ________

h 49 100 ________

i 9 ________

19 Divide by 1000.

a 1 ________

b 34 ________

c 387 ________

d 6640 ________

e 91 113 ________

f 8756 ________

g 4600 ________

h 82 403 ________

i 3 ________

20 Fill in the table.

	Number	÷ 10	÷ 100	÷ 1000
a	4			
b	78			
c	469			
d	371			

HALVES – FRACTIONS AND COLLECTIONS

When an object or group of objects is shared into equal parts, each part is a fraction. When an object or group of objects is divided into two equal parts, each part is one-half ($\frac{1}{2}$).

This square is divided into two equal parts.

Half ($\frac{1}{2}$) of this square is orange.

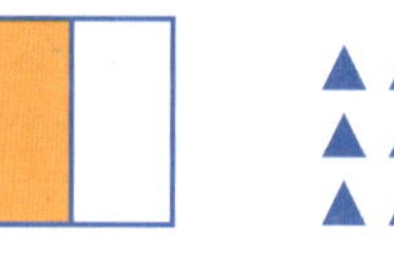

Half of the group of triangles is circled. 6 out of the 12 triangles are circled.

Half of 12 is 6. $\frac{1}{2}$ of 12 = 6

Example 1:
Colour $\frac{1}{2}$.

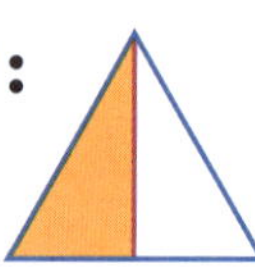

Example 4:
Divide into two equal parts.

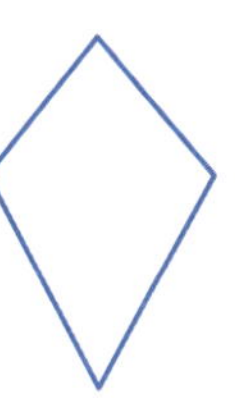

You can find $\frac{1}{2}$ by dividing by 2.

Example 2:
Divide into halves.

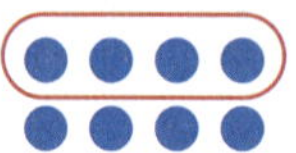

Example 5:
Circle $\frac{1}{2}$.

Check your answer on the video!

Example 3:
How many in one-half? 3

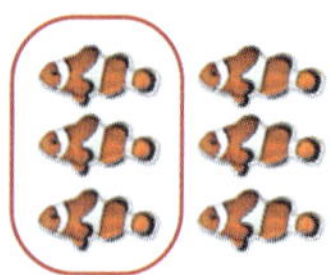

Example 6:
How many in one-half? ___

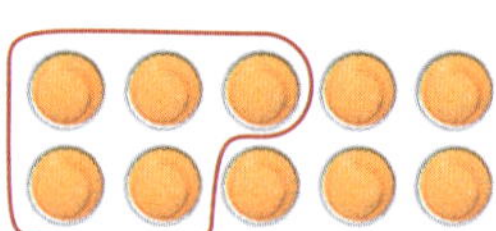

1 Circle the objects that are divided into 2 equal parts.

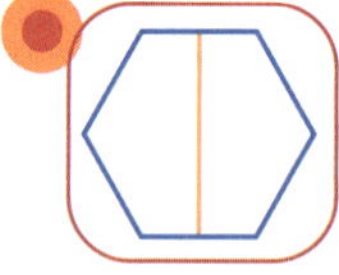

a 

b 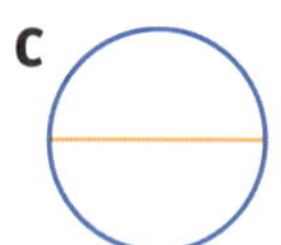

c

d

e 

2 Circle half of each group and then write the missing numbers.

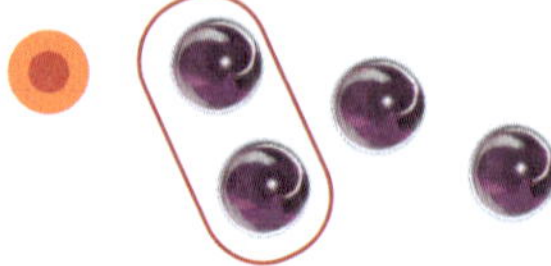

Half of 4 = 2

a 

Half of ___ = ___

b 

Half of ___ = ___

SELF CHECK Tick how you feel		
Got it!	Need help...	I don't get it

Check your answers
How many did you get correct?

CATCH UP MATHS YEAR 6 BOOK A © PASCAL PRESS ISBN: 9781925726183

1 Cross the shapes that are not divided in half.

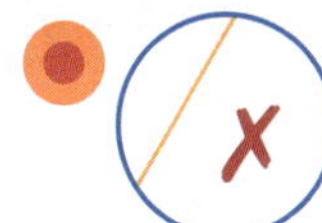

a b c d e

2 Cut these shapes into halves and colour $\frac{1}{2}$ of each shape.

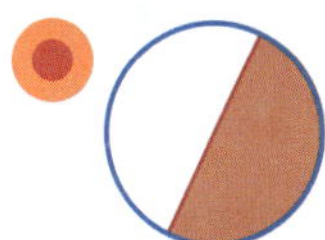

a 

b 

c 

d 

e 

3 Circle half of each group and then write the missing numbers.

Each group has 3 of the 6 bears.

Half of 6 is 3 $\frac{1}{2}$ of 6 = 3

b 

Each group has ___ of the ___ bears.

Half of ___ is ___ $\frac{1}{2}$ of ___ = ___

a 

Each group has ___ of the ___ bears.

Half of ___ is ___ $\frac{1}{2}$ of ___ = ___

c 

Each group has ___ of the ___ bears.

Half of ___ is ___ $\frac{1}{2}$ of ___ = ___

4 Circle half of each collection and then fill in the missing numbers.

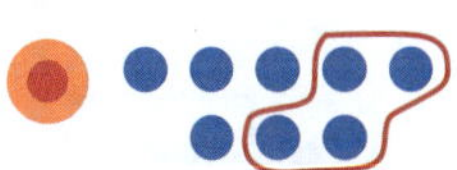

Half of 8 = 4

$\frac{1}{2}$ × 8 = 4

a 

Half of ___ = ___

$\frac{1}{2}$ × ___ = ___

b 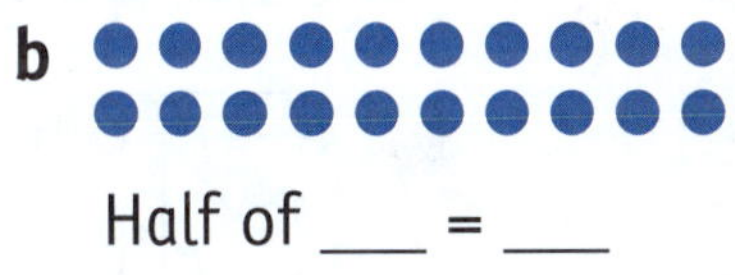

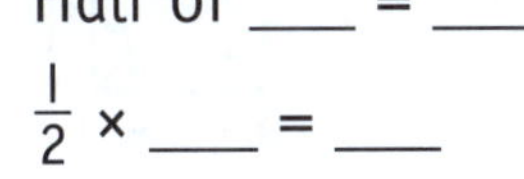

Half of ___ = ___

$\frac{1}{2}$ × ___ = ___

5 Complete.

$\frac{1}{2}$ × 36 = 18

a $\frac{1}{2}$ × 2 = ___

b $\frac{1}{2}$ × 6 = ___

c $\frac{1}{2}$ × 100 = ___

d $\frac{1}{2}$ × 48 = ___

e $\frac{1}{2}$ × 26 = ___

f $\frac{1}{2}$ × 42 = ___

g $\frac{1}{2}$ × 30 = ___

h $\frac{1}{2}$ × 22 = ___

i $\frac{1}{2}$ × 50 = ___

j $\frac{1}{2}$ × 56 = ___

k $\frac{1}{2}$ × 60 = ___

l $\frac{1}{2}$ × 66 = ___

m $\frac{1}{2}$ × 64 = ___

n $\frac{1}{2}$ × 120 = ___

o $\frac{1}{2}$ × 144 = ___

p $\frac{1}{2}$ × 124 = ___

q $\frac{1}{2}$ × 130 = ___

r $\frac{1}{2}$ × 400 = ___

s $\frac{1}{2}$ × 96 = ___

QUARTERS AND EIGHTHS – FRACTIONS AND COLLECTIONS

When a whole is cut into 4 equal parts, each part is one-quarter ($\frac{1}{4}$).

One-quarter ($\frac{1}{4}$) is yellow.

Example 1:
Here, 16 balls are divided into quarters. There are four equal parts. Each part has 4 balls.

One-quarter of 16 is 4. $\frac{1}{4} \times 16 = 4$

Example 2:
Divide into quarters.

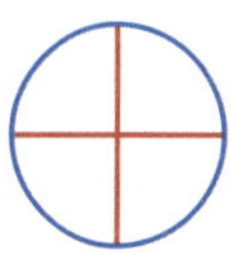

Example 3:
Circle ($\frac{1}{4}$).

When a whole is cut into 8 equal parts, each part is one-eighth ($\frac{1}{8}$).

One-eighth ($\frac{1}{8}$) is orange.

SCAN to watch video

Example 4:
Here, 16 balls are divided into eighths. There are eight equal parts. Each part has 2 balls.

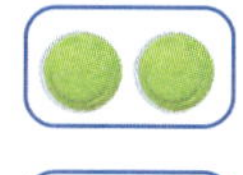 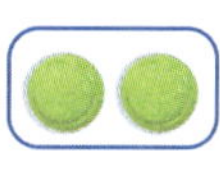 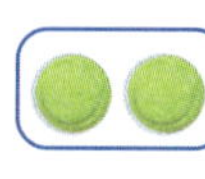 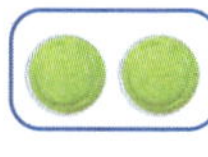
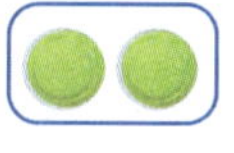 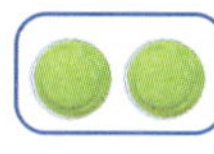 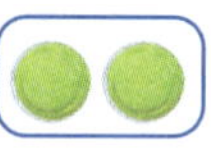 

One-eighth of 16 is 2. $\frac{1}{8} \times 16 = 2$

Example 5:
Divide into eighths.

Example 6:
Circle ($\frac{1}{8}$).

1 Use red to circle the objects divided into eighths and use green to circle the objects divided into quarters.

a 

b 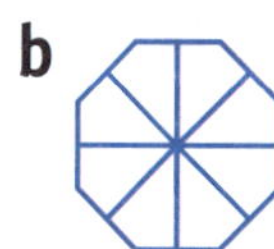

c 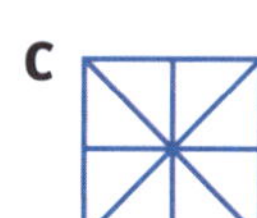

d 

e 

2 Circle and then answer.

 $\frac{1}{8}$ of 24 = 3

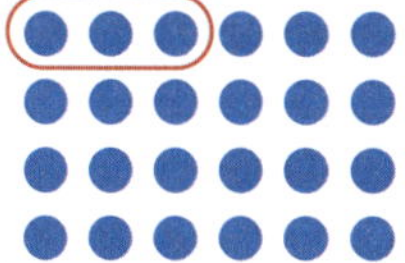

a $\frac{1}{4}$ of ___ = __

b $\frac{1}{8}$ of ___ = __

SELF CHECK Tick how you feel

Got it!	Need help...	I don't get it
 ☐	 ☐	 ☐

Check your answers
How many did you get correct? ☐

CATCH UP MATHS YEAR 6 BOOK A © PASCAL PRESS ISBN: 9781925726183

PRACTICE

1 Colour $\frac{3}{8}$ of each shape.

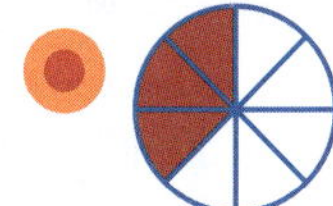

a 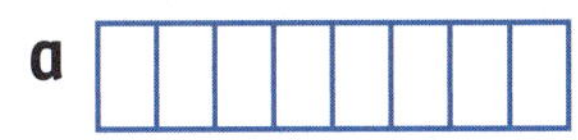

b 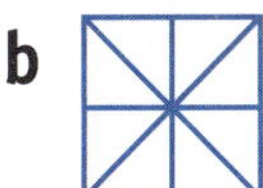

c 

d 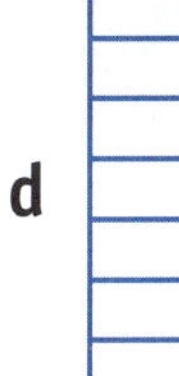

2 Colour three-quarters ($\frac{3}{4}$) of each shape.

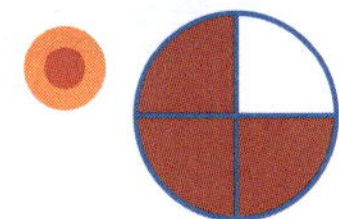

a

b 

c

d 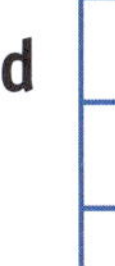

3 Complete the table.

	Fraction name	Fraction	Diagram
●	two-quarters	$\frac{2}{4}$	
a		$\frac{5}{8}$	
b	one-quarter		
c	three-eighths		
d			
e		$\frac{4}{8}$	
f	four-quarters		
g			

4 What is one-eighth? (Divide by 8.)

● 16 <u>2</u> b 8 ___ d 80 ___ f 40 ___

a 32 ___ c 56 ___ e 24 ___ g 64 ___

5 What is one-quarter? (Divide by 4.)

● 4 <u>1</u> b 24 ___ d 48 ___ f 32 ___

a 20 ___ c 44 ___ e 100 ___ g 36 ___

6 Solve.

● $\frac{1}{4} \times 8 =$ <u>2</u> b $\frac{1}{4} \times 12 =$ ___ d $\frac{1}{8} \times 72 =$ ___ f $\frac{1}{4} \times 28 =$ ___

a $\frac{1}{8} \times 48 =$ ___ c $\frac{1}{4} \times 16 =$ ___ e $\frac{1}{8} \times 80 =$ ___ g $\frac{1}{4} \times 40 =$ ___

THIRDS AND FIFTHS – FRACTIONS AND COLLECTIONS

When a whole is cut into 3 equal parts, each part is one-third ($\frac{1}{3}$).

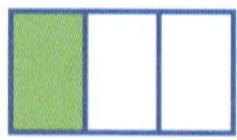 One-third ($\frac{1}{3}$) is green.

Example 1:
Here, 9 balls are divided into thirds.
There are three equal parts.
Each part has 3 balls.

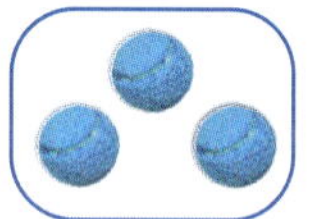 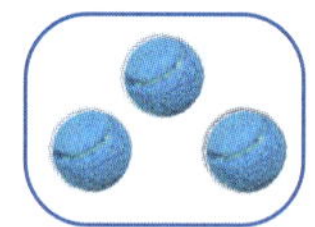 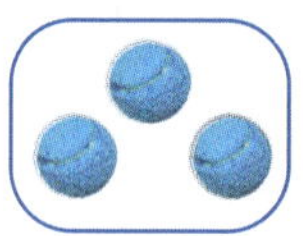

One-third of 9 is 3. $\frac{1}{3} \times 9 = 3$

When a whole is cut into 5 equal parts, each part is one-fifth ($\frac{1}{5}$).

 One-fifth ($\frac{1}{5}$) is pink.

SCAN to watch video

Example 4:
Here, 10 balls are divided into fifths.
There are 5 equal parts.
Each part has 2 balls.

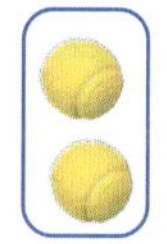   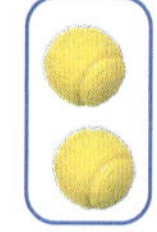 

One-fifth of 10 is 2. $\frac{1}{5} \times 10 = 2$

Divide by 3 to find $\frac{1}{3}$.

Divide by 5 to find $\frac{1}{5}$.

Example 2:
Divide into thirds.

Example 5:
Divide into fifths.

Example 3:
Circle ($\frac{1}{3}$). 

Example 6:
Circle ($\frac{1}{5}$). 

Check your answer on the video!

Your turn

1 Use green to circle the objects divided into fifths and use red to circle the objects divided into thirds.

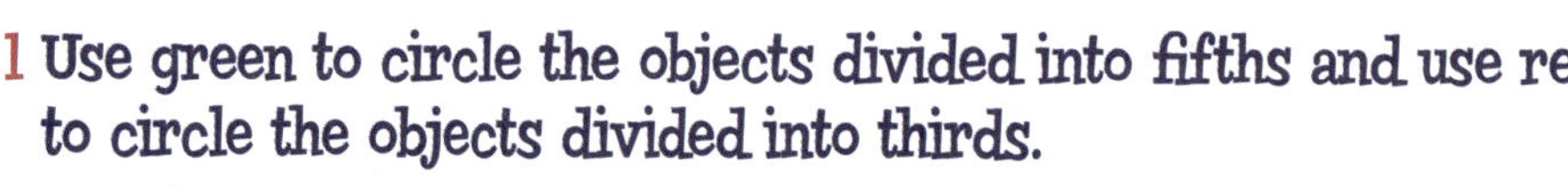

 a  b  c 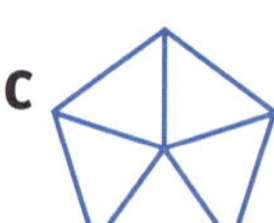 d  e 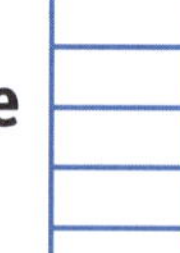

2 Circle and then answer.

$\frac{1}{5}$ of 15 = 3

a $\frac{1}{3}$ of ___ = ___

b $\frac{1}{5}$ of ___ = ___

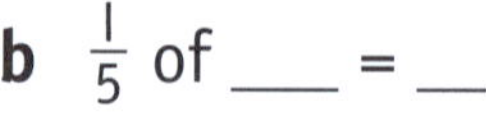

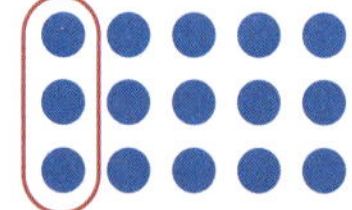  

SELF CHECK Tick how you feel

Got it!	Need help...	I don't get it
☐	☐	☐

Check your answers
How many did you get correct? ☐

CATCH UP MATHS YEAR 6 BOOK A © PASCAL PRESS ISBN: 9781925726183

1 Complete the table.

	Fraction name	Fraction	Diagram
●	one-third	$\frac{1}{3}$	
a		$\frac{2}{5}$	
b			
c	two-thirds		
d		$\frac{3}{3}$	
e			

2 What fraction is orange? Write your answer in the box.

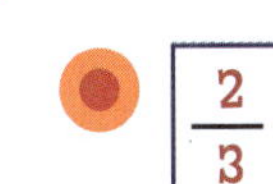

b  

d 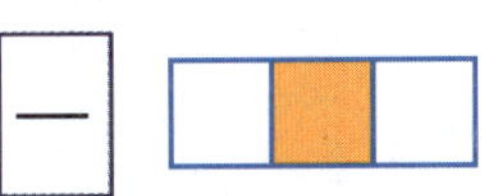

f 

a 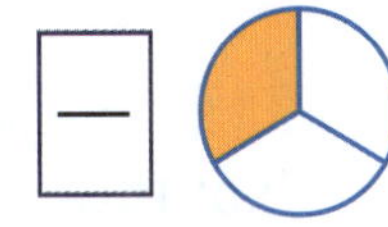

c 

e 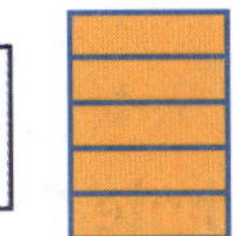

g 

3 What fraction is not orange in each shape in Question 2?

 $\frac{1}{3}$ b ___ d ___ f ___

a ___ c ___ e ___ g ___

4 What is one-fifth? (Divide by 5.)

● 20 4 b 30 ___ d 60 ___ f 5 ___ h 40 ___

a 50 ___ c 25 ___ e 100 ___ g 10 ___ i 15 ___

5 What is one-third? (Divide by 3.)

● 6 2 b 24 ___ d 21 ___ f 90 ___ h 33 ___

a 12 ___ c 66 ___ e 30 ___ g 36 ___ i 9 ___

6 Solve the following.

● $\frac{1}{3} \times 15 = 5$ b $\frac{1}{5} \times 35 =$ ___ d $\frac{1}{3} \times 27 =$ ___ f $\frac{1}{3} \times 69 =$ ___

a $\frac{1}{3} \times 18 =$ ___ c $\frac{1}{5} \times 45 =$ ___ e $\frac{1}{5} \times 55 =$ ___ g $\frac{1}{5} \times 75 =$ ___

EQUIVALENT FRACTIONS

Equivalent means equal or the same.
Equivalent fractions are equal or the same value.

$\frac{1}{2} = \frac{1}{4}$

Two of these four parts are exactly the same size as one-half.

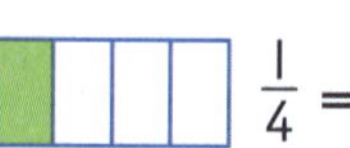 $\frac{1}{4} = \frac{2}{8}$ 

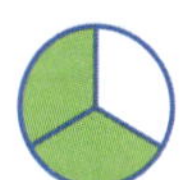 $\frac{2}{3} = \frac{4}{6}$

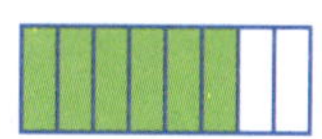 $\frac{6}{8} = \frac{3}{4}$ 

To make an equivalent fraction, multiply or divide the numerator and the denominator by the same number.

$\frac{1}{2} \xrightarrow{\times 4} \frac{4}{8}$ (× 4)

$\frac{1}{2} \xrightarrow{\times 3} \frac{3}{6}$ (× 3)

$\frac{4}{6} \xrightarrow{\div 2} \frac{2}{3}$ (÷ 2)

If you have the denominator, work out the relationship between the denominators to find the number you need to multiply or divide by.

$\frac{3}{5} \xrightarrow{\times 2} \frac{6}{10}$ (× 2)

$\frac{3}{4} \xrightarrow{\times 2} \frac{6}{8}$ (× 2)

$\frac{20}{100} \xrightarrow{\div 20} \frac{1}{5}$ (÷ 20)

Example 1: × 3: $\frac{1}{5} = \frac{3}{15}$ × 3

Example 2: × 4: $\frac{3}{5} = \frac{12}{20}$ × 4

Example 3: ÷ 2: $\frac{2}{16} = \frac{}{8}$ ÷ 2

Example 4: ÷ 2: $\frac{6}{16} = \frac{}{8}$ ÷ 2

Example 5: ÷ 2: $\frac{6}{8} = \frac{}{4}$ ÷ 2

Check your answer on the video!

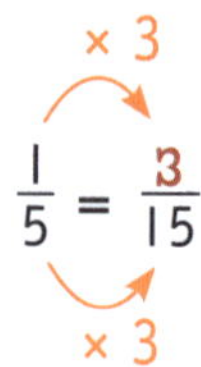

Colour to show the equivalent fraction.

● $\frac{2}{3} = \frac{4}{6}$

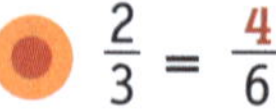
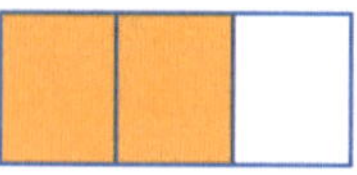
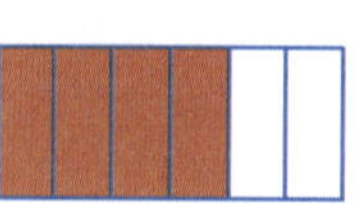

b $\frac{4}{5} = \frac{}{10}$

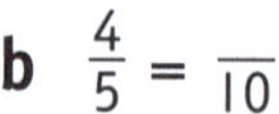
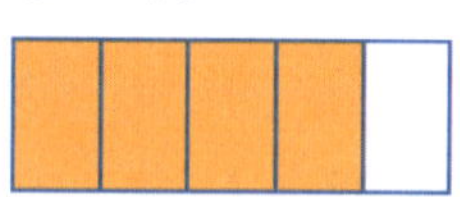

a $\frac{3}{4} = \frac{}{8}$

c $\frac{8}{12} = \frac{}{6}$

SELF CHECK Tick how you feel

Got it!	Need help...	I don't get it
 ☐	 ☐	☐

Check your answers
How many did you get correct? ☐

CATCH UP MATHS YEAR 6 BOOK A © PASCAL PRESS ISBN: 9781925726183

1 Work out the equivalent fractions.

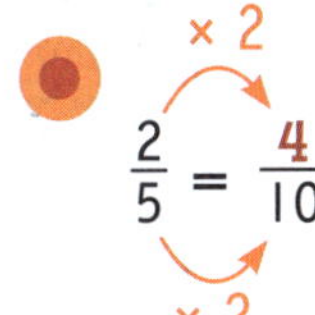

$\frac{2}{5} = \frac{4}{10}$ (× 2)

b $\frac{3}{5} = \frac{}{15}$ (× 3)

d $\frac{16}{20} = \frac{}{5}$ (÷ 4)

f $\frac{3}{4} = \frac{}{20}$ (× 5)

a $\frac{2}{8} = \frac{}{16}$ (× 2)

c $\frac{4}{12} = \frac{}{3}$ (÷ 4)

e $\frac{4}{5} = \frac{}{20}$ (× 4)

g $\frac{7}{10} = \frac{}{100}$ (× 10)

2 Work out the equivalent fractions.

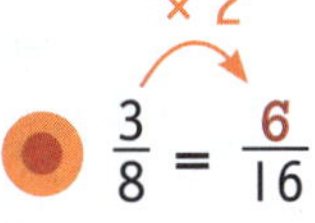

$\frac{3}{8} = \frac{6}{16}$ (× 2)

b $\frac{2}{3} = \frac{}{12}$

d $\frac{3}{8} = \frac{}{16}$

f $\frac{3}{5} = \frac{}{10}$

h $\frac{5}{20} = \frac{}{100}$

a $\frac{4}{10} = \frac{}{100}$

c $\frac{6}{12} = \frac{}{2}$

e $\frac{10}{12} = \frac{}{6}$

g $\frac{6}{10} = \frac{}{100}$

i $\frac{3}{10} = \frac{}{20}$

3 Write equivalent fractions.

$\frac{1}{2} = \frac{3}{6} = \frac{4}{8} = \frac{5}{10}$

b $\frac{2}{5} = \frac{}{10} = \frac{}{15} = \frac{}{100}$

d $\frac{1}{4} = \frac{}{8} = \frac{}{16} = \frac{}{100}$

a $\frac{3}{4} = \frac{}{8} = \frac{}{12} = \frac{}{20}$

c $\frac{3}{5} = \frac{}{10} = \frac{}{20} = \frac{}{100}$

e $\frac{1}{5} = \frac{}{10} = \frac{}{20} = \frac{}{100}$

4 Circle the equivalent fractions in each set.

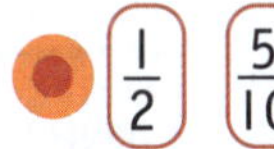

$\frac{1}{2}$ $\frac{5}{10}$ $\frac{3}{4}$ $\frac{4}{10}$ $\frac{4}{8}$

b $\frac{2}{5}$ $\frac{2}{6}$ $\frac{1}{3}$ $\frac{3}{9}$ $\frac{3}{4}$

d $\frac{4}{5}$ $\frac{4}{8}$ $\frac{6}{12}$ $\frac{3}{2}$ $\frac{1}{2}$

a $\frac{1}{4}$ $\frac{2}{16}$ $\frac{2}{8}$ $\frac{4}{16}$ $\frac{5}{20}$

c $\frac{1}{10}$ $\frac{3}{15}$ $\frac{5}{10}$ $\frac{1}{5}$ $\frac{2}{10}$

e $\frac{10}{100}$ $\frac{1}{50}$ $\frac{1}{10}$ $\frac{20}{200}$ $\frac{10}{50}$

5 Write T for true or F for false.

F One-half is equivalent to two-thirds

a ☐ One-quarter is equivalent to $\frac{2}{5}$

b ☐ One-third is equal to two-sixths

c ☐ Two-eighths is equivalent to one-quarter

d ☐ Six-eighths is equal to $\frac{3}{4}$

e ☐ Three-fifths is equivalent to six-tenths

f ☐ $\frac{1}{4}$ is equal to $\frac{2}{5}$

g ☐ $\frac{2}{3}$ is equivalent to $\frac{4}{6}$

h ☐ $\frac{3}{5}$ is equivalent to $\frac{6}{10}$

i ☐ $\frac{4}{5}$ is equal to $\frac{8}{10}$

COMPARING FRACTIONS

A fraction wall makes it easier to compare fractions.

SCAN to watch video

$\frac{1}{10}$ is the smallest fraction and $\frac{1}{2}$ is the largest fraction.

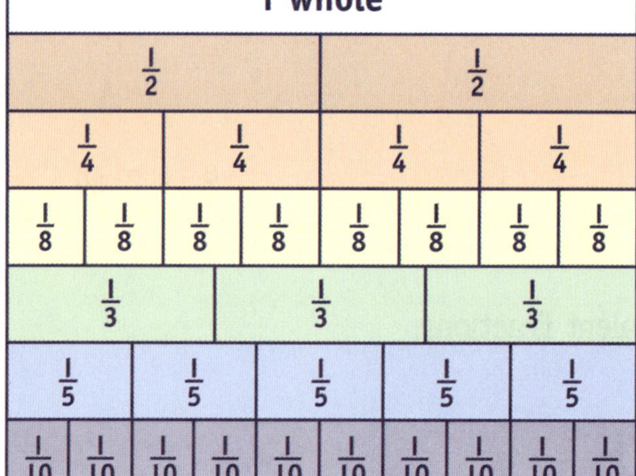

A fraction wall can help you work out which fractions are equal (equivalent).

When the numerator is 1, the larger the denominator, the smaller the fraction.

Example 1:
Circle the smaller fraction: $\frac{1}{3}$ (circled) $\frac{1}{2}$

Example 2:
Circle the larger fraction: $\frac{3}{8}$ (circled) $\frac{1}{3}$

Example 3:
Write a fraction equivalent to $\frac{4}{10}$. $\frac{2}{5}$

Example 4:
Circle the smaller fraction: $\frac{3}{4}$ $\frac{1}{2}$

Example 5:
Circle the larger fraction: $\frac{4}{10}$ $\frac{3}{5}$

Example 6:
Write a fraction equivalent to $\frac{1}{2}$. ____

Check your answer on the video!

Your turn

1 Write a smaller fraction.

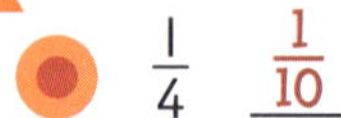

● $\frac{1}{4}$ $\frac{1}{10}$ **a** $\frac{2}{3}$ ____ **b** $\frac{8}{10}$ ____ **c** $\frac{3}{5}$ ____ **d** $\frac{1}{2}$ ____

2 Write a larger fraction.

● $\frac{1}{3}$ $\frac{3}{8}$ **a** $\frac{1}{2}$ ____ **b** $\frac{3}{8}$ ____ **c** $\frac{3}{4}$ ____ **d** $\frac{2}{10}$ ____

3 Write an equivalent fraction.

● $\frac{1}{4}$ $\frac{3}{12}$ **a** $\frac{4}{10}$ ____ **b** $\frac{4}{8}$ ____ **c** $\frac{6}{8}$ ____ **d** $\frac{3}{5}$ ____

SELF CHECK Tick how you feel		
Got it! ☐	Need help... ☐	I don't get it ☐

Check your answers
How many did you get correct? ☐

CATCH UP MATHS YEAR 6 BOOK A © PASCAL PRESS ISBN: 9781925726183

PRACTICE

1 Number the boxes to order these fractions from largest (1) to smallest (6).

● $\frac{1}{4}$ $\frac{2}{3}$ $\frac{3}{5}$ $\frac{5}{8}$ $\frac{3}{4}$ $\frac{1}{3}$

$\frac{1}{4}$	$\frac{2}{3}$	$\frac{3}{5}$	$\frac{5}{8}$	$\frac{3}{4}$	$\frac{1}{3}$
6	3	4	2	1	5

a $\frac{1}{4}$ $\frac{1}{2}$ $\frac{2}{3}$ $\frac{4}{5}$ $\frac{1}{8}$ $\frac{5}{8}$

b $\frac{6}{8}$ $\frac{4}{5}$ $\frac{3}{3}$ $\frac{2}{3}$ $\frac{2}{4}$ $\frac{1}{8}$

c $\frac{7}{10}$ $\frac{2}{5}$ $\frac{3}{4}$ $\frac{4}{8}$ $\frac{6}{10}$ $\frac{2}{3}$

2 Write the fractions on the number lines.

● $\frac{1}{8}$, $\frac{3}{4}$, $\frac{2}{4}$, $\frac{3}{8}$, $\frac{7}{8}$

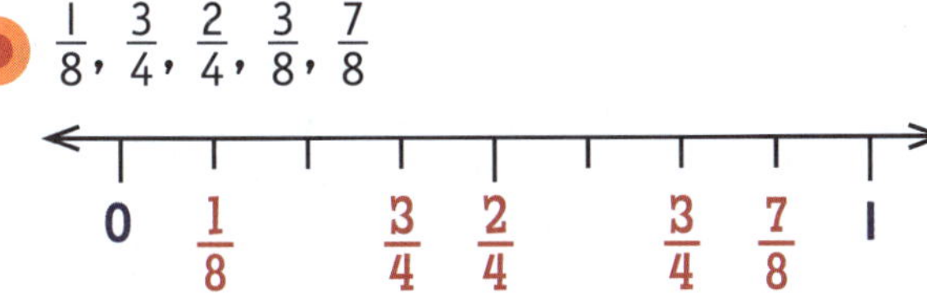

b $\frac{1}{3}$, $\frac{3}{10}$, $\frac{1}{2}$, $\frac{3}{4}$, $\frac{8}{10}$

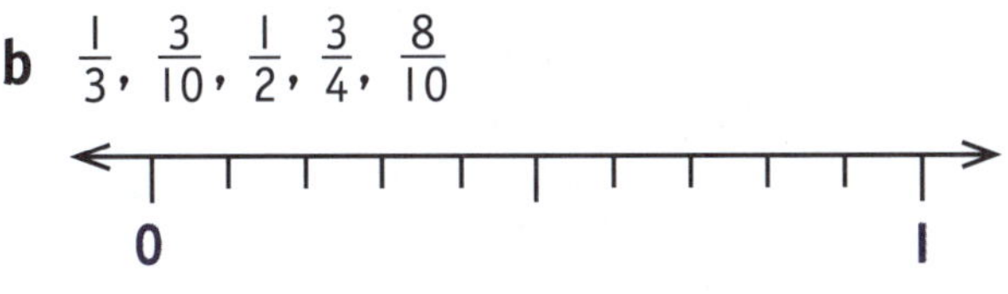

a $\frac{1}{5}$, $\frac{4}{10}$, $\frac{1}{2}$, $\frac{4}{5}$, $\frac{1}{4}$

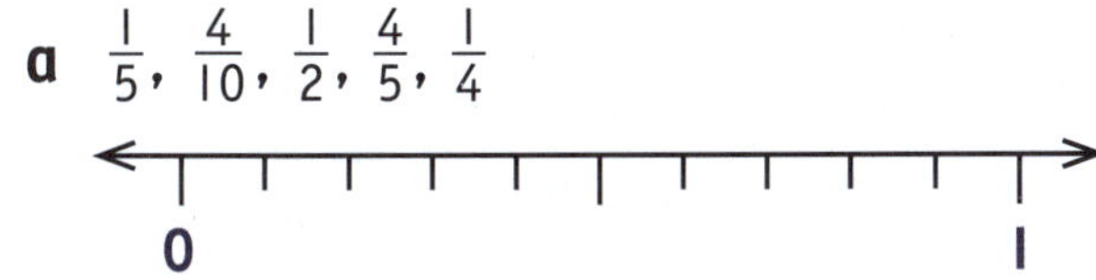

c $\frac{1}{12}$, $\frac{1}{8}$, $\frac{1}{3}$, $\frac{3}{4}$, $\frac{1}{2}$

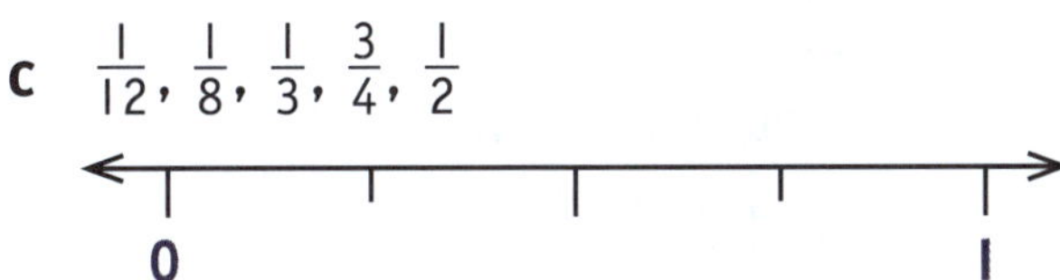

3 Write T for true or F for false.

● $\frac{1}{10} < \frac{1}{5}$ T

a $\frac{3}{4} > \frac{1}{3}$ ☐

b $\frac{1}{3} < \frac{1}{5}$ ☐

c $\frac{7}{8} = \frac{3}{4}$ ☐

d $\frac{3}{4} = \frac{6}{8}$ ☐

e $\frac{1}{2} < \frac{4}{8}$ ☐

f $\frac{2}{5} = \frac{4}{12}$ ☐

g $\frac{7}{10} > \frac{6}{8}$ ☐

h $\frac{1}{2} > \frac{3}{8}$ ☐

i $\frac{2}{3} < \frac{5}{5}$ ☐

j $\frac{3}{10} < \frac{1}{2}$ ☐

k $\frac{3}{4} > \frac{2}{5}$ ☐

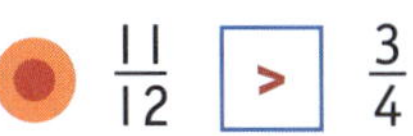

4 Write >, < or = to make the statements true.

● $\frac{11}{12}$ > $\frac{3}{4}$

a $\frac{1}{2}$ ☐ $\frac{3}{5}$

b $\frac{5}{6}$ ☐ $\frac{1}{3}$

c $\frac{10}{12}$ ☐ $\frac{9}{10}$

d $\frac{2}{3}$ ☐ $\frac{9}{12}$

e 1 ☐ $\frac{4}{4}$

f $\frac{10}{100}$ ☐ $\frac{1}{10}$

g $\frac{3}{5}$ ☐ $\frac{2}{3}$

h $\frac{7}{8}$ ☐ $\frac{9}{10}$

i $\frac{4}{6}$ ☐ $\frac{2}{3}$

j $\frac{8}{8}$ ☐ $\frac{12}{12}$

k $\frac{2}{3}$ ☐ $\frac{5}{6}$

PROPER AND IMPROPER FRACTIONS

A **proper fraction** has a numerator that is smaller than the denominator.

$\frac{3}{4}$ ← The numerator is smaller. ← The denominator is larger.

These fractions are proper fractions: $\frac{1}{4}$, $\frac{7}{12}$, $\frac{75}{100}$

An **improper fraction** has a numerator that is larger than the denominator.

$\frac{10}{7}$ ← The numerator is larger. ← The denominator is smaller.

These fractions are improper fractions: $\frac{4}{2}$, $\frac{20}{17}$, $\frac{258}{53}$

SCAN to watch video

Example 1:
Complete to show a proper fraction. $\frac{10}{15}$

Example 2:
Circle the proper fraction.

$\frac{8}{2}$ $\frac{16}{10}$ $\frac{9}{5}$ $\frac{1}{5}$ (circled) $\frac{125}{100}$

Example 3:
Write three proper fractions.

$\frac{1}{7}$ $\frac{2}{25}$ $\frac{97}{100}$

Example 4:
Complete to show an improper fraction.

$\frac{\_\_}{10}$

Example 5:
Circle the improper fraction.

$\frac{1}{5}$ $\frac{2}{9}$ $\frac{7}{10}$ $\frac{2}{5}$ $\frac{500}{100}$

Example 6:
Write three improper fractions.

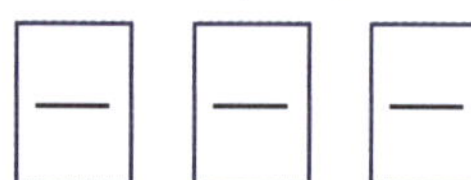

Use green to circle the proper fractions and use blue to circle the improper fractions.

$\frac{2}{3}$ (green) $\frac{5}{4}$ (blue) $\frac{6}{1}$ $\frac{10}{12}$ $\frac{8}{5}$ $\frac{10}{9}$ $\frac{3}{4}$ $\frac{4}{10}$ $\frac{89}{100}$ $\frac{2}{1}$

$\frac{1}{8}$ $\frac{100}{97}$ $\frac{9}{10}$ $\frac{5}{2}$ $\frac{7}{2}$ $\frac{2}{6}$ $\frac{5}{3}$ $\frac{16}{8}$ $\frac{3}{2}$ $\frac{5}{12}$ $\frac{12}{7}$

$\frac{3}{1}$ $\frac{1}{5}$ $\frac{3}{5}$ $\frac{15}{12}$ $\frac{4}{3}$ $\frac{4}{7}$ $\frac{9}{12}$ $\frac{8}{6}$ $\frac{4}{5}$ $\frac{25}{20}$ $\frac{4}{8}$

Check your answers
How many did you get correct?

CATCH UP MATHS YEAR 6 BOOK A © PASCAL PRESS ISBN: 9781925726183

PRACTICE

1 Circle the improper fraction in each set.

● $\frac{2}{4}$ $\frac{3}{5}$ $\frac{2}{1}$ (circled) $\frac{3}{10}$ $\frac{8}{12}$

a $\frac{1}{2}$ $\frac{2}{3}$ $\frac{20}{25}$ $\frac{8}{3}$ $\frac{2}{10}$

b $\frac{7}{10}$ $\frac{10}{12}$ $\frac{10}{9}$ $\frac{3}{4}$ $\frac{2}{5}$

c $\frac{1}{3}$ $\frac{3}{5}$ $\frac{4}{8}$ $\frac{5}{12}$ $\frac{12}{10}$

d $\frac{6}{3}$ $\frac{2}{4}$ $\frac{9}{10}$ $\frac{12}{14}$ $\frac{16}{18}$

e $\frac{2}{3}$ $\frac{3}{2}$ $\frac{1}{2}$ $\frac{3}{3}$ $\frac{1}{3}$

2 Circle the proper fraction in each set.

● $\frac{50}{27}$ $\frac{53}{25}$ $\frac{24}{30}$ (circled) $\frac{51}{28}$ $\frac{63}{60}$

a $\frac{1}{3}$ $\frac{3}{1}$ $\frac{4}{3}$ $\frac{3}{2}$ $\frac{4}{2}$

b $\frac{27}{30}$ $\frac{30}{29}$ $\frac{60}{30}$ $\frac{81}{18}$ $\frac{27}{20}$

c $\frac{17}{10}$ $\frac{14}{20}$ $\frac{20}{15}$ $\frac{70}{63}$ $\frac{20}{14}$

d $\frac{30}{24}$ $\frac{83}{43}$ $\frac{18}{14}$ $\frac{16}{35}$ $\frac{90}{63}$

e $\frac{6}{3}$ $\frac{9}{13}$ $\frac{9}{3}$ $\frac{8}{3}$ $\frac{4}{3}$

3 Write five proper fractions that match the description.

a 6 is the denominator

● $\frac{1}{6}$ ___ ___ ___ ___

b 20 is the denominator

c 12 is the denominator

d 10 is the denominator

4 Write five improper fractions that match the description.

a 10 is the numerator

● $\frac{10}{3}$ ___ ___ ___ ___

b 36 is the numerator

c 24 is the numerator

d 5 is the numerator

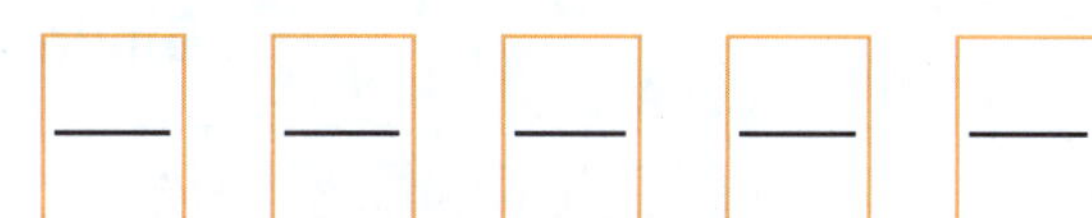

e 12 is the numerator

f 8 is the numerator

MIXED NUMBERS

A mixed number has a whole number and a proper fraction written together.

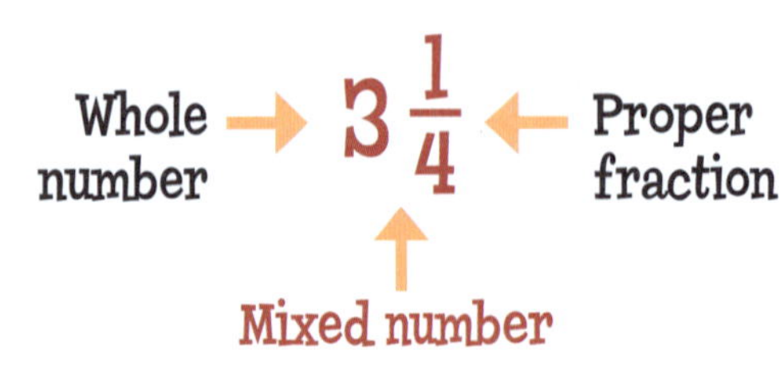	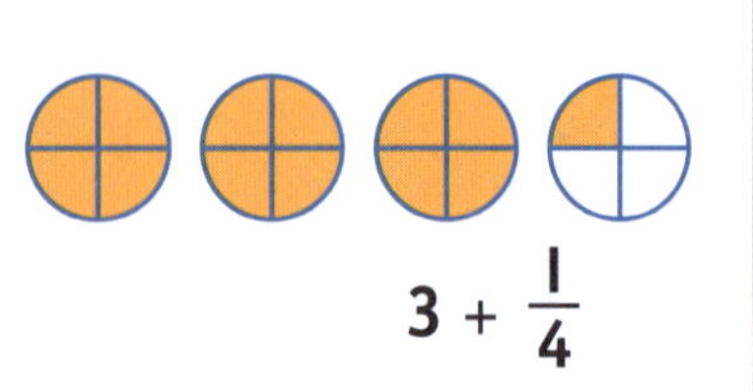
Whole number → $3\frac{1}{4}$ ← Proper fraction; Mixed number	$3 + \frac{1}{4}$

Change a mixed number to an improper fraction

- Multiply the denominator by the whole number.
- Add the answer to the numerator.
- Keep the same denominator.

Example 1:

$2\frac{1}{3} = \frac{7}{3}$ $\quad 3 \times 2 + 1 = 7$

Example 2:

$4\frac{1}{2} = \frac{\_\_}{2}$ $\quad 2 \times \_\_ + \_\_ = \_\_$

Change an improper fraction to a mixed number

- Divide the numerator by the denominator to get the whole number.
- Write the remainder as the numerator.
- Keep the same denominator.

Example 3:

$\frac{9}{4} = 2\frac{1}{4}$ $\quad 9 \div 4 = 2$ remainder 1

Example 4:

$\frac{6}{5} = \_\_\frac{\_\_}{5}$ $\quad 6 \div \_\_ = \_\_$ remainder __

Complete the table.

	Diagram	Improper Fraction	Mixed number
●		$\frac{11}{3}$	$3\frac{2}{3}$
a			__ $\frac{\_\_}{\_\_}$
b			__ $\frac{\_\_}{\_\_}$
c			__ $\frac{\_\_}{\_\_}$

SELF CHECK Tick how you feel

Got it!	Need help...	I don't get it
☐	☐	☐

Check your answers
How many did you get correct? ☐

CATCH UP MATHS YEAR 6 BOOK A © PASCAL PRESS ISBN: 9781925726183

PRACTICE

1 Change these improper fractions into mixed numbers.

- $\frac{12}{5} = 2\frac{2}{5}$
- c $\frac{13}{2} =$ ___
- f $\frac{59}{50} =$ ___
- i $\frac{43}{7} =$ ___
- a $\frac{10}{4} =$ ___
- d $\frac{25}{20} =$ ___
- g $\frac{19}{4} =$ ___
- j $\frac{52}{12} =$ ___
- b $\frac{13}{3} =$ ___
- e $\frac{47}{40} =$ ___
- h $\frac{17}{6} =$ ___
- k $\frac{36}{5} =$ ___

2 Change these mixed numbers into improper fractions.

- $3\frac{1}{3} = \frac{10}{3}$
- c $2\frac{3}{5} =$ ___
- f $12\frac{8}{10} =$ ___
- i $9\frac{3}{8} =$ ___
- a $5\frac{3}{4} =$ ___
- d $3\frac{4}{5} =$ ___
- g $13\frac{4}{7} =$ ___
- j $7\frac{4}{9} =$ ___
- b $3\frac{5}{6} =$ ___
- e $10\frac{4}{7} =$ ___
- h $20\frac{4}{6} =$ ___
- k $6\frac{11}{12} =$ ___

3 Write T for true or F for false.

- $3\frac{2}{3} = \frac{11}{3}$ T
- c $7\frac{3}{4} = \frac{32}{4}$ ☐
- f $6\frac{3}{5} = \frac{33}{5}$ ☐
- a $4\frac{1}{5} = \frac{22}{5}$ ☐
- d $12\frac{10}{12} = \frac{154}{12}$ ☐
- g $8\frac{1}{3} = \frac{23}{3}$ ☐
- b $3\frac{5}{8} = \frac{29}{8}$ ☐
- e $5\frac{2}{6} = \frac{31}{6}$ ☐
- h $10\frac{7}{10} = \frac{170}{10}$ ☐

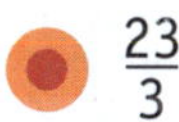

4 Circle the mixed number that matches the improper fraction.

- $\frac{23}{3}$: $7\frac{1}{2}$ (circled: $7\frac{2}{3}$) $7\frac{1}{3}$
- c $\frac{13}{3}$: $3\frac{2}{3}$ $4\frac{1}{3}$ $3\frac{1}{3}$
- a $\frac{31}{4}$: $7\frac{3}{4}$ $7\frac{1}{4}$ $7\frac{2}{4}$
- d $\frac{24}{10}$: $2\frac{12}{10}$ $2\frac{3}{10}$ $2\frac{4}{10}$
- b $\frac{29}{2}$: $14\frac{1}{2}$ $14\frac{1}{4}$ $14\frac{1}{9}$
- e $\frac{19}{4}$: $4\frac{3}{4}$ $5\frac{3}{4}$ $4\frac{1}{4}$

5 Circle the improper fraction that matches the mixed number.

- $3\frac{3}{8}$: $\frac{25}{8}$ (circled: $\frac{27}{8}$) $\frac{26}{8}$
- b $9\frac{7}{9}$: $\frac{88}{9}$ $\frac{90}{9}$ $\frac{98}{9}$
- d $8\frac{6}{8}$: $\frac{72}{8}$ $\frac{70}{8}$ $\frac{69}{8}$
- a $4\frac{7}{8}$: $\frac{36}{8}$ $\frac{38}{8}$ $\frac{39}{8}$
- c $7\frac{3}{5}$: $\frac{38}{5}$ $\frac{39}{5}$ $\frac{33}{5}$
- e $10\frac{3}{4}$: $\frac{44}{4}$ $\frac{43}{4}$ $\frac{46}{4}$

ADD AND SUBTRACT FRACTIONS WITH THE SAME DENOMINATOR

If the denominators are the same, you can add or subtract the numerators to get the answer.

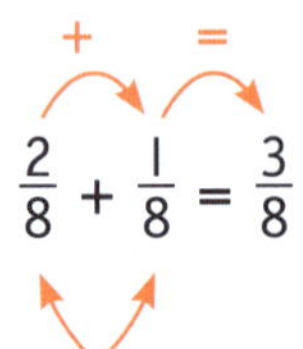

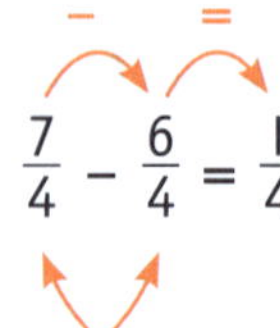

Adding fractions

Example 1:

$\frac{2}{8} + \frac{1}{8} = \frac{3}{8}$

The denominators are the same.

Example 2: $\frac{3}{4} + \frac{2}{4} = \frac{5}{4}$

$= 1\frac{1}{4}$

Example 3: $\frac{3}{6} + \frac{1}{6} = \frac{\quad}{6}$

Example 4: $\frac{2}{4} + \frac{1}{4} = \frac{\quad}{4}$

Subtracting fractions

Example 5:

$\frac{7}{4} - \frac{6}{4} = \frac{1}{4}$

The denominators are the same.

Example 6: $\frac{10}{4} - \frac{3}{4} = \frac{7}{4}$

$= 1\frac{3}{4}$

Example 7: $\frac{7}{8} - \frac{3}{8} = \frac{\quad}{8}$

Example 8: $\frac{8}{10} - \frac{2}{10} = \frac{\quad}{10}$

Solve these additions and subtractions.
Write your answer as a mixed number if you need to.

● $\frac{6}{10} - \frac{1}{10} = \frac{5}{10}$

a $\frac{11}{12} - \frac{5}{12} =$ ____

b $\frac{2}{15} + \frac{4}{15} =$ ____

c $\frac{12}{20} + \frac{3}{20} =$ ____

d $\frac{17}{10} - \frac{4}{10} =$ ____

= ____

e $\frac{5}{6} + \frac{7}{6} =$ ____

= ____

f $\frac{19}{8} - \frac{7}{8} =$ ____

= ____

g $\frac{3}{5} + \frac{3}{5} =$ ____

= ____

h $\frac{8}{12} - \frac{6}{12} =$ ____

i $\frac{10}{15} - \frac{7}{15} =$ ____

SELF CHECK Tick how you feel		
Got it! ☐	Need help... ☐	I don't get it ☐

Check your answers
How many did you get correct? ☐

CATCH UP MATHS YEAR 6 BOOK A © PASCAL PRESS ISBN: 9781925726183

1 Solve these additions.

- ● $\frac{1}{3} + \frac{1}{3} = \frac{2}{3}$
- b $\frac{4}{6} + \frac{1}{6} =$ ____
- d $\frac{6}{10} + \frac{2}{10} =$ ____
- f $\frac{4}{8} + \frac{3}{8} =$ ____
- a $\frac{3}{10} + \frac{4}{10} =$ ____
- c $\frac{7}{10} + \frac{2}{10} =$ ____
- e $\frac{5}{12} + \frac{5}{12} =$ ____
- g $\frac{5}{10} + \frac{1}{10} =$ ____

2 Add and write your answer as a mixed number.

- ● $\frac{6}{10} + \frac{6}{10} = \frac{12}{10} = 1\frac{2}{10}$
- b $\frac{7}{10} + \frac{4}{10} =$ ____ = ____
- d $\frac{3}{4} + \frac{5}{4} =$ ____ = ____
- a $\frac{3}{6} + \frac{5}{6} =$ ____ = ____
- c $\frac{10}{12} + \frac{8}{12} =$ ____ = ____
- e $\frac{3}{5} + \frac{3}{5} =$ ____ = ____

3 Solve these subtractions.

- ● $\frac{7}{10} - \frac{1}{10} = \frac{6}{10}$
- b $\frac{3}{2} - \frac{2}{2} =$ ____
- d $\frac{5}{9} - \frac{2}{9} =$ ____
- f $\frac{36}{100} - \frac{21}{100} =$ ____
- a $\frac{9}{12} - \frac{4}{12} =$ ____
- c $\frac{10}{12} - \frac{8}{12} =$ ____
- e $\frac{15}{20} - \frac{10}{20} =$ ____
- g $\frac{17}{20} - \frac{3}{20} =$ ____

4 Subtract and write your answer as a mixed number.

- ● $\frac{15}{10} - \frac{2}{10} = \frac{13}{10} = 1\frac{3}{10}$
- c $\frac{19}{10} - \frac{5}{10} =$ ____ = ____
- f $\frac{16}{8} - \frac{5}{8} =$ ____ = ____
- a $\frac{23}{12} - \frac{10}{12} =$ ____ = ____
- d $\frac{13}{8} - \frac{4}{8} =$ ____ = ____
- g $\frac{5}{3} - \frac{1}{3} =$ ____ = ____
- b $\frac{17}{5} - \frac{12}{5} =$ ____ = ____
- e $\frac{11}{6} - \frac{3}{6} =$ ____ = ____
- h $\frac{21}{12} - \frac{7}{12} =$ ____ = ____

5 Complete the tables.

Remember to subtract the smaller mixed number from the larger mixed number.

	Mixed Numbers		Add	Subtract
●	$1\frac{1}{8}$	$1\frac{4}{8}$	$2\frac{5}{8}$	$\frac{3}{8}$
a	$1\frac{3}{10}$	$1\frac{2}{10}$		
b	$1\frac{2}{12}$	$2\frac{5}{12}$		
c	$3\frac{7}{8}$	$1\frac{3}{8}$		
d	$3\frac{24}{100}$	$5\frac{32}{100}$		

	Mixed Numbers		Add	Subtract
e	$2\frac{11}{12}$	$1\frac{4}{12}$		
f	$3\frac{5}{6}$	$1\frac{3}{6}$		
g	$2\frac{3}{8}$	$5\frac{5}{8}$		
h	$3\frac{1}{4}$	$5\frac{3}{4}$		
i	$7\frac{3}{5}$	$5\frac{2}{5}$		

ADD AND SUBTRACT FRACTIONS WITH DIFFERENT DENOMINATORS

To add or subtract fractions that have different denominators, first make the denominators the same by finding equivalent fractions.

Example 1:

$\frac{1}{4} + \frac{1}{8}$

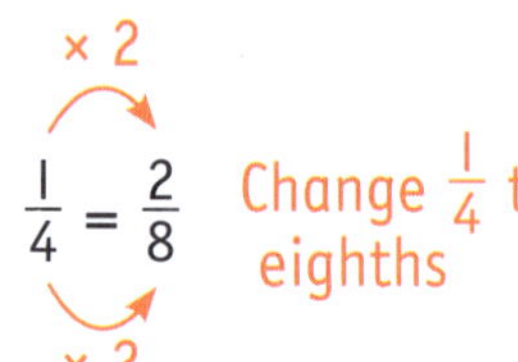

× 2

$\frac{1}{4} = \frac{2}{8}$ Change $\frac{1}{4}$ to eighths

× 2

$\frac{2}{8} + \frac{1}{8} = \frac{3}{8}$

Example 2:

$\frac{3}{10} + \frac{2}{5}$

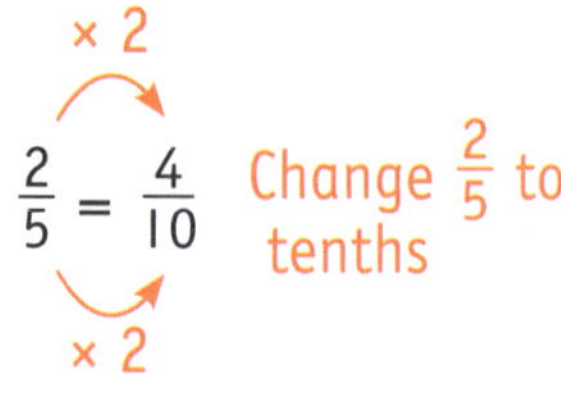

× 2

$\frac{2}{5} = \frac{4}{10}$ Change $\frac{2}{5}$ to tenths

× 2

$\frac{3}{10} + \frac{4}{10} = \frac{7}{10}$

Example 3:

$\frac{3}{4} + \frac{1}{8}$

× 2

$\frac{3}{4} = \frac{6}{8}$

× 2

$\frac{6}{8} + \frac{1}{8} = \frac{7}{8}$

Example 4:

$\frac{5}{6} - \frac{1}{2}$

× 3

$\frac{1}{2} = \frac{3}{6}$

× 3

$\frac{5}{6} - \frac{3}{6} = \frac{2}{6}$

Example 5:

$\frac{1}{2} + \frac{1}{8}$

× 4

$\frac{1}{2} = \frac{\_\_}{8}$

× 4

$\frac{\_\_}{8} + \frac{\_\_}{8} = \frac{\_\_}{8}$

Example 6:

$\frac{4}{5} - \frac{3}{10}$

× 2

$\frac{4}{5} = \frac{\_\_}{10}$

× 2

$\frac{\_\_}{10} - \frac{\_\_}{10} = \frac{\_\_}{10}$

Check your answer on the video!

Solve.

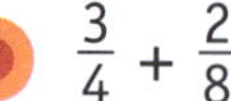

● $\frac{3}{4} + \frac{2}{8}$

× 2

$\frac{3}{4} = \frac{6}{8}$

× 2

$\frac{6}{8} + \frac{2}{8} = \frac{8}{8}$

a $\frac{7}{10} - \frac{2}{5}$

$\frac{2}{5} = \frac{\_\_}{10}$

____ − ____ = ____

b $\frac{3}{5} - \frac{7}{20}$

$\frac{3}{5} = \frac{\_\_}{20}$

____ − ____ = ____

SELF CHECK Tick how you feel

Got it!	Need help...	I don't get it

Check your answers

How many did you get correct?

CATCH UP MATHS YEAR 6 BOOK A © PASCAL PRESS ISBN: 9781925726183

Solve the following additions.

Example: $\frac{1}{2} + \frac{3}{8} = \frac{7}{8}$

$\frac{1}{2} = \frac{4}{8}$ (×4)

$\frac{4}{8} + \frac{3}{8} = \frac{7}{8}$

b $\frac{1}{2} + \frac{2}{10}$

___ = ___

___ + ___ = ___

d $\frac{4}{5} + \frac{1}{20}$

___ = ___

___ + ___ = ___

a $\frac{3}{4} + \frac{3}{8}$

___ = ___

___ + ___ = ___

c $\frac{3}{4} + \frac{1}{8}$

___ = ___

___ + ___ = ___

e $\frac{3}{5} + \frac{5}{20}$

___ = ___

___ + ___ = ___

Solve the following subtractions.

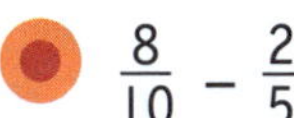

Example: $\frac{8}{10} - \frac{2}{5}$

$\frac{2}{5} = \frac{4}{10}$ (×2)

$\frac{8}{10} - \frac{4}{10} = \frac{4}{10}$

b $\frac{6}{8} - \frac{1}{2}$

___ = ___

___ − ___ = ___

d $\frac{5}{6} - \frac{5}{12}$

___ = ___

___ − ___ = ___

a $\frac{3}{10} - \frac{2}{20}$

___ = ___

___ − ___ = ___

c $\frac{2}{3} - \frac{7}{12}$

___ = ___

___ − ___ = ___

e $\frac{3}{4} - \frac{3}{8}$

___ = ___

___ − ___ = ___

Solve these additions and subtractions of mixed numbers.

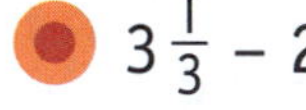

Example: $3\frac{1}{3} - 2\frac{1}{12}$

$\frac{1}{3} = \frac{4}{12}$ (×4)

$3\frac{4}{12} - 2\frac{1}{12} = 1\frac{3}{12}$

a $5\frac{3}{5} - 1\frac{2}{10}$

___ = ___

___ − ___ = ___

b $1\frac{3}{5} + 2\frac{1}{10}$

___ = ___

___ + ___ = ___

FRACTIONS REVIEW

Cut these shapes into halves and colour $\frac{1}{2}$ of each shape.

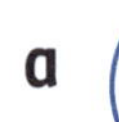

a 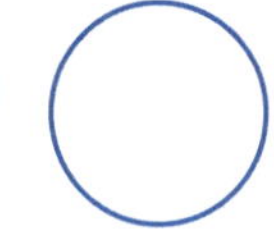

c 

e 

g 

b 

d 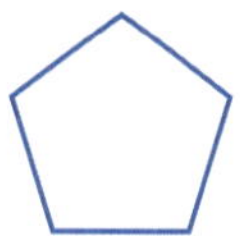

f 

h 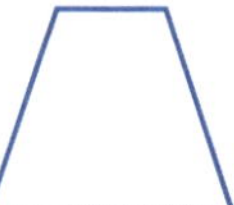

Circle half of each group and then write the missing numbers.

a 

Each group has ___ of the ___ balls.
Half of ___ is ___ $\frac{1}{2}$ of ___ = ___

c 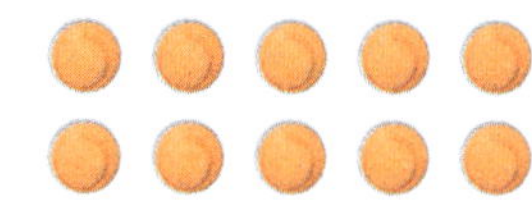

Each group has ___ of the ___ balls.
Half of ___ is ___ $\frac{1}{2}$ of ___ = ___

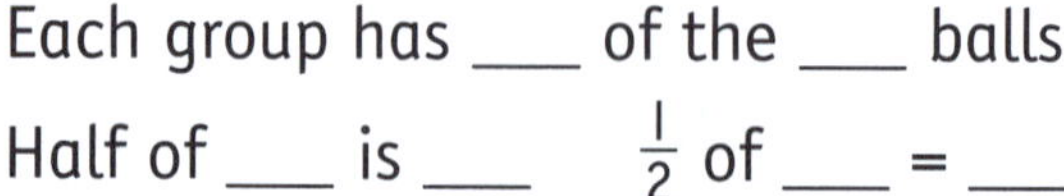

b 

Each group has ___ of the ___ balls.
Half of ___ is ___ $\frac{1}{2}$ of ___ = ___

d 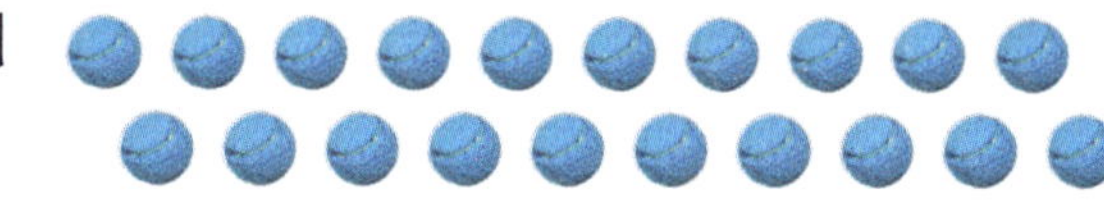

Each group has ___ of the ___ balls.
Half of ___ is ___ $\frac{1}{2}$ of ___ = ___

Write the answer.

a $\frac{1}{2} \times 14 =$ ___

b $\frac{1}{2} \times 30 =$ ___

c $\frac{1}{2} \times 120 =$ ___

d $\frac{1}{2} \times 42 =$ ___

e $\frac{1}{2} \times 66 =$ ___

f $\frac{1}{2} \times 2 =$ ___

g $\frac{1}{2} \times 126 =$ ___

h $\frac{1}{2} \times 148 =$ ___

Use red to colour the shapes that have been cut into eighths and green to colour the shapes that have been cut into quarters.

a 

c 

e 

g 

b 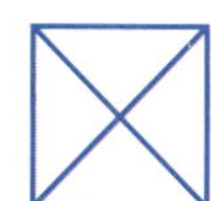

d 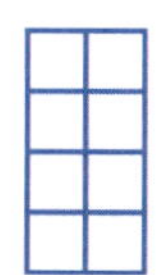

f 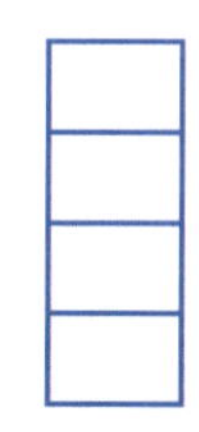

  ISBN: 9781925726183

5 Circle one-eighth of each group and then write the missing numbers.

a 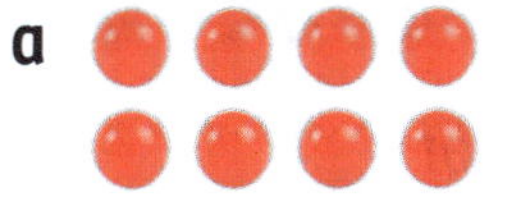

$\frac{1}{8}$ of ___ = ___

b 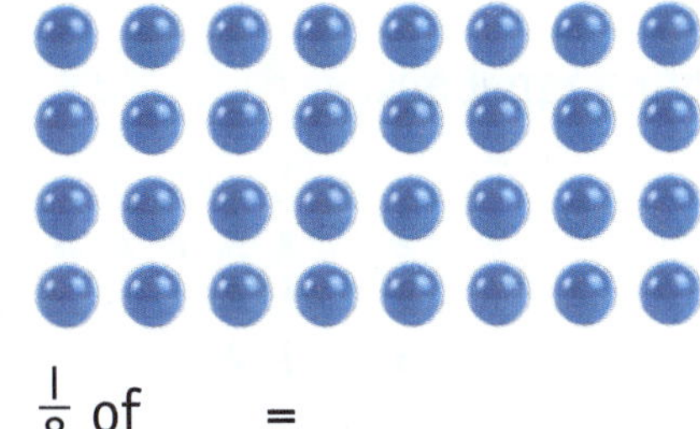

$\frac{1}{8}$ of ___ = ___

c 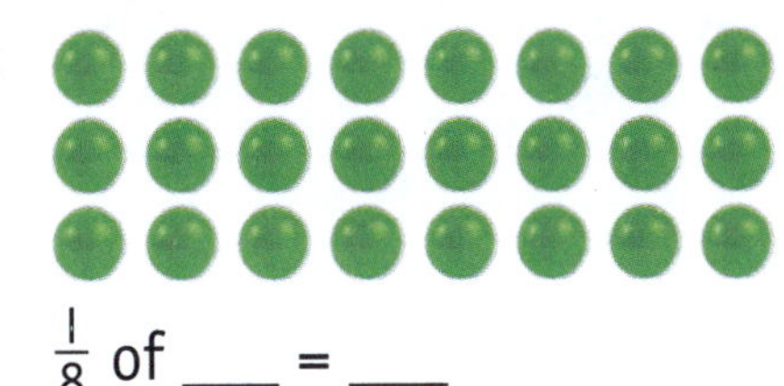

$\frac{1}{8}$ of ___ = ___

d 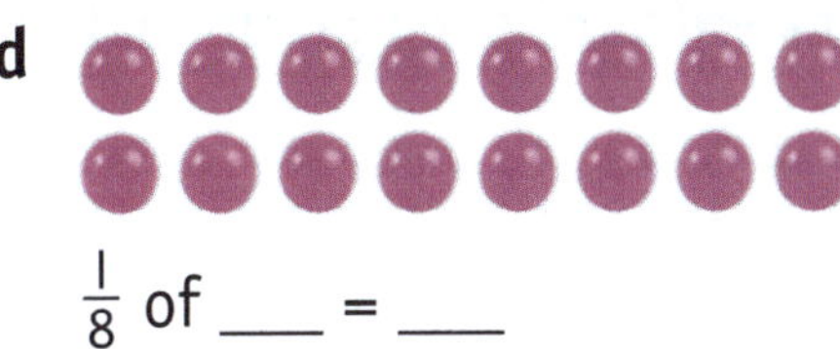

$\frac{1}{8}$ of ___ = ___

6 Circle one-quarter of each group and then write the missing numbers.

a 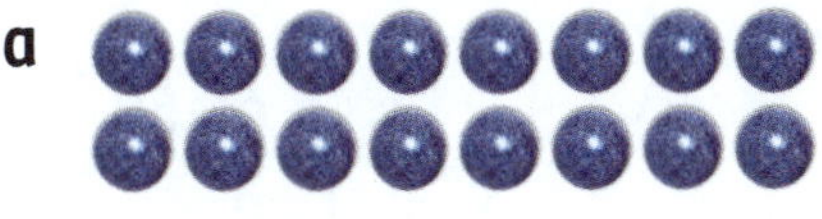

$\frac{1}{4}$ of ___ = ___

b 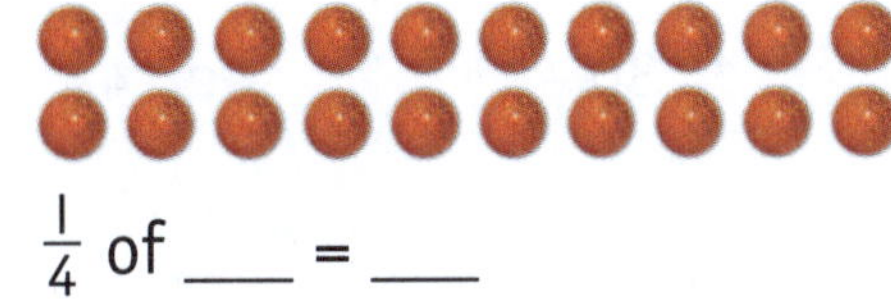

$\frac{1}{4}$ of ___ = ___

c 

$\frac{1}{4}$ of ___ = ___

d 

$\frac{1}{4}$ of ___ = ___

7 What is one-eighth?

a 80 ___ b 72 ___ c 56 ___ d 32 ___ e 64 ___

8 What is one-quarter?

a 8 ___ b 24 ___ c 36 ___ d 44 ___ e 52 ___

9 Write the answer.

a $\frac{1}{4} \times 28 =$ ___ c $\frac{1}{8} \times 80 =$ ___ e $\frac{1}{4} \times 48 =$ ___ g $\frac{1}{8} \times 48 =$ ___

b $\frac{1}{8} \times 40 =$ ___ d $\frac{1}{4} \times 16 =$ ___ f $\frac{1}{4} \times 40 =$ ___ h $\frac{1}{8} \times 96 =$ ___

10 Use red to colour the shapes divided into fifths and blue to colour the shapes divided into thirds.

a  b  c d  e f 

REVIEW

11 Circle one-third of each group and then write the missing numbers.

a $\frac{1}{3}$ of ___ = ___

b $\frac{1}{3}$ of ___ = ___

c $\frac{1}{3}$ of ___ = ___

12 Circle one-fifth of each group and then write the missing numbers.

a

$\frac{1}{5}$ of ___ = ___

b

$\frac{1}{5}$ of ___ = ___

c $\frac{1}{5}$ of ___ = ___

13 What is one-fifth?

a 30 ___ b 5 ___ c 20 ___ d 40 ___ e 60 ___

14 What is one-third?

a 30 ___ b 3 ___ c 21 ___ d 36 ___ e 48 ___

15 Write the answer.

a $\frac{1}{3} \times 6 =$ ___
b $\frac{1}{3} \times 18 =$ ___
c $\frac{1}{5} \times 50 =$ ___
d $\frac{1}{5} \times 75 =$ ___
e $\frac{1}{3} \times 27 =$ ___
f $\frac{1}{5} \times 45 =$ ___
g $\frac{1}{3} \times 24 =$ ___
h $\frac{1}{5} \times 55 =$ ___
i $\frac{1}{3} \times 30 =$ ___
j $\frac{1}{3} \times 12 =$ ___
k $\frac{1}{5} \times 60 =$ ___
l $\frac{1}{3} \times 60 =$ ___

16 Write equivalent fractions.

a $\frac{3}{4} = \frac{}{8}$
b $\frac{1}{2} = \frac{}{10}$
c $\frac{2}{6} = \frac{}{18}$
d $\frac{3}{5} = \frac{}{10}$
e $\frac{1}{4} = \frac{}{8}$
f $\frac{7}{12} = \frac{}{24}$
g $\frac{5}{6} = \frac{}{12}$
h $\frac{2}{3} = \frac{}{6}$
i $\frac{3}{8} = \frac{}{16}$
j $\frac{1}{2} = \frac{}{8}$
k $\frac{10}{12} = \frac{}{6}$
l $\frac{4}{12} = \frac{}{3}$
m $\frac{16}{20} = \frac{}{5}$
n $\frac{12}{20} = \frac{}{5}$
o $\frac{4}{12} = \frac{}{3}$
p $\frac{8}{10} = \frac{}{5}$

CATCH UP MATHS YEAR 6 BOOK A © PASCAL PRESS ISBN: 9781925726183

17 Number the boxes to order these fractions from smallest (1) to largest (5).

a $\frac{2}{4}$ $\frac{2}{3}$ $\frac{3}{5}$ $\frac{5}{8}$ $\frac{2}{2}$

b $\frac{1}{2}$ $\frac{1}{4}$ $\frac{4}{5}$ $\frac{2}{3}$ $\frac{1}{8}$

c $\frac{6}{10}$ $\frac{2}{5}$ $\frac{4}{8}$ $\frac{8}{10}$ $\frac{1}{3}$

d $\frac{3}{4}$ $\frac{1}{5}$ $\frac{2}{3}$ $\frac{7}{8}$ $\frac{5}{10}$

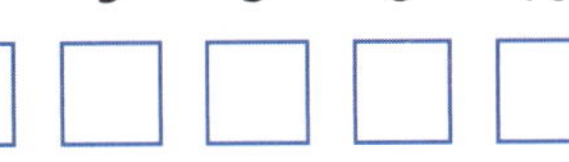

18 Mark and label the fractions on the number lines.

a $\frac{1}{5}$, $\frac{1}{2}$, $\frac{1}{4}$, $\frac{3}{8}$, $\frac{4}{4}$

0 1

b $\frac{1}{10}$, $\frac{2}{5}$, $\frac{7}{8}$, $\frac{3}{4}$, $\frac{2}{3}$

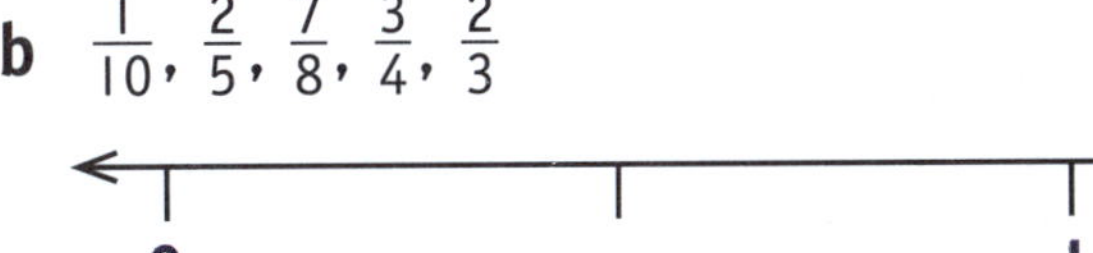

19 Write five fractions that match each description.

a improper fractions with 3 as the denominator

b proper fractions with 4 as the numerator

c improper fractions with 10 as the numerator

d proper fractions with 12 as the denominator

20 Change these improper fractions into mixed numbers.

a $\frac{13}{3}$ = ___

b $\frac{24}{7}$ = ___

c $\frac{109}{10}$ = ___

d $\frac{17}{4}$ = ___

e $\frac{22}{5}$ = ___

f $\frac{23}{2}$ = ___

g $\frac{64}{10}$ = ___

h $\frac{38}{3}$ = ___

i $\frac{91}{2}$ = ___

j $\frac{53}{4}$ = ___

k $\frac{38}{4}$ = ___

l $\frac{59}{7}$ = ___

m $\frac{63}{5}$ = ___

n $\frac{29}{4}$ = ___

o $\frac{39}{10}$ = ___

p $\frac{42}{8}$ = ___

REVIEW

21 Change these mixed numbers into improper fractions.

a $1\frac{2}{3}$ = ___

b $4\frac{5}{12}$ = ___

c $8\frac{5}{9}$ = ___

d $3\frac{1}{4}$ = ___

e $7\frac{6}{8}$ = ___

f $5\frac{10}{12}$ = ___

g $9\frac{3}{5}$ = ___

h $2\frac{7}{12}$ = ___

i $11\frac{4}{5}$ = ___

j $4\frac{3}{8}$ = ___

k $10\frac{2}{5}$ = ___

l $6\frac{8}{10}$ = ___

m $12\frac{2}{3}$ = ___

n $7\frac{3}{10}$ = ___

o $9\frac{3}{4}$ = ___

p $11\frac{4}{10}$ = ___

22 Find the sum.

a $\frac{2}{8} + \frac{2}{8}$ = ___

b $\frac{3}{4} + \frac{1}{4}$ = ___

c $\frac{7}{8} + \frac{1}{8}$ = ___

d $\frac{3}{10} + \frac{2}{10}$ = ___

e $\frac{5}{10} + \frac{2}{10}$ = ___

f $\frac{3}{12} + \frac{4}{12}$ = ___

g $\frac{4}{8} + \frac{3}{8}$ = ___

h $\frac{1}{10} + \frac{2}{10}$ = ___

23 Find the difference.

a $\frac{10}{12} - \frac{2}{12}$ = ___

b $\frac{7}{10} - \frac{1}{10}$ = ___

c $\frac{3}{8} - \frac{1}{8}$ = ___

d $\frac{6}{12} - \frac{2}{12}$ = ___

e $\frac{9}{10} - \frac{2}{10}$ = ___

f $\frac{11}{12} - \frac{5}{12}$ = ___

g $\frac{7}{8} - \frac{6}{8}$ = ___

h $\frac{4}{5} - \frac{2}{5}$ = ___

24 Solve the following.

a $3\frac{5}{8} - 2\frac{1}{8}$ = ______

b $4\frac{10}{12} - 2\frac{5}{12}$ = ______

c $6\frac{3}{4} - 2\frac{1}{4}$ = ______

d $8\frac{9}{10} - 1\frac{4}{10}$ = ______

e $10\frac{9}{12} - 5\frac{6}{12}$ = ______

f $7\frac{3}{5} - 2\frac{1}{5}$ = ______

g $1\frac{1}{4} + 1\frac{2}{4}$ = ______

h $3\frac{3}{5} + 1\frac{1}{5}$ = ______

i $5\frac{4}{10} + 2\frac{2}{10}$ = ______

j $7\frac{4}{12} + 3\frac{5}{12}$ = ______

k $6\frac{1}{5} + 2\frac{2}{5}$ = ______

l $4\frac{1}{8} + 3\frac{5}{8}$ = ______

25 Answer the following.

a $\frac{4}{5} - \frac{6}{20}$

___ = ___

___ − ___ = ___

b $\frac{8}{10} - \frac{3}{5}$

___ = ___

___ − ___ = ___

c $\frac{3}{5} - \frac{1}{10}$

___ = ___

___ − ___ = ___

CATCH UP MATHS YEAR 6 BOOK A © PASCAL PRESS ISBN: 9781925726183

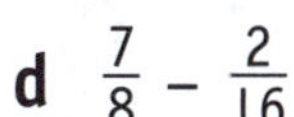

d $\frac{7}{8} - \frac{2}{16}$

___ = ___

___ − ___ = ___

e $\frac{7}{10} - \frac{2}{5}$

___ = ___

___ − ___ = ___

f $\frac{6}{8} - \frac{1}{4}$

___ = ___

___ − ___ = ___

g $\frac{4}{5} + \frac{3}{20}$

___ = ___

___ + ___ = ___

h $\frac{3}{4} + \frac{2}{8}$

___ = ___

___ + ___ = ___

i $\frac{4}{10} + \frac{3}{20}$

___ = ___

___ + ___ = ___

j $\frac{1}{2} + \frac{3}{10}$

___ = ___

___ + ___ = ___

k $\frac{2}{10} + \frac{1}{5}$

___ = ___

___ + ___ = ___

l $\frac{3}{4} + \frac{5}{8}$

___ = ___

___ + ___ = ___

26 Solve the following.

a $4\frac{1}{3} - 2\frac{2}{12}$

___ = ___

___ − ___ = ___

b $4\frac{3}{5} - 1\frac{3}{10}$

___ = ___

___ − ___ = ___

c $9\frac{3}{4} - 2\frac{2}{8}$

___ = ___

___ − ___ = ___

d $2\frac{3}{5} + 1\frac{1}{10}$

___ = ___

___ + ___ = ___

e $3\frac{1}{2} + 2\frac{1}{8}$

___ = ___

___ + ___ = ___

f $5\frac{3}{5} + 3\frac{2}{10}$

___ = ___

___ + ___ = ___

  ISBN: 9781925726183

WRITING DECIMALS

You can write decimals in fraction form and in words.

Example 1: 32 out of 100 = $\frac{32}{100}$ = 0.32 = zero point three two

No whole numbers ↗ ↑ three-tenths ↖ two-hundredths

Example 2: 423 out of 100 = $\frac{423}{100}$ = 4.23 = four point two three

Example 3: 64 out of 100 = $\frac{}{100}$ = __.__ __ = zero point six ______

Example 4: ___ out of 100 = $\frac{72}{100}$ = 0.72 = ____________________

Example 5: 898 out of 100 = $\frac{}{100}$ = __.__ __ = eight point nine eight

Remember, 1 whole = 10 tenths = 100 hundredths

1 tenth = 0.1
1 hundredth = 0.01

1 Use green to circle the whole numbers and blue to circle the decimal fraction.

- 27.42   **a** 1.46   **b** 2.37   **c** 0.43   **d** 18.62

2 Write how many tenths.

- 33.24 <u>2</u>   **a** 1.39 ___   **b** 14.83 ___   **c** 78.09 ___

3 Write how many hundredths.

- 124.15 <u>5</u>   **a** 7.56 ___   **b** 93.02 ___   **c** 8.7 ___

4 Write how many ones.

- 8.34 <u>8</u>   **a** 6.73 ___   **b** 12.20 ___   **c** 10.09 ___

SELF CHECK Tick how you feel		
Got it! ☐	Need help... ☐	I don't get it ☐

Check your answers
How many did you get correct? ☐

CATCH UP MATHS YEAR 6 BOOK A © PASCAL PRESS ISBN: 9781925726183

PRACTICE

1 Write each decimal fraction in words.

- 5.62 five point six two
- a 12.07 ______
- b 63.45 ______
- c 19.3 ______
- d 0.36 ______
- e 78.59 ______

2 Write the decimal fractions in Question 1 in ascending order.

0.36, ______, ______, ______, ______, ______

3 Write the decimal numbers in the table.

- 9 ones, 43 hundredths
- a 6 tens, 5 ones, 9 hundredths
- b 3 tens, 72 hundredths
- c 8 ones, 6 hundredths
- d 7 tens, 4 ones, 12 hundredths
- e 5 tens, 9 hundredths
- f 1 ten, 3 ones, 5 tenths, 8 hundredths
- g 4 tens, 2 ones, 49 hundredths

	Tens	Ones	.	Tenths	Hundredths
●		9	.	4	3
a					
b					
c					
d					
e					
f					
g					

4 Write the decimal fractions in Question 3 in descending order.

74.12, ______, ______, ______, ______, ______, ______, ______

5 Write the decimals using the values of each digit.

- 5.64 5 ones + 6 tenths + 4 hundredths
- a 2.17 ______
- b 4.05 ______
- c 13.89 ______
- d 17.50 ______
- e 0.85 ______

THOUSANDTHS

After tenths and hundredths, the next decimal place value is thousandths. There are 1000 thousandths in 1 whole.

Example 1:

37.462 has:

3 tens
7 ones
4 tenths
6 hundredths
2 thousandths.

3 tens, 7 ones, 4 tenths, 6 hundredths, 2 thousandths

37.462

Tens	Ones	.	Tenths	Hundredths	Thousandths
3	7	.	4	6	2

Example 2:

143.261 has:

1 hundred
4 tens
3 ones
2 tenths
6 hundredths
1 thousandth.

Example 3:

4593.270 has:

__ thousands
__ hundreds
__ tens
__ ones
__ tenths
__ hundredths
__ thousandths.

Example 4:

3978.642 has:

__ thousands
__ hundreds
__ tens
__ ones
__ tenths
__ hundredths
__ thousandths.

1 Use yellow to trace over the tens.

● 432.3541 **a** 206.1573 **b** 35.9824 **c** 5962.1374

2 Use green to trace over the ones.

● 905.4371 **a** 6493.1275 **b** 147.2495 **c** 8753.6219

3 Use blue to trace over the tenths.

● 193.4872 **a** 2468.1357 **b** 201.5946 **c** 67.3258

4 Use red to trace over the hundredths.

● 1493.5680 **a** 145.6032 **b** 34.9876 **c** 9082.4137

5 Use orange to trace over the thousandths.

● 70.524 **a** 950.134 **b** 5490.376 **c** 4156.372

SELF CHECK Tick how you feel

Got it!	Need help...	I don't get it
☐	☐	☐

Check your answers
How many did you get correct? ☐

CATCH UP MATHS YEAR 6 BOOK A © PASCAL PRESS ISBN: 9781925726183

1 What is the place value of 6?

- 45.936 thousandths
- a 36.547 ____________
- b 63.297 ____________
- c 1497.637 ____________
- d 682.794 ____________
- e 6593.123 ____________
- f 2.063 ____________
- g 163.824 ____________

2 Complete the table.

	Decimal	Thousands	Hundreds	Tens	Ones	.	Tenths	Hundredths	Thousandths
	1436.209	1	4	3	6	.	2	0	9
a	247.376								
b	5891.423								
c	149.380								
d	62.413								
e	8.625								
f	4963.820								

3 Write the numbers.

- 8 in the thousandths place, 3 in the ones place, 5 in the tenths place, 6 in the hundredths place

  3.568

- a 2 in the ones place, 6 in the tenths place, 7 in the hundreds place, 3 in the thousandths place

  ____

- b 5 in the thousands place, 6 in the thousandths place, 2 in the hundredths place, 7 in the hundreds place, 3 in the ones place, 4 in the tens place

  ____

- c 3 in the hundreds place, 4 in the thousandths place

  ____

4 Write as decimal fractions.

- 6735 thousandths = 6.735
- a 14 926 thousandths = ______
- b 5493 thousandths = ______
- c 73 498 thousandths = ______
- d 826 thousandths = ______
- e 492 thousandths = ______
- f 109 thousandths = ______
- g 660 thousandths = ______

PARTITIONING DECIMALS

A decimal is made up of parts. These parts can be whole numbers and fractions of whole numbers.

There are many different ways to partition and write decimals.

Example 1:

$4.78 = 4\frac{78}{100}$

4.78 is a mixed number that has 4 wholes + $\frac{7}{10}$ + $\frac{8}{100}$

It can be written as $4 + \frac{78}{100}$

Example 2:

$8.425 = 8\frac{425}{1000}$

8.425 is a mixed number that has 8 wholes + $\frac{4}{10}$ + $\frac{2}{100}$ + $\frac{5}{1000}$

It can be written as $8 + \frac{425}{1000}$

Example 3:

$4.625 = 4\frac{625}{1000}$ = 4 wholes + $\frac{6}{10}$ + $\frac{2}{100}$ + $\frac{5}{1000}$

Example 4:

$7.49 = 7\frac{49}{100}$ = ___ wholes + $\frac{\ }{10}$ + $\frac{\ }{100}$

Example 5:

$2.546 = 2\frac{546}{1000}$ = ___ wholes + $\frac{\ }{10}$ + $\frac{\ }{100}$ + $\frac{\ }{1000}$

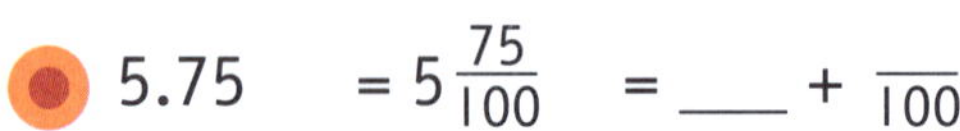

Complete the fractions.

Check your answer on the video!

- ● 5.75 $= 5\frac{75}{100}$ = ___ + $\frac{\ }{100}$
- **a** 2.56 $= 2\frac{56}{100}$ $= 2 + \frac{\ }{10} + \frac{\ }{100}$
- **b** 10.254 $= 10\frac{254}{1000}$ = ___ + $\frac{\ }{10}$ + $\frac{\ }{100}$ + $\frac{\ }{1000}$
- **c** 6.21 $= 6\frac{21}{100}$ $= 6 + \frac{\ }{10} + \frac{\ }{100}$
- **d** 9.34 $= 9\frac{34}{100}$ = ___ + $\frac{\ }{10}$ + $\frac{\ }{100}$
- **e** 4.19 $= 4\frac{19}{100}$ $= 4 + \frac{\ }{100}$
- **f** 8.354 $= 8\frac{354}{1000}$ = ___ + $\frac{\ }{10}$ + $\frac{\ }{100}$ + $\frac{\ }{1000}$

SELF CHECK Tick how you feel		
Got it! ☐	Need help... ☐	I don't get it ☐

Check your answers

How many did you get correct? ☐

ISBN: 9781925726183

1 Write the missing numbers.

7	5 • 2	4	9	thousandths

= 75 + 249 thousandths

= 75 + $\frac{249}{1000}$

a

4	9	3 • 6	tenths	2	hundredths	1	thousandths

= ______ + __ tenths + __ hundredths + __ thousandth

= ______ + $\frac{\quad}{10}$ + $\frac{\quad}{100}$ + $\frac{\quad}{1000}$

b

5	2	4	3 • 7	8	1	thousandths

= _______ + ______ thousandths

= _______ + $\frac{\quad}{1000}$

c

6 • 7	3	hundredths	8	thousandths

= ___ + ___ hundredths + __ thousandths

= ___ + $\frac{\quad}{100}$ + $\frac{\quad}{1000}$

2 Complete the table.

	Decimal	Mixed Number	Wholes	Tenths	Hundredths	Thousandths
●	7.314	$7\frac{314}{1000}$	7	3	1	4
a			8	0	4	3
b		$16\frac{475}{1000}$				
c			9	7	4	2
d	5.037					
e			6	2	0	1
f		$11\frac{230}{1000}$				
g	23.573					

  ISBN: 9781925726183

3 Complete the table.

	Decimal	Mixed Number	Wholes	Thousandths
●	4.072	$4\frac{72}{1000}$	4	$\frac{72}{1000}$
a			9	$\frac{342}{1000}$
b		$15\frac{642}{1000}$		
c			10	$\frac{999}{1000}$
d			37	$\frac{109}{1000}$
e	5.337			
f			16	$\frac{673}{1000}$
g		$1\frac{541}{1000}$		

4 Write the decimal.

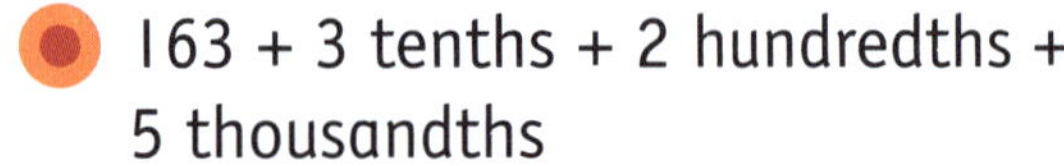

● 163 + 3 tenths + 2 hundredths + 5 thousandths

163.325

a $24 + \frac{3}{10} + \frac{5}{100} + \frac{7}{1000}$ ________

b $4937 + \frac{5}{10} + \frac{6}{100} + \frac{9}{1000}$ ________

c 53 + 2 hundredths + 7 thousandths

____

d $702 + \frac{6}{100} + \frac{4}{1000}$ ________

e 836 + 4 tenths + 3 hundredths

____

f $5 + \frac{3}{10} + \frac{5}{100}$ ________

g 17 + 8 thousandths ________

h $19 + \frac{4}{10} + \frac{7}{100}$ ________

i 27 + 9 thousandths ________

j 86 + 3 tenths + 1 thousandth

____

k 93 + 4 hundredths ________

l $184 + \frac{4}{10} + \frac{3}{100} + \frac{8}{1000}$ ________

m $109 + \frac{2}{10} + \frac{5}{100} + \frac{7}{1000}$ ________

n 246 + 2 tenths + 2 hundredths + 8 thousandths

____

o $6290 + \frac{4}{10} + \frac{7}{100} + \frac{8}{1000}$ ________

p 18 + 8 thousandths ________

q 124 + 5 hundredths ________

r $72 + \frac{7}{100} + \frac{7}{1000}$ ________

s $150 + \frac{8}{10} + \frac{9}{1000}$ ________

CATCH UP MATHS YEAR 6 BOOK A © PASCAL PRESS ISBN: 9781925726183

ADDING AND SUBTRACTING DECIMALS

When you add or subtract decimals, make sure the decimal points line up. If you need to, write zeros at the end so they have the same number of decimal places.

Example 1: 0.5 + 0.28

```
  0 . 5 0  ← Write zero here so both numbers have two decimal places.
+ 0 . 2 8
  -------
  0 . 7 8
```

Example 2: 36.02 + 1.7

```
  3 6 . 0 2
+   1 . 7 0
  ---------
  3 7 . 7 2
```

Example 3: 8.724 − 1.2

```
  8 . 7 2 4
− 1 . 2 0 0
  ---------
  7 . 5 2 4
```

Start by lining up the decimal points.

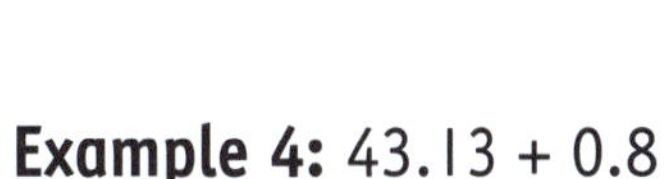

Example 4: 43.13 + 0.8

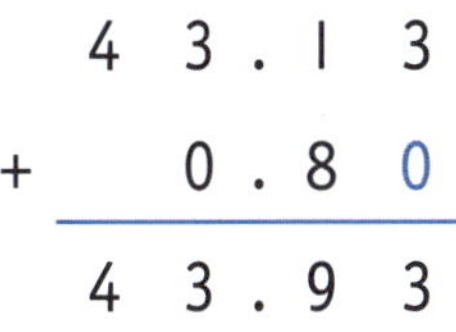

```
  4 3 . 1 3
+   0 . 8 0
  ---------
  4 3 . 9 3
```

Example 5: 3.25 + 15.736

```
  0 3 . 2 5 0
+ 1 5 . 7 3 6
  -----------

```

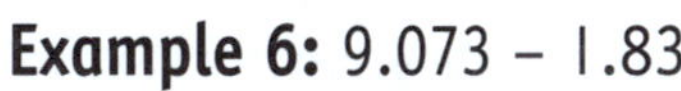

Example 6: 9.073 − 1.83

```
  9 . 0 7 3
− 1 . 8 3 0
  ---------

```

Your turn Solve these additions and subtractions.

```
  7 5 . 0 0 0
+ 2 1 . 5 5 0
  -----------
  9 6 . 5 5 0
```

a
```
  5 1 . 6
− 3 2 . 4 5 2
  -----------

```

b
```
  8 2 . 1 0
+ 1 4 . 9 3 8
  -----------

```

c
```
  1 . 8 9 5
− 0 . 3 7
  ---------

```

d
```
  8 1 . 3 7 4
− 1 4 . 8
  -----------

```

e
```
  9 3 . 9 2 9
+   9 . 8 7 7
  -----------

```

Check your answers
How many did you get correct?

PRACTICE

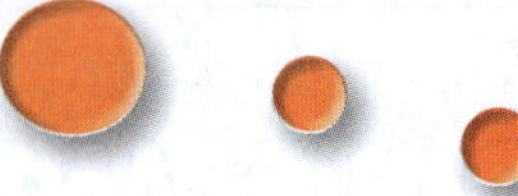

1 Complete the addition algorithms.

●
```
   ¹3 ¹7 ¹6 . ¹8 4
 +  7  4  6 .  5 8
 11 2  3  . 4  2
```

a
```
   9 5 7 . 6 8
 + 9 5 8 . 1
         .
```

b
```
   3 5 4 9 3 . 7 0
 +   2 1 9 5 . 3 6
             .
```

c
```
       6 4 3 . 7
   2 5 8 4 . 6 3
 +     7 2 2 . 1
         8 5 . 6 2
             .
```

d
```
   3 5 4 . 3 5
       1 7 . 8 9 6
 + 1 0 3 . 1 2
       6 2 . 4
           .
```

e
```
   1 5 9 0 . 2 4 5
       6 2 4 . 1
 +       1 4 . 7 3
       1 9 7 . 7 1
               .
```

f
```
   8 4 9 3 . 4 9
         7 1 . 1 5 3
 +         3 . 5 6
     1 2 4 . 9 9 5
             .
```

2 Solve these additions. Use the working space below.

● 62.3 + 12.496 + 1.56 + 2.073 = 78.429

a 12.5 + 72.59 + 36.891 + 35.741 = ____________

b 71.35 + 0.2 + 1.4 + 143.899 = ____________

c 864.1 + 1.43 + 5.637 + 0.4 = ____________

d 7.1 + 3.25 + 14.369 + 283.1 = ____________

Working space

●
```
   6 ¹2 . ²3 0 0
   1  2 .  4 9 6
 + 0  1 .  5 6 0
   0  2 .  0 7 3
   7  8 .  4 2 9
```


CATCH UP MATHS YEAR 6 BOOK A © PASCAL PRESS ISBN: 9781925726183

3 Complete these subtraction algorithms.

● 5 ⁶7̸ ¹¹2̸ ¹4 . 9
 − 1 2 4 9 . 6
 4 4 7 5 . 3

a 7 4 9 5 . 1 2 3
 − 1 0 3 . 2
 .

b 7 4 2 . 1
 − 9 4 . 1 5 9
 .

c 2 4 7 . 8 1
 − 6 . 7 9 4
 .

d 9 0 9 . 0 3 6
 − 9 9 . 7 9 3
 .

e 3 2 1 . 0 7 9
 − 2 . 9
 .

f 5 9 3 4 . 1
 − 1 2 . 7 6 5
 .

g 1 7 3 . 5 7 9
 − 0 . 3 5
 .

h 1 9 2 9 . 6 8
 − 3 7 1 . 1
 .

4 Solve. Use the working space below.

● 64.2 − 1.897 = 62.303

a 1.4 − 0.379 = ______

b 136.1 − 49.38 = ______

c 4834.1 − 124.751 = ______

d 849.372 − 14.5 = ______

e 4483.157 − 1.2 = ______

f 7493.2 − 186.437 = ______

Working space

 6 ³4̸ . ¹¹2̸ ⁹0̸ ¹0
 − 1 . 8 9 7
 6 2 . 3 0 3

MULTIPLYING DECIMALS

When you multiply a decimal by a whole number, the answer will have the same number of digits after the decimal point as the decimal being multiplied.

$$\begin{array}{r} 486.04 \\ \times \quad 8 \\ \hline 3888.32 \\ \hline \end{array}$$

(carries: +6 +4 +3)

2 decimal places after the decimal point

$$\begin{array}{r} 348.426 \\ \times \quad 7 \\ \hline 2438.982 \\ \hline \end{array}$$

(carries: +3 +5 +2 +1 +4)

3 decimal places after the decimal point

Example 1:
The answer to 319.724 × 3 has 3 decimal places.

Example 2:
The answer to 14.2 × 7 has 1 decimal place.

Example 3:
The answer to 1.42 × 7 has 2 decimal places.

Example 4:
The answer to 1.098 × 95 has 3 decimal places.

Example 5:
The answer to 847.14 × 16 has __ decimal places.

Example 6:
The answer to 7.49 × 5 has __ decimal places.

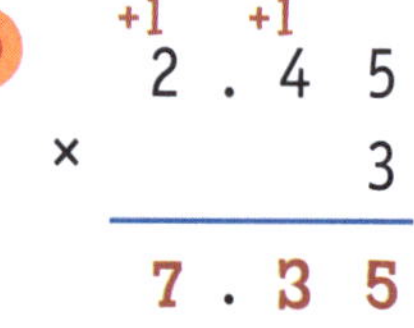

1 Solve.

● 2.45 × 3 = 7.35 (carries: +1 +1)

a 7.4 × 8 = ___.___

b 1.795 × 4 = ___.___

c 14.2 × 7 = ___.___

2 Solve.

● 1.937 × 16 (carries: +5 +2 +4)

$$\begin{array}{r} 1.937 \\ \times \quad 16 \\ \hline 11622 \\ +\ 19370 \\ \hline 30.992 \\ \hline \end{array}$$

a 2.435 × 15

b 7.483 × 22

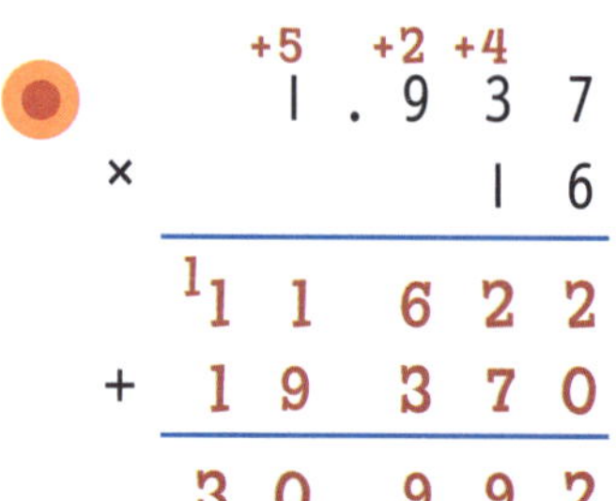

Check your answers
How many did you get correct?

CATCH UP MATHS YEAR 6 BOOK A © PASCAL PRESS ISBN: 9781925726183

Complete the table.

×	3	2	9	12	6	8	7	13
0.4	1.2	0.8	3.6	4.8	2.4	3.2	2.8	5.2
a 1.3								
b 2.25								
c 4.32								
d 5.535								
e 1.116								

Use the prices below to work out the following costs.

Grapes $23.45/kg

Mangoes $17.99/kg

Oranges $6.99/kg

Apples $5.28/kg

Pineapples $6.49/kg

- 3 kg of grapes and 2 kg of oranges $70.35 + $13.98 = $84.33

a 2 kg of mangoes and 2 kg of pineapples ____________

b 7 kg of apples and 4 kg of grapes ____________

c 3 kg of grapes, 3 kg of mangoes and 3 kg of pineapples ____________

d 3 kg of oranges, 4 kg of mangoes and 4 kg of apples ____________

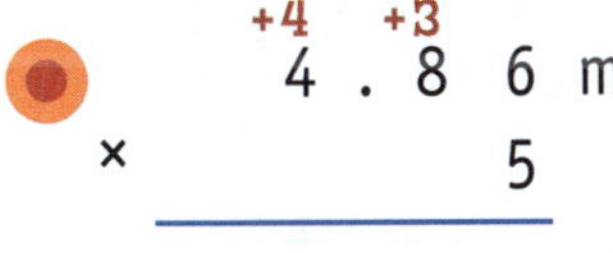

3 Solve these multiplications of metres.

- (carries +4, +3) 4.86 m × 5 = 24.30 m

a 3.73 m × 4 = ____

b 15.32 m × 6 = ____

c 15.36 m × 15 = ____

d 27.59 m × 42 = ____

e 47.34 m × 28 = ____

f 73.64 m × 78 = ____

DIVIDING DECIMALS

When you divide a decimal by a whole number, the answer will have the same number of digits after the decimal point as the decimal being divided.

$$5\overline{)28.^{3}4^{4}5} = 5.69$$

2 decimal places after the decimal point

$$7\overline{)17.^{3}5} = 2.5$$

1 decimal place after the decimal point

Example 1:

Add the decimal point.

$$9\overline{)74.^{2}0^{2}7} \quad 8.23$$

Example 2:

Add the decimal point.

$$4\overline{)88.44} \quad 2\,2\,1\,1$$

Example 3:

Add the decimal point.

$$2\overline{)3^{1}04.6} \quad 1\,5\,2\,3$$

Example 4:

Add the decimal point.

$$5\overline{)6^{1}2^{2}5.55} \quad 1\,2\,5\,1\,1$$

Your turn

Calculate the answer.

- ● $4\overline{)6^{2}5.^{1}0^{2}04} = 16.251$
- **a** $5\overline{)131.5}$
- **b** $8\overline{)17.232}$
- **c** $7\overline{)364.98}$
- **d** $2\overline{)23.0}$
- **e** $6\overline{)92.82}$
- **f** $9\overline{)23.283}$
- **g** $3\overline{)19.584}$
- **h** $4\overline{)26.092}$
- **i** $5\overline{)4.370}$
- **j** $6\overline{)7.548}$
- **k** $8\overline{)1006.4}$

Check your answers

How many did you get correct?

CATCH UP MATHS YEAR 6 BOOK A © PASCAL PRESS ISBN: 9781925726183

PRACTICE

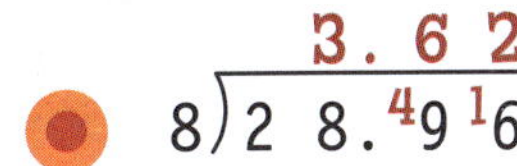

1 Calculate the answer.

● $8\overline{)28.^{4}9^{1}6}$ = 3.62

a $5\overline{)37.50}$

b $9\overline{)156.15}$

c $5\overline{)2510.75}$

d $7\overline{)4025.70}$

e $8\overline{)88.888}$

f $8\overline{)4553.60}$

g $7\overline{)253.40}$

h $6\overline{)1065.36}$

i $6\overline{)219.12}$

j $9\overline{)2120.40}$

k $7\overline{)67.55}$

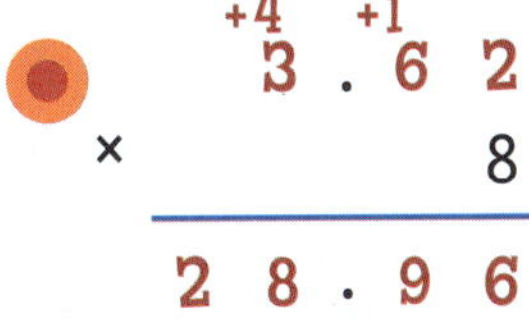

2 Use multiplication to check your answers for Question 1.

● 3.62 × 8 = 28.96 (carries +4, +1)

a ___ . ___ × 5 = ___ . ___

b ___ . ___ × 9 = ___ . ___

c ___ . ___ × 5 = ___ . ___

d ___ . ___ × 7 = ___ . ___

e ___ . ___ × 8 = ___ . ___

f ___ . ___ × 8 = ___ . ___

g ___ . ___ × 7 = ___ . ___

h ___ . ___ × 6 = ___ . ___

i ___ . ___ × 6 = ___ . ___

j ___ . ___ × 9 = ___ . ___

k ___ . ___ × 7 = ___ . ___

DIVIDING DECIMALS BY POWERS OF 10

When you multiply, you move the decimal point to the right.
When you divide, move the decimal point to the left.

SCAN to watch video

Multiplying decimals by powers of 10

× 10 — 10 has 1 zero. Move the decimal point 1 place to the right.

× 100 — 100 has 2 zeros. Move the decimal point 2 places to the right.

× 1000 — 1000 has 3 zeros. Move the decimal point 3 places to the right.

× → Move . to the right

Example 1: 0.36 × 10 = 3.6

Example 2: 2.85 × 100 = 285.

Example 3: 31.895 × 1000 = 31 895.

Example 4:
3.2 × 10 = ________

Example 5:
2.093 × 100 = ________

Example 6:
0.695 × 1000 = ________

Dividing decimals by powers of 10

÷ 10 — 10 has 1 zero. Move the decimal point 1 place to the left.

÷ 100 — 100 has 2 zeros. Move the decimal point 2 places to the left.

÷ 1000 — 1000 has 3 zeros. Move the decimal point 3 places to the left.

÷ ← Move . to the left

Example 7: 0.36 ÷ 10 = 0.036

Example 8: 2.85 ÷ 100 = 0.0285

Example 9: 31.895 ÷ 1000 = 0.031 895

Example 10:
3.2 ÷ 10 = ________

Example 11:
2.093 ÷ 100 = ________

Example 12:
0.695 ÷ 1000 = ________

Your turn

Write the answer.

- 0.0375 × 1000 = 37.5
- a 2.79 × 10 = ________
- b 8.62 ÷ 10 = ________
- c 1.932 ÷ 1000 = ________
- d 6.3 × 1000 = ________
- e 8.493 × 100 = ________
- f 0.9 ÷ 1000 = ________
- g 1.825 ÷ 1000 = ________

SELF CHECK Tick how you feel

Got it!	Need help...	I don't get it
☐	☐	☐

Check your answers
How many did you get correct? ☐

CATCH UP MATHS YEAR 6 BOOK A © PASCAL PRESS ISBN: 9781925726183

PRACTICE

1 Multiply each decimal by 10, 100 and 1000.

	Decimal	× 10	× 100	× 1000
●	0.3	3.0	30	300
a	31.4			
b	122.7			
c	9.9			
d	0.54			
h	1.246			
i	38.720			
j	0.329			
k	179.855			

2 Divide each decimal by 10, 100 and 1000.

	Decimal	÷ 10	÷ 100	÷ 1000
●	7.2	0.72	0.072	0.0072
a	21.3			
b	802.4			
c	0.9			
d	0.37			
h	0.994			
i	14.816			
j	5.049			
k	736.863			

3 Answer.

- ● 6.037 × 10 = 60.37
- a 9.02 × 100 = ________
- b 0.3 × 1000 = ________
- c 17.3 ÷ 100 = ________
- d 830.205 ÷ 1000 = ________
- e 6.75 ÷ 10 = ________
- f 6.203 × 10 = ________
- g 3.603 × 100 = ________
- h 703.24 ÷ 1000 = ________
- i 609.909 ÷ 100 = ________

PERCENTAGES

Per cent (%) means out of 100.

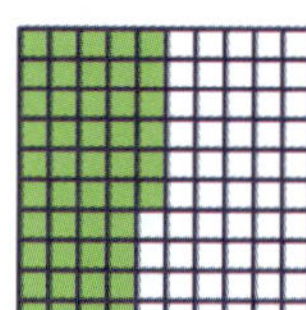

Fraction	Decimal	Percentage
$\frac{46}{100}$	0.46	46%

46 hundredths are coloured green

SCAN to watch video

Example 1:
60% means 60 out of 100

Example 2:
3% means ___ out of ___

Example 3:
98% means ___ out of ___

Example 4:
59% means ___ out of ___

Example 5:
14% means ___ out of ___

Example 6:
28% means ___ out of ___

Check your answer on the video!

Your turn

Colour each grid to show the percentage and then write the equivalent fraction and decimal.

● 62%

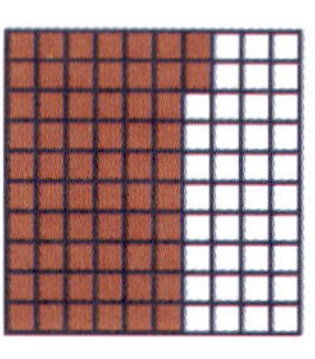

0.62 | $\frac{62}{100}$

b 94%

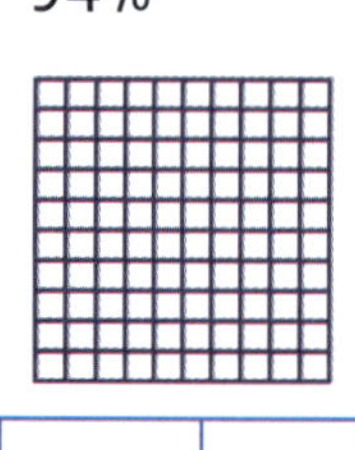

0.___ | $\frac{\quad}{100}$

d 2%

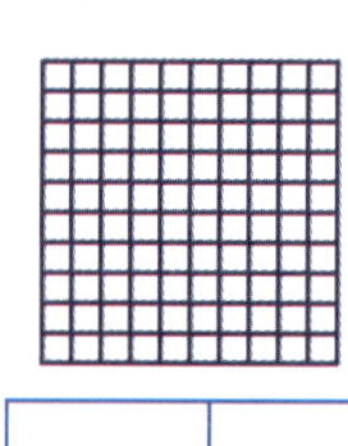

0.___ | $\frac{\quad}{100}$

f 36%

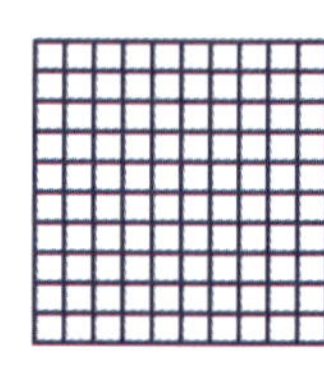

0.___ | $\frac{\quad}{100}$

a 15%

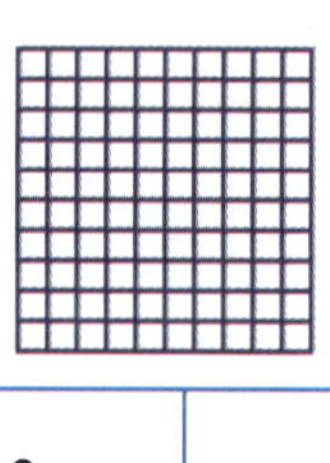

0.___ | $\frac{\quad}{100}$

c 57%

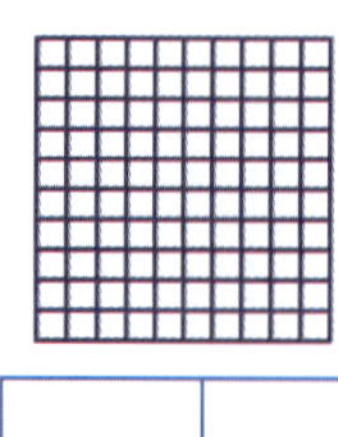

0.___ | $\frac{\quad}{100}$

e 21%

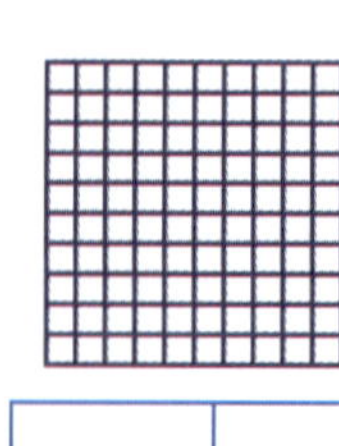

0.___ | $\frac{\quad}{100}$

g 83%

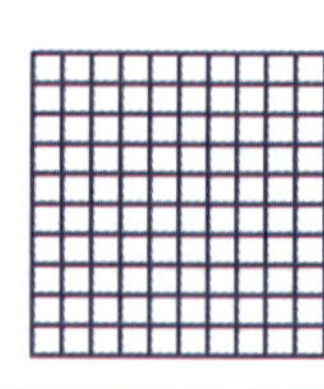

0.___ | $\frac{\quad}{100}$

SELF CHECK Tick how you feel

Got it!	Need help...	I don't get it
 ☐	 ☐	 ☐

Check your answers
How many did you get correct? ☐

CATCH UP MATHS YEAR 6 BOOK A © PASCAL PRESS ISBN: 9781925726183

PRACTICE

1 Convert each fraction to a percentage.

- ● $\frac{31}{100}$ = 31%
- a $\frac{62}{100}$ = ____
- b $\frac{90}{100}$ = ____
- c $\frac{14}{100}$ = ____
- d $\frac{27}{100}$ = ____
- e $\frac{44}{100}$ = ____
- f $\frac{53}{100}$ = ____
- g $\frac{79}{100}$ = ____

2 Convert each decimal to a percentage.

- ● 0.38 = 38%
- a 0.64 = ____
- b 0.99 = ____
- c 0.17 = ____
- d 0.22 = ____
- e 0.41 = ____
- f 0.56 = ____
- g 0.77 = ____

3 Complete the tables.

	Fraction	Decimal	Percentage
●	$\frac{1}{10}$	0.10	10%
a	$\frac{2}{10}$		
b	$\frac{3}{10}$		
c	$\frac{4}{10}$		
d	$\frac{5}{10}$		
e	$\frac{6}{10}$		
f	$\frac{7}{10}$		
g	$\frac{8}{10}$		

	Fraction	Decimal	Percentage
h	$\frac{9}{10}$		
i	$\frac{10}{10}$		
j	$\frac{1}{4}$		
k	$\frac{1}{2}$		
l	$\frac{3}{4}$		
m	$\frac{1}{5}$		
n	$\frac{1}{3}$		
o	$\frac{1}{8}$		

4 How much?

- ● 30% of $1.00 = $0.30
- a 20% of $1.00 = ______
- b 25% of $1.00 = ______
- c 50% of $2.00 = ______
- d 75% of $2.00 = ______
- e 90% of $2.00 = ______
- f 10% of $5.00 = ______
- g 40% of $10.00 = ______
- h 60% of $20.00 = ______
- i 80% of $20.00 = ______
- j 100% of $120.00 = ______
- k $33\frac{1}{3}$% of $100.00 = ______
- l $66\frac{1}{3}$% of $2.00 = ______
- m 12.5% of $10.00 = ______

  ISBN: 9781925726183

EQUIVALENT FRACTIONS, DECIMALS AND PERCENTAGES

You can write the same value as a fraction, a decimal and a percentage. For example, $\frac{32}{100}$ is the same value as 0.32 and 32%.

Example 1:

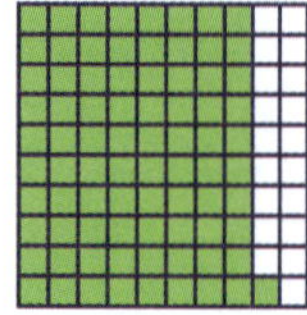

Fraction	Decimal	Percentage
$\frac{81}{100}$	0.81	81%

Example 2:

Fraction	Decimal	Percentage
$\frac{163}{100}$	1.63	163%

Example 3:

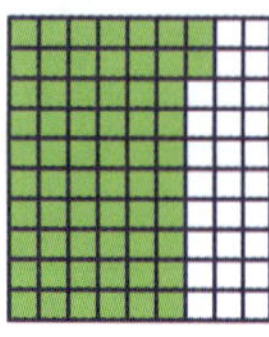

Fraction	Decimal	Percentage
	0.62	

Example 4:

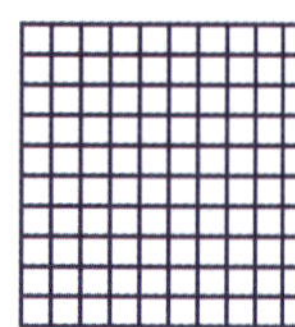

Fraction	Decimal	Percentage
$\frac{71}{100}$		

Example 5:

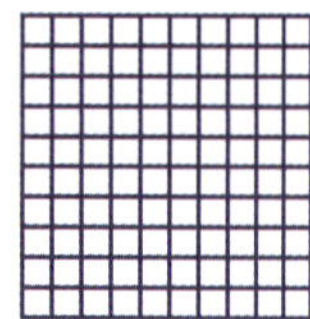

Fraction	Decimal	Percentage
		50%

1 Write as a decimal.

- 147% 1.47 a 24% ______ b $\frac{15}{1000}$ ______

2 Write as a fraction.

- 2.48 $\frac{248}{100}$ a 127% ______ b 0.61 ______

3 Write as a percentage.

- 3.76 376% a $\frac{103}{100}$ ______ b 8.01 ______

SELF CHECK Tick how you feel

Got it!	Need help...	I don't get it
☐	☐	☐

Check your answers

How many did you get correct? ☐

CATCH UP MATHS YEAR 6 BOOK A © PASCAL PRESS ISBN: 9781925726183

PRACTICE

1 Complete the table.

	Fraction	Decimal	Percentage
●	$\frac{24}{100}$	0.24	24%
a		0.83	
b			76%
c	$\frac{3}{100}$		
d			49%
e		0.50	
f	$\frac{10}{100}$		
g			13%

	Fraction	Decimal	Percentage
h	$\frac{149}{100}$		
i		2.62	
j			501%
k	$\frac{830}{100}$		
l		4.87	
m			990%
n		1.09	
o	$\frac{747}{100}$		

2 Convert to percentages.

● 4.37 437%

a 2.95 ______

b $\frac{436}{100}$ ______

c 0.80 ______

d 7.0 ______

e $\frac{310}{100}$ ______

f $\frac{895}{100}$ ______

g 6.06 ______

3 Convert to decimals.

● 344% 3.44

a $\frac{717}{100}$ ______

b $\frac{124}{100}$ ______

c 33% ______

d 1% ______

e $\frac{606}{100}$ ______

f $\frac{7}{100}$ ______

g 736% ______

4 Convert to fractions.

● 5.49 $\frac{549}{100}$

a 2.29 ______

b 43% ______

c 6% ______

d 4.4 ______

e 3.86 ______

f 5.91 ______

g 16.8 ______

h 72% ______

i 7.80 ______

j 6.0 ______

k 23.95 ______

l 1.88 ______

m 99% ______

n 63% ______

o 4.89 ______

  ISBN: 9781925726183

DISCOUNTS

A discount is a reduction in price.
Here are two ways to work out a percentage discount.

SCAN to watch video

Multiply the cost by the percentage on a calculator.

20% of $40

= $ 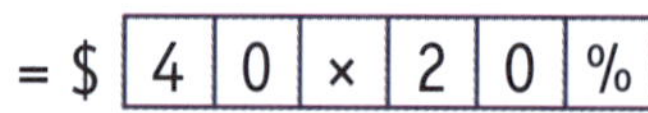

= $8

Write the percentage as a decimal and then multiply the cost.

20% of $40

= 0.2 × 40

= $8

This means the shirt is 20% cheaper.

	Item and cost	Discount	Calculator	Decimal
Example 1:	$180	30%	30% of $180 180 × 30% = $54	30% of $180 0.3 × 180 = $54
Example 2:	$75	60%	60% of $75 = 75 × 60% = ______	60% of $75 = 0.6 × 75 = ______

Check your answer on the video!

Find the discounted price of each item.

	Item and cost	Discount	Calculator	Decimal	Discounted Price
●	$55	10%	10% of $55 55 × 10% = $5.50	10% of $55 0.1 × 55 = $5.50	$55.00 – $ 5.50 $49.50
a	$170	25%	25% of $170 ____ × ____ = ______	25% of $170 ____ × ____ = ______	$170.00 – $ ______
b	$15	20%	20% of $15 ____ × ____ = ______	20% of $15 ____ × ____ = ______	$15.00 – $ ______

SELF CHECK Tick how you feel

Got it!	Need help...	I don't get it
☐	☐	☐

Check your answers
How many did you get correct? ☐

CATCH UP MATHS YEAR 6 BOOK A © PASCAL PRESS ISBN: 9781925726183

PRACTICE

Write the discount in dollars and the discounted price.

	Item	Price	% Discount	$ Discount	Discounted Price
●	Toaster	$89	20%	$17.80	$71.20
a	Kettle	$125	10%		
b	Griller	$45	25%		
c	Computer	$1524	30%		
d	Microwave	$399	50%		
e	Blender	$155	40%		
f	Juicer	$49	20%		

Johnny found the jacket he wanted to buy at three different stores. Circle the store that has the cheapest price.

STAR SHIRTS

20% off $120

T-SHIRT CITY

10% off $110

DRESS ME

30% off $130

Dom and Livvy went shopping. Work out the discounts of the items they bought.

Dom

Item	Original price	Discount	Discounted price
T-shirt	$45.00	10%	
Jeans	$50.00	20%	
Shoes	$150.00	30%	
Total:		Total:	

Livvy

Item	Original price	Discount	Discounted price
T-shirt	$50.00	20%	
Jeans	$60.00	30%	
Shoes	$170.00	20%	
Total:		Total:	

a Who spent more money after all the discounts? _______

b What was the value of Dom's discount? _______

c What was the value of Livvy's discount? _______

DECIMALS REVIEW

1 Write each decimal fraction in words.

a 0.58 ______________________

b 2.63 ______________________

c 8.495 ______________________

2 Write the decimal numbers in the table below.

a 8 ones, 7 hundredths

b 3 tens, 4 ones, 5 thousands, 6 thousandths, 23 hundredths

c 8 thousandths, 4 ones, 5 tenths, 6 hundredths

d 3 tenths, 4 thousandths, 3 hundredths

e 5 hundredths, 5 thousands

	Thousands	Hundreds	Tens	Ones	.	Tenths	Hundredths	Thousandths
a								
b								
c								
d								
e								

3 Write the decimals using the values of each digit.

a 2.72 ______________________

b 0.39 ______________________

c 9.1 ______________________

d 57.485 ______________________

4 What is the place value of the 9?

a 45.936 __________

b 96.574 __________

c 36.792 __________

d 9563.12 __________

e 9.375 __________

f 190.375 __________

g 1953.46 __________

h 124.859 __________

i 9143.78 __________

j 94.27 __________

CATCH UP MATHS YEAR 6 BOOK A © PASCAL PRESS ISBN: 9781925726183

5 Write the decimal.

a 146 + 4 tenths + 3 hundredths + 7 thousandths = ______

b 33 + 3 hundredths = ______

c $302 + \frac{6}{10} + \frac{5}{100} + \frac{7}{1000}$ = ______

d 16 + 8 thousandths = ______

e $857 + \frac{6}{100} + \frac{9}{1000}$ = ______

f $9 + \frac{2}{10} + \frac{5}{1000}$ = ______

g 8 + 7 tenths + 9 thousandths = ______

h $41\frac{362}{1000}$ = ______

6 Solve these additions.

a
```
  428.34
+ 131.12
  ______
```

b
```
  756.12
+ 241.01
  ______
```

c
```
  430.25
+  29.43
  ______
```

d
```
   634.8
     3.94
+ 1431.2
    25.17
  _______
```

e
```
   43.759
    2.13
+ 125.9
   50.302
  _______
```

f
```
     2.91
+  517.5
    32.815
  ________
```

g
```
  4849.1
    57.593
+    6.24
   345.4
  _______
```

7 Solve these additions. Use the working space at the right.

a 32.1 + 15.892 + 2.47 + 1.037 = ______

b 49.37 + 0.9 + 3.4 + 522.185 = ______

Working space

8 Solve these subtractions.

a
```
  3825.8
- 1413.1
  ______
```

b
```
  6954.321
-  450.1
  ________
```

c
```
  352.1
-   0.345
  _______
```

REVIEW

9 Solve these subtractions.
Use the working space at the right.

a 84.2 – 2.869 = ________

b 5836.1 – 124.326 = ____________

c 746.3 – 49.625 = ___________

Working space

10 Solve these decimal multiplications.

a 1.873 × 4 = ____.____

b 8.754 × 3 = ____.____

c 2.073 × 8 = ____.____

d 9.154 × 7 = ____.____

e 4.715 × 6 = ____.____

f 3.563 × 7 = ____.____

11 Solve these decimal divisions.

a 9)5755.77

b 4)2609.20

c 7)1109.92

d 6)883.50

e 5)1846.290

f 7)876.792

g 3)1564.20

h 8)1178.064

i 9)1111.104

12 Use multiplication to check your answers for Question 11.

a ____.____ × 9 = ____.____

b ____.____ × 4 = ____.____

c ____.____ × 7 = ____.____

CATCH UP MATHS YEAR 6 BOOK A © PASCAL PRESS ISBN: 9781925726183

d
× 6

f
× 7

h
× 8

e
× 5

g
× 3

i
× 9

13 Complete the tables.

	Decimal	× 10	× 100	× 1000
a	0.6			
b	15.4			
c	184.59			
d	4432.573			
e	0.895			

	Decimal	÷ 10	÷ 100	÷ 1000
f	0.8			
g	36.47			
h	164.86			
i	5981.246			
j	0.458			

14 Solve.

a 9.31 × 100 = ________

b 731.16 × 10 = ________

c 8.097 × 1000 = ________

d 2.45 ÷ 10 = ________

e 584.251 ÷ 100 = ________

f 0.953 ÷ 1000 = ________

g 17.569 × 100 = ________

h 1495.231 ÷ 1000 = ________

i 7.02 ÷ 100 = ________

j 0.2 × 1000 = ________

k 59.24 ÷ 10 = ________

l 0.064 × 10 = ________

REVIEW

15 Convert to percentages.

a 7.49 ______ c $\frac{895}{100}$ ______ e 8.0 ______ g $\frac{637}{100}$ ______

b 5.2 ______ d 0.60 ______ f $\frac{410}{100}$ ______ h 7.04 ______

16 Convert to decimals.

a 543% ______ c $\frac{560}{100}$ ______ e $\frac{910}{100}$ ______ g $\frac{8}{100}$ ______

b $\frac{104}{100}$ ______ d 856% ______ f 790% ______ h 17% ______

17 Convert to fractions.

a 4.95 ______ c 8.2 ______ e 7.11 ______ g 4.37 ______

b 519% ______ d 15% ______ f 2% ______ h 48% ______

18 Complete the tables.

	Fraction	Decimal	Percentage
a		0.21	
b	$\frac{34}{100}$		
c			57%
d		0.40	

	Fraction	Decimal	Percentage
e	$\frac{79}{100}$		
f			69%
g			157%
h		5.06	

19 Write the missing numbers.

a 31% means ____ out of ___

b 72% means ____ out of ___

c 544% means _____ out of ___

d 801% means _____ out of ___

20 Complete the tables.

	Fraction	Decimal	Percentage
a	$\frac{3}{10}$		
b	$\frac{1}{4}$		
c	$\frac{3}{4}$		
d	$\frac{2}{5}$		

	Fraction	Decimal	Percentage
e	$\frac{2}{3}$		
f	$\frac{1}{8}$		
g	$\frac{10}{10}$		
h	$\frac{1}{5}$		

  ISBN: 9781925726183

21 How much?

a 20% of $1.00 = ________

b 50% of $5.00 = ________

c 75% of $10.00 = ________

d 100% of $60.00 = ________

e 10% of $70.00 = ________

f 10% of $85.00 = ________

g 40% of $50.00 = ________

h 80% of $110.00 = ________

i $33\frac{1}{3}$% of $10.00 = ________

j 30% of $240.00 = ________

22 Write the discount in dollars and the discounted price.

	Item	Price	% Discount	$ Discount	Discounted Price
a	Shoes	$190	10%		
b	T-shirt	$60	50%		
c	Shorts	$120	30%		
d	Jeans	$140	60%		
e	Pullover	$75	20%		
f	Trackpants	$90	70%		
g	Belt	$45	40%		

23 Max and Zoe went shopping. Work out the discounts of the items they bought.

Max

Item	Original price	Discount	Discounted price
Dress	$120.00	40%	
Shoes	$140.00	30%	
Hat	$55.00	20%	
Jacket	$190.00	30%	
Total:		Total:	

Zoe

Item	Original price	Discount	Discounted price
Dress	$150.00	50%	
Shoes	$160.00	20%	
Hat	$40.00	10%	
Jacket	$180.00	20%	
Total:		Total:	

a Who spent more money after all the discounts? ________

b What was Max's total before discounts? ________

c What was Zoe's total before discounts? ________

d How much money did Max save? ________

e How much money did Zoe save? ________

NUMBER PATTERNS

A number pattern is a series of numbers that follows a rule.
The rule tells you what to do with each number to continue the pattern.

	Rule	Number Pattern
Example 1	Add 5	9, 14, 19, 24, 29, 34, 39, 44
Example 2	Subtract 2	27, 25, 23, 21, 19, 17, 15, 13
Example 3	Multiply by 3	6, 18, 54, 162, 486, 1458, 4374, 13 122
Example 4	Divide by 2	128, 64, 32, 16, 8, 4, 2, 1
Example 5	Add 6	4, ___, 16, 22, ___, ___
Example 6	Subtract 3	20, 17, ___, ___, 8, 5
Example 7	Multiply by 2	7, ___, ___, 56, 112, ___
Example 8	Divide by 2	512, 256, ___, ___, 32, ___

1 Use the rule to continue the pattern.

- 57, 60, 63, 66, 69, 72 — Rule: + 3
- **a** 70, 63, ___, ___, 42, ___ — Rule: − 7
- **b** 29, ___, ___, 17, 13, ___ — Rule: − 4
- **c** 1, ___, 25, ___, 625, ___ — Rule: × 5

2 Work out the rule for each pattern.

- 13, 18, 23, 28, 33, 38 — Rule: + 5
- **a** 72, 63, 54, 45, 36, 27 — Rule: ___
- **b** 102, 106, 110, 114, 118, 122 — Rule: ___
- **c** 300, 150, 75, $37\frac{1}{2}$, $18\frac{3}{4}$, $9\frac{3}{8}$ — Rule: ___
- **d** 8, 24, 72, 216, 648, 1944 — Rule: ___

Check your answers
How many did you get correct?

CATCH UP MATHS YEAR 6 BOOK A © PASCAL PRESS ISBN: 9781925726183

PRACTICE

1 Complete the table.

	Rule	Number Pattern
●	× 2	4, 8, 16, 24, 64, 128
a	Add 4	___, 13, ___, 21, ___, 29
b	Subtract 2	101, ___, 97, ___, ___, 91
c	+ 8	___, 1364, ___, ___, 1388, 1396
d	− 16	___, 12 487, ___, 12 455, 12 439, ___

2 Continue the pattern and write the rule.

● 7, 13, 19, 25, 31, 37 Rule: + 6

a 80, 71, 62, ___, ___, ___ Rule: ___

b 5, 25, 125, ___, ___, ___ Rule: ___

c 195, 191, 187, ___, ___, ___ Rule: ___

d 4096, 2048, ___, ___, ___ Rule: ___

e ___, 12, ___, 48, 96, ___ Rule: ___

3 Write a number sentence to describe the pattern.

● D = E + F

D	E	F
21	9	12
43	5	38
18	9	9
16	11	5

a ___

G	H	I
3	24	8
9	36	4
2	10	5
11	110	10

b ___

A	B	C
12	3	4
20	5	4
81	9	9
108	12	9

c ___

J	K	L
10	7	3
39	24	15
34	16	18
146	125	21

4 Write the missing numbers in these sequences with two patterns.

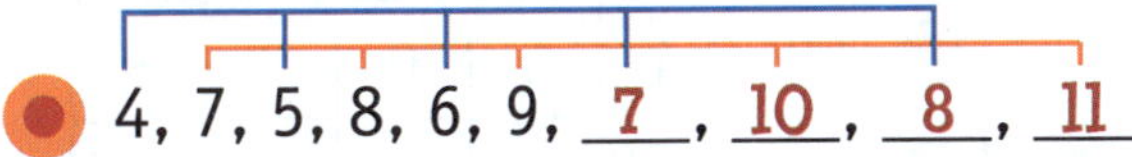

● 4, 7, 5, 8, 6, 9, 7, 10, 8, 11

a 10, 4, 9, 5, 8, 6, ___, ___, ___, ___

b 2, 3, 4, 6, 6, 9, ___, ___, ___, ___

c 3, 4, 6, 7, 9, 10, ___, ___, ___, ___

d 4, 5, 8, 7, 12, 9, ___, ___, ___, ___

e 25, 5, 20, 10, 15, 15, ___, ___, ___

PATTERN GRIDS

You can use tables to show patterns.
The rows of numbers are related by a rule.

SCAN to watch video

Example 1: ■ = ● + 8

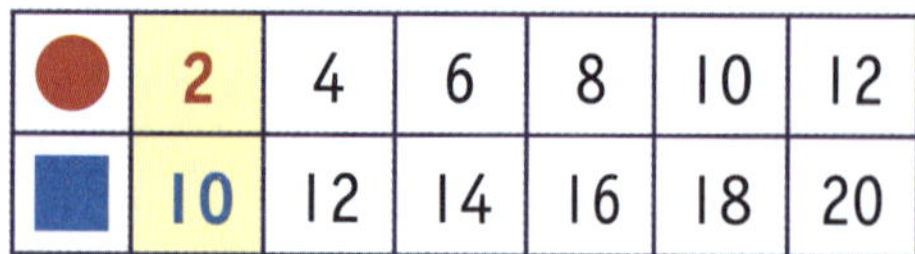

●	2	4	6	8	10	12
■	10	12	14	16	18	20

■ = 2 + 8
= 10

Example 2: ★ = ▲ – 5

▲	100	99	98	97	96	95
★	95	94	93	92	91	90

★ = 100 – 5
= 95

Example 3: ● × 2 + 1 = ▲

●	1	2	3	4	5	6
▲	3	5	7	9	11	13

▲ = 1 × 2 + 1
= 3

Example 4: (F + 10) × 2 = E

F	5	4	3	2	1	0
E	30	28	26	24	22	20

E = (5 + 10) × 2
= 30

Example 5: ▲ = ⬢ ÷ 3

⬢	36	33	30	27	24	21
▲	12	11				7

▲ = 36 ÷ 3
= 12

Example 6: A = B + 7

B	5	10	15	20	25	30
A	12					

A = 5 + 7
= 12

Your turn

Complete the pattern grids.

● ▲ × 3 – 1 = ■

▲	1	3	5	7	9	11
■	2	8	14	20	26	32

a ◆ = (● + 5) × 2

●	5	10	15	20	25	30
◆						

b ▲ = ★ × 3 + 2

★	2	4	6	8	10	12
▲						

c (● – 3) × 4 = ⬢

●	3	4	5	6	7	8
⬢						

SELF CHECK Tick how you feel

Got it! ☐ Need help... ☐ I don't get it ☐

Check your answers
How many did you get correct? ☐

CATCH UP MATHS YEAR 6 BOOK A © PASCAL PRESS ISBN: 9781925726183

1 Complete the pattern grids.

■ × 7 + 3 = ●

■	3	6	9	12	15	18
●	24	45	66	87	108	129

a ■ + 16 = ●

■	5	10	15	20	25	30
●						

b ▲ = ■ − 6 + 34

■	10	12	14	16	18	20
▲						

c (▲ + 10) × 3 = ■

▲	8	7	6	5	4	3
■						

2 Write the rule for each pattern grid.

Rule: ★ × 2 + 1 = ●

★	1	2	3	4	5	6
●	3	5	7	9	11	13

a Rule: ▲ = ■ × __ + __

■	1	2	3	4	5	6
▲	6	10	14	18	22	26

b Rule: ★ = (▲ − __) × __

▲	10	9	8	7	6	5
★	28	24	20	16	12	8

c Rule: ● × __ − __ = ◆

●	2	4	6	8	10	12
◆	4	10	16	22	28	34

3 Use the rules to write the numbers.

Rule	2	4	6	8	10	12
+ 6	8	10	12	14	16	18
× 3	6	12	18	24	30	36

a

Rule	1	3	5	7	9	11
− 5						
× 2						

b

Rule	5	6	7	8	9	10
× 2						
+ 4						
− 6						
× 11						

c

Rule	5	6	7	8	9	10
× 1						
+ 8						
− 10						
÷ 2						

d

Rule	6	8	10	12	14	16
× 3						
+ 15						
÷ 2						
× 10						
− 6						

EQUIVALENT NUMBER SENTENCES

Number sentences that are equal to each other are equivalent. One side of the number sentence is equal to or equivalent to the other side.

Example 1:

4 + 6 = 5 + 5

Both sides equal 10.

Example 2:

22 – 10 = 2 × 6

Both sides equal 12.

Example 3:

9 × 2 = 1 × 18

Both sides equal 18.

Example 4:

7 × 4 = 10 + 18

Both sides equal 28.

Example 5:

36 ÷ 12 = 12 – 9

Both sides equal 3.

Example 6:

8 + 8 = 4 × 4

Both sides equal ___.

Example 7:

35 – 5 = ☐ × 6

Both sides equal ___.

Example 8:

7 × ☐ = 42 – 21

Both sides equal ___.

Equivalent means 'equal in value'.

Check your answer on the video!

Your turn

Tick the number sentences that are equivalent.

- 2 × 4 = 10 – 2 ☑
- **a** 7 × 2 = 20 – 6 ☐
- **b** 10 × 3 = 36 – 35 ☐
- **c** 40 + 2 = 7 × 6 ☐
- **d** 12 + 0 = 24 ÷ 2 ☐
- **e** 7 × 7 = 56 – 7 ☐
- **f** 10 × 12 = 119 + 2 ☐
- **g** 4 × 11 = 48 – 10 ☐
- **h** 8 × 2 = 4 × 3 ☐
- **i** 24 – 3 = 6 × 4 ☐

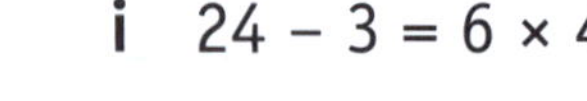

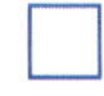

SELF CHECK Tick how you feel

Got it!	Need help...	I don't get it
☐	☐	☐

Check your answers

How many did you get correct? ☐

CATCH UP MATHS YEAR 6 BOOK A © PASCAL PRESS ISBN: 9781925726183

PRACTICE

1 Complete the equivalent number sentences.

- (example) 8 × 8 = [60] + 4
- a 120 ÷ 10 = 6 × [ ]
- b 9 × [ ] = 53 + 10
- c [ ] × 4 = 20 + 12
- d 12 ÷ 4 = 20 − [ ]
- e [ ] + 10 = 5 × 11
- f 3 × 5 = 30 ÷ [ ]
- g 45 − 7 = 40 − [ ]
- h 12 + 12 = [ ] ÷ 2
- i 56 ÷ [ ] = 20 − 13
- j 24 ÷ [ ] = 50 − 42
- k 63 ÷ 7 = 12 − [ ]

2 Write two equivalent number sentences.

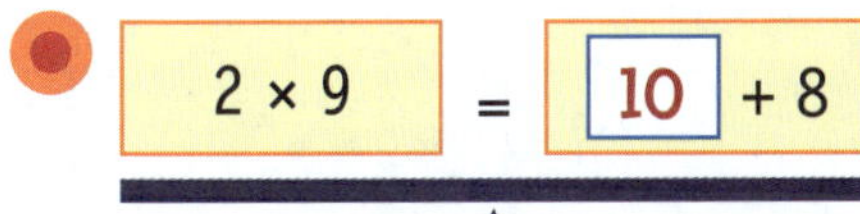

- (example) 12 ÷ 3 = 10 − 6 = 2 × 2
- a 4 × 9 = ________ = ________
- b 55 ÷ 5 = ________ = ________
- c 5 × 2 = ________ = ________
- d 7 × 7 = ________ = ________
- e 18 ÷ 2 = ________ = ________
- f 12 × 2 = ________ = ________
- g 64 ÷ 8 = ________ = ________
- h 6 × 8 = ________ = ________
- i 36 ÷ 3 = ________ = ________

3 Write numbers that balance the number sentences.

- (example) 2 × 9 = [10] + 8
- a 35 − 7 = 20 + [ ]
- b 3 × [ ] = 21 + 6
- c 36 ÷ 12 = 9 ÷ [ ]
- d 7 × [ ] = 73 − 10
- e [ ] × 2 = 110 ÷ 11
- f 42 + 8 = 5 × [ ]
- g 7 × [ ] = 23 + 33
- h 3 × [ ] = 57 − 21
- i [ ] + 25 = 9 × 7
- j 11 × [ ] = 61 − 17
- k 29 − 23 = 24 ÷ [ ]

PATTERNS WITH FRACTIONS AND DECIMALS

These number sentences and patterns are like other number sentences and patterns you have seen and solved before, but they have fractions and decimals in them.

Pattern with fractions

Example 1:

$\frac{1}{4}$, $\frac{1}{2}$, $\frac{3}{4}$, 1 Rule: Add $\frac{1}{4}$

$+\frac{1}{4}$ $+\frac{1}{4}$ $+\frac{1}{4}$

Example 2:

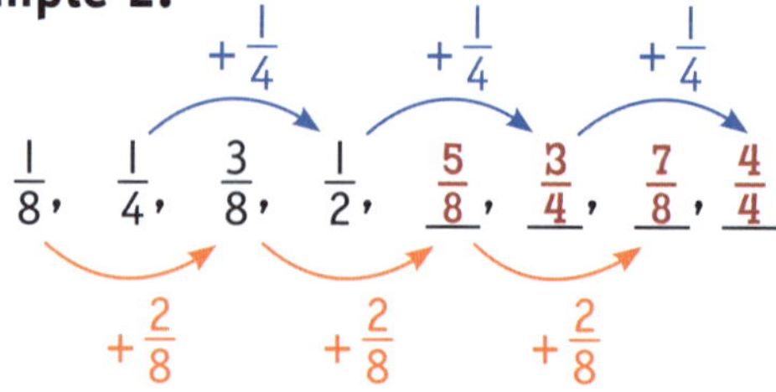

$\frac{1}{8}$, $\frac{1}{4}$, $\frac{3}{8}$, $\frac{1}{2}$, $\frac{5}{8}$, $\frac{3}{4}$, $\frac{7}{8}$, $\frac{4}{4}$

$+\frac{1}{4}$ $+\frac{1}{4}$ $+\frac{1}{4}$

$+\frac{2}{8}$ $+\frac{2}{8}$ $+\frac{2}{8}$

Example 3:

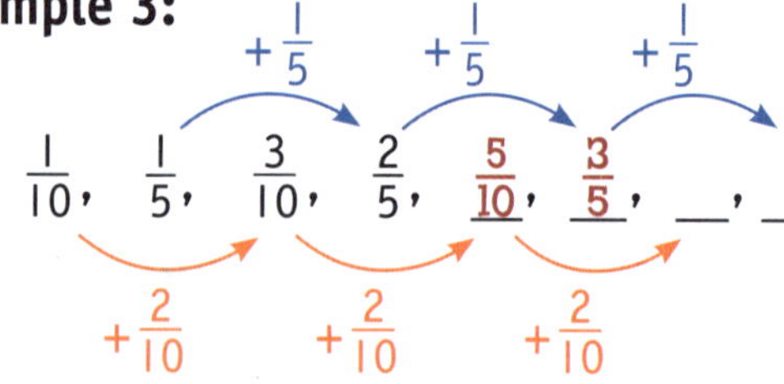

$\frac{1}{10}$, $\frac{1}{5}$, $\frac{3}{10}$, $\frac{2}{5}$, $\frac{5}{10}$, $\frac{3}{5}$, ___, ___

$+\frac{1}{5}$ $+\frac{1}{5}$ $+\frac{1}{5}$

$+\frac{2}{10}$ $+\frac{2}{10}$ $+\frac{2}{10}$

Pattern with decimals

Example 4:

2.1, 1.9, 1.7, 1.5, 1.3

Rule: Subtract 0.2

Example 5:

0.8, 1.0, 1.2, 1.4, 1.6, ___, ___, ___

+0.2 +0.2 +0.2 +0.2 +0.2 +0.2 +0.2

Example 6:

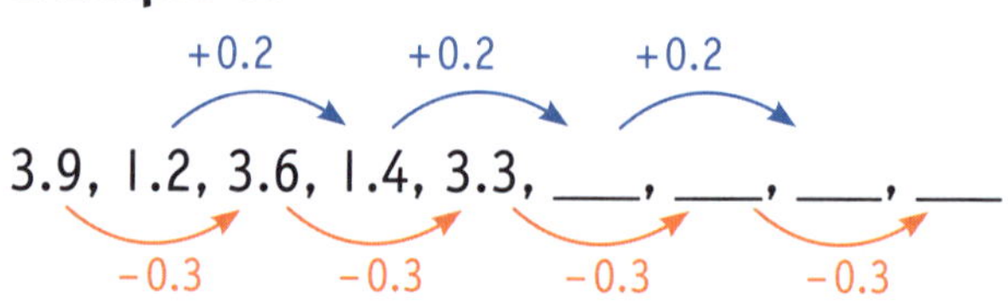

3.9, 1.2, 3.6, 1.4, 3.3, ___, ___, ___, ___

+0.2 +0.2 +0.2

−0.3 −0.3 −0.3 −0.3

Check your answer on the video!

Continue the patterns.

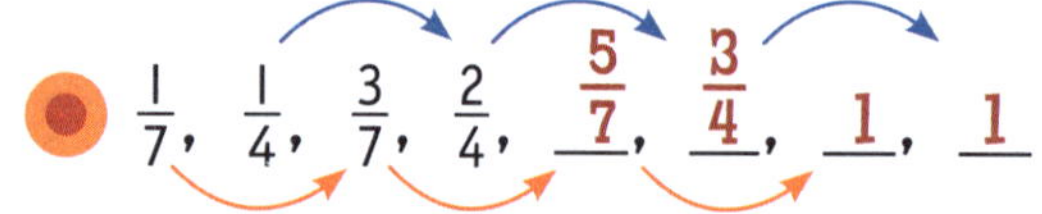

● $\frac{1}{7}$, $\frac{1}{4}$, $\frac{3}{7}$, $\frac{2}{4}$, $\frac{5}{7}$, $\frac{3}{4}$, 1, 1

a $\frac{1}{8}$, $1\frac{1}{4}$, $\frac{1}{16}$, 1, ___, ___, ___, ___

b 2.6, 1, 2.4, 1.3, 2.2, ___, ___, ___, ___

c 10.3, 10.5, 10.7, 10.9, ___, ___, ___, ___

Sometimes there is more than one pattern in a sequence of numbers.

SELF CHECK Tick how you feel

Got it!	Need help...	I don't get it
 ☐	 ☐	 ☐

Check your answers

How many did you get correct? ☐

CATCH UP MATHS YEAR 6 BOOK A © PASCAL PRESS ISBN: 9781925726183

PRACTICE

1 Continue these patterns with fractions.

- $0, \frac{1}{3}, \frac{2}{3}, \frac{3}{3}, \underline{\frac{4}{3}}, \underline{\frac{5}{3}}, \underline{\frac{6}{3}}, \underline{\frac{7}{3}}$

a $\frac{1}{4}, \frac{2}{4}, \frac{3}{4}, \frac{4}{4}$, ___, ___, ___, ___

b $\frac{1}{8}, \frac{2}{8}, \frac{3}{8}, \frac{4}{8}$, ___, ___, ___, ___

c $\frac{1}{6}, \frac{2}{6}, \frac{3}{6}, \frac{4}{6}$, ___, ___, ___, ___

d $\frac{1}{12}, \frac{3}{12}, \frac{5}{12}, \frac{7}{12}$, ___, ___, ___, ___

e $1\frac{1}{3}, 1\frac{2}{3}, 2, 2\frac{1}{3}$, ___, ___, ___, ___

2 Continue these sequences with two patterns.

- $\frac{2}{5}, \frac{1}{3}, \frac{3}{5}, \frac{1}{9}, \frac{4}{5}, \frac{1}{27}, \underline{\frac{5}{5}}, \underline{\frac{1}{81}}, \underline{\frac{6}{5}}, \underline{\frac{1}{243}}$

a $\frac{6}{5}, \frac{8}{12}, \frac{5}{5}, \frac{7}{12}, \frac{4}{5}, \frac{6}{12}$, ___, ___, ___, ___

b $\frac{9}{8}, 1, \frac{8}{8}, \frac{3}{4}, \frac{7}{8}, \frac{2}{4}$, ___, ___, ___, ___

3 Use the rule to continue the pattern.

- $\frac{1}{4}, \frac{2}{4}, \frac{3}{4}, \underline{\frac{4}{4}}, \underline{\frac{5}{4}}, \underline{\frac{6}{4}}, \underline{\frac{7}{4}}$ Rule: $+ \frac{1}{4}$

a $\frac{1}{3}, \frac{2}{3}, \frac{3}{3}$, ___, ___, ___, ___ Rule: $+ \frac{1}{3}$

b $\frac{8}{7}, \frac{7}{7}, \frac{6}{7}$, ___, ___, ___, ___ Rule: $- \frac{1}{7}$

c $\frac{1}{2}, \frac{1}{4}, \frac{1}{8}$, ___, ___, ___, ___ Rule: $\times \frac{1}{2}$

4 Continue these patterns with decimals.

- 1.4, 1.5, 1.6, <u>1.7</u>, <u>1.8</u>, <u>1.9</u>, <u>2.0</u>

a 32.3, 32.1, 31.9, ____, ____, ____, ____

b 8.9, 9.3, 9.7, ____, ____, ____, ____

c 8.6, 8.0, 7.4, ____, ____, ____, ____

d 8.2, 16.4, 32.8, ____, ____, ____, ____

e 81.9, 72.9, 63.9, ____, ____, ____, ____

5 Use the rule to continue the pattern.

- 5, 5.2, 5.4, 5.6, <u>5.8</u>, <u>6.0</u>, <u>6.2</u>, <u>6.4</u> Rule: Increase by 0.2

a 10.5, 9.1, 7.7, ____, ____, ____, ____ Rule: Decrease by 1.4

b 3.2, 9.6, 28.8, ____, ____, ____, ____ Rule: Multiply by 3

TERMS IN SEQUENCES

A term is one of the parts or members of a pattern.

Example 1:
This sequence has 5 terms:

5, 10, 15, 20, 25

1st term, 2nd term, 3rd term, 4th term, 5th term

Each number in a pattern is a term. Count the terms from left to right.

Example 2:
This sequence has 6 terms:

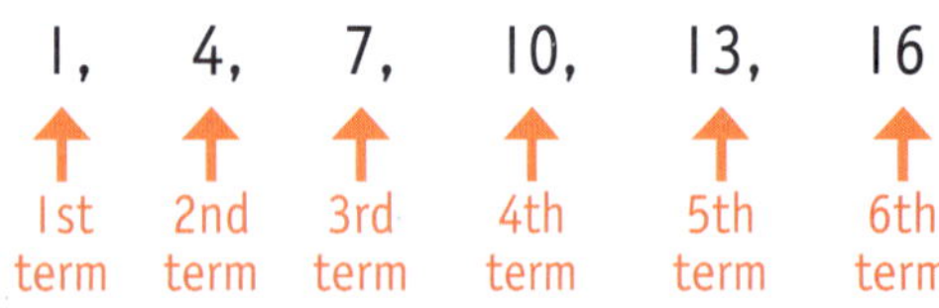

Example 3:
2, 4, 6, 8, 10

This sequence has 5 terms.

Example 4:
3, 6, 9, 12, 15, 18, 21

This sequence has ___ terms.

Example 5: 1, 4, 9

This sequence has ___ terms.

Example 6: 6, 12, 18, 24

This sequence has ___ terms.

Example 7:
110, 120, 130, 140, 150, 160

This sequence has ___ terms.

Your turn

1 Circle the first and third term in each sequence.

a 10, 20, 30, 40, 50

b 6, 12, 18, 24

c 9, 18, 27, 36, 45, 54

d 20, 25, 30, 35, 40

e 1, 2, 3, 4, 5

f 7, 14, 21, 28, 35

g 11, 22, 33, 44, 55

2 Write the second term of each sequence in Question 1.

a ____ b ____ c ____ d ____ e ____ f ____ g ____

SELF CHECK Tick how you feel

Got it! ☐ Need help... ☐ I don't get it ☐

Check your answers
How many did you get correct? ☐

CATCH UP MATHS YEAR 6 BOOK A © PASCAL PRESS ISBN: 9781925726183

PRACTICE

1 What term has been circled in each number sequence?

- ● 2, 4, 6, (8), 10 — 4th
- a 2, 3, 4, 5, (6) ____
- b 5, (10), 15, 20, 25 ____
- c (12), 10, 8, 6, 4 ____
- d 4, 8, (12), 16, 20 ____
- e 15, 12, 9, (6), 3 ____

2 Write the missing terms.

- ● 108, 106, 104, 102, 100, 98
- a 24, 28, ____, ____, 40, 44
- b 20, 25, 30, ____, ____, ____
- c 108, 99, ____, 81, ____, 63, ____
- d 72, 84, ____, ____, 120, ____
- e 56, ____, ____, ____, 84, 91

3 Describe the pattern of the terms in Question 2.

- ● The terms decrease by 2.
- a The terms ____________ by ____.
- b The terms ____________ by ____.
- c The terms ____________ by ____.
- d The terms ____________ by ____.
- e The terms ____________ by ____.

4 Continue each sequence of numbers.

- ● Add 0.2
  0.7, 0.9, 1.1, 1.3, 1.5
- a Subtract 4
  27, ____, ____, ____, ____
- b Subtract 0.4
  4.3, ____, ____, ____, ____
- c Add 0.6
  2, ____, ____, ____, ____
- d Add 3
  7, ____, ____, ____, ____
- e Subtract 0.7
  5.2, ____, ____, ____, ____

5 Complete the tables.

●

Order of term	1	2	3	4	5	6
Term	4	8	12	16	20	24

a

Order of term	1	2	3	4	5	6
Term	56	48	40			

b

Order of term	1	2	3	4	5	6
Term	72	63	54			

c

Order of term	1	2	3	4	5	6
Term	17	27	37			

6

Order of term	1	2	3	4	5	6	7	8	9
Term	3	6	9						

- a Complete the table.
- b What is the 10th term? ____
- c What is the 15th term? ____
- d What is the 20th term? ____

THE CARTESIAN NUMBER PLANE

This is a Cartesian number plane.

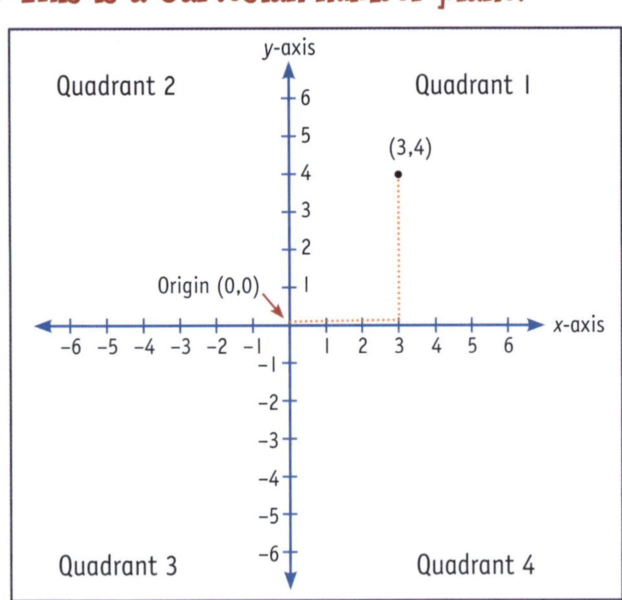

(3,4) is a set of coordinates. It is also called an ordered pair.

The first coordinate is for the *x*-axis. The second coordinate is for the *y*-axis. The *x*-axis and *y*-axis intersect at the origin, (0,0).

SCAN to watch video

How to plot a point on the number plane

To plot (3,4):

- Start at (0,0) on the *x*–axis. Move 3 units to the right.
- Then move 4 units up the *y*–axis.

(3,4) is in Quadrant 1.

Example 1:

Circle the ordered pair that is not in Quadrant 1.

(1,1) (2,3) (4,5)  (5,–1)

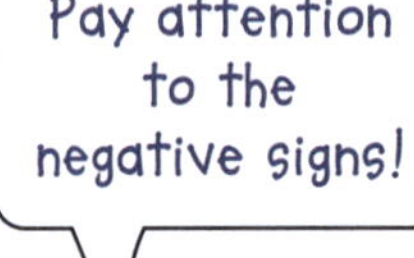

Example 2:

Which ordered pairs are in Quadrant 2?

(–1,1) (–1,–1) (–1,2) (4,2)

Example 3:

Circle the coordinates that are not in Quadrant 3.

(–1,–3) (–4,–2) (–4,2) (5,5)

Example 4:

Which coordinates are in Quadrant 4?

(–2,–4) (1,–4) (3,–4) (5,1)

Plot these ordered pairs on the Cartesian number plane. Use the letter to label each point.

 A (5,3) G (4,3)

B (–1,4) H (–2,–5)

C (3,–5) I (6,4)

D (–4,–4) J (–6,–6)

E (1,3) K (6,–6)

F (–3,5) L (4,–5)

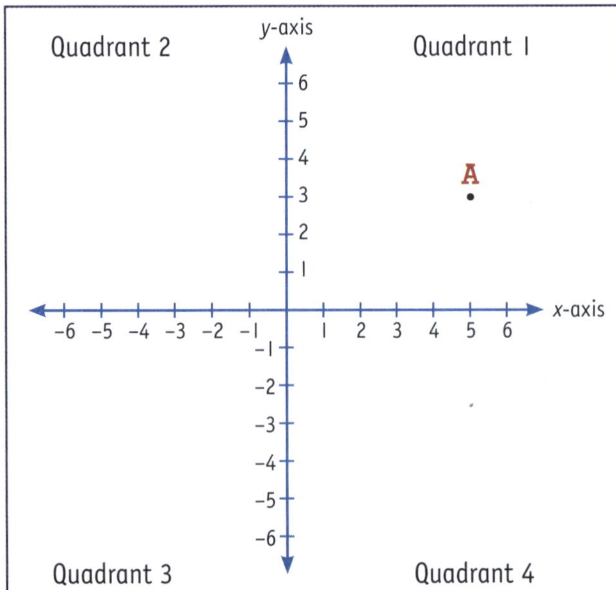

Check your answers
How many did you get correct?

CATCH UP MATHS YEAR 6 BOOK A © PASCAL PRESS ISBN: 9781925726183

PRACTICE

1 Write the ordered pair and quadrant for each point.

	Ordered pair	Quadrant		Ordered pair	Quadrant
A	(1,6)	1	K		
B			L		
C			M		
D			N		
E			O		
F			P		
G			Q		
H			R		
I			S		
J			T		

Quadrant 2
y-axis
Quadrant 1
A
P
N
Q
E
M
O
C
L
D
x-axis
H
G
R
I
S
B
F
K
T
J
Quadrant 3
Quadrant 4

2 Plot the points on the Cartesian plane.

	x	y
B	3	5
Z	−1	2
M	−2	−4
T	6	−3
F	−10	1
J	9	9
N	7	10
P	4	1
D	3	−2
C	−2	−2
A	3	3
Q	5	−1
R	−6	2
U	−4	3
E	−3	6
S	−8	3
V	10	−7
W	−4	−10
I	−2	2

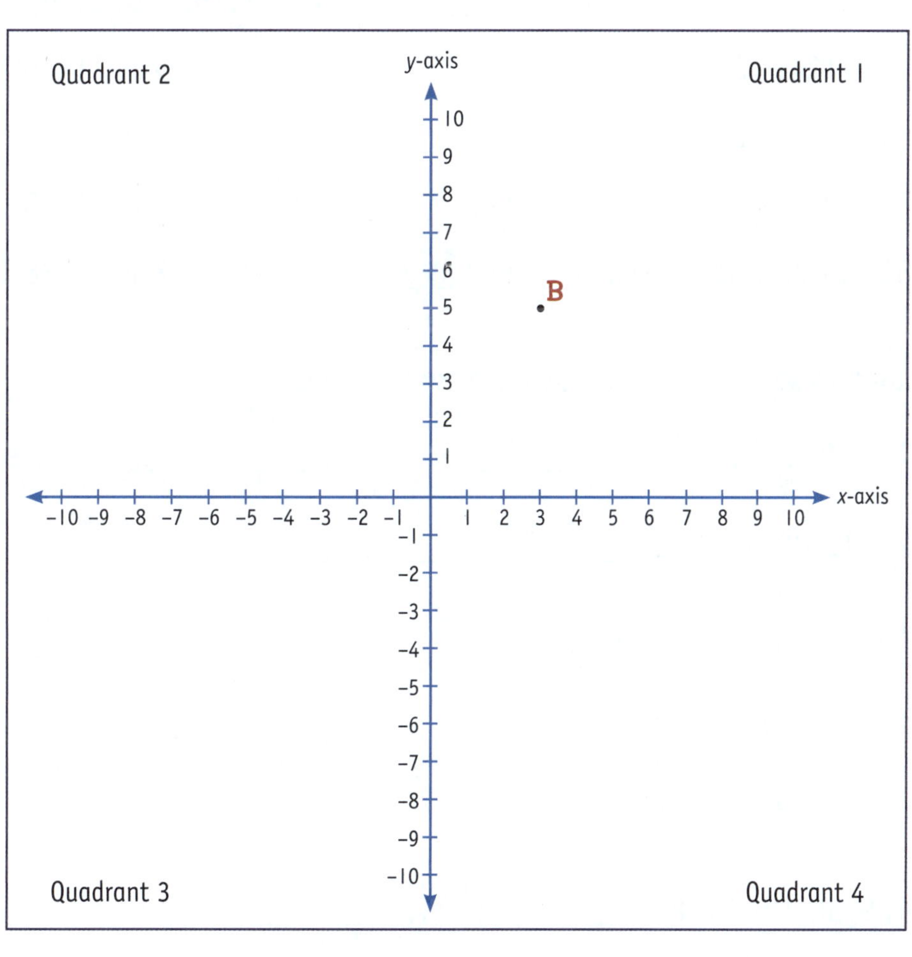

PATTERNS AND ALGEBRA REVIEW

1 Use the rules to continue the patterns.

a 7, ____, ____, ____, ______, ______ Rule: × 5

b 100, ____, ____, ____, ____, ______ Rule: ÷ 10

c 14, ____, ____, ____, ____, ____ Rule: + 6

d 49, ____, ____, ____, ____, ____ Rule: − 7

e 3, ____, ____, ____, ____, ____ Rule: × 2

f 3, ____, ____, ____, ____, ______ Rule: × 4

g 2, ____, ____, ____, ____, ______ Rule: × 6

h 16, _____, _____, _____, _____, _____ Rule: + 105

2 Write the rule for each pattern.

a 8, 15, 22, 29, 36 Rule: _____

b 127, 124, 121, 118, 115 Rule: _____

c 64, 128, 256, 512 Rule: _____

d 77, 72, 67, 62, 57 Rule: _____

e 63.3, 62.8, 62.3, 61.8 Rule: _____

f 2.25, 2.5, 2.75, 3 Rule: _____

g 226, 220, 214, 208, 202 Rule: _____

h 84, 91, 98, 105, 112 Rule: _____

i 9.36, 9.21, 9.06, 8.91 Rule: _____

j 7.08, 7.18, 7.28, 7.38 Rule: _____

3 Continue these sequences that have two patterns.

a 7, 7, 8, 6, 9, 5, ____, ____, ____, ____, ____, ____

b 21, 1, 23, 3, 25, 5, ____, ____, ____, ____, ____, ____

c 15, 7, 14, 10, 13, 13, ____, ____, ____, ____, ____, ____

d 2, 6, 7, 10, 12, 14, ____, ____, ____, ____, ____, ____

e $1\frac{1}{2}$, $\frac{1}{8}$, $1\frac{3}{4}$, $\frac{3}{8}$, 2, $\frac{5}{8}$, ____, ____, ____, ____, ____, ____

f 4, 9, 9, 18, 14, 27, _____, _____, _____, _____, _____, _____

g 20, 28, 24, 24, 28, 20, ____, ____, ____, ____, ____, ____

CATCH UP MATHS YEAR 6 BOOK A © PASCAL PRESS ISBN: 9781925726183

4 Write the operation to complete the rule.

a A ☐ B = C

A	B	C
32	8	4
50	10	5
63	9	7
48	12	4

b B = C ☐ A

C	B	A
3	21	7
7	35	5
7	49	7
12	108	9

c A = C ☐ B

B	A	C
15	5	20
7	25	32
12	24	36
18	32	50

d C = B ☐ A

C	B	A
24	9	15
37	20	17
54	40	14
72	62	10

5 Complete the pattern grids.

a ▲ = ● × 3 + 1

●	2	4	6	8	10	12	14	16	18	20	22
▲											

b ◆ = (▲ + 6) × 3

▲	3	5	7	9	11	13	15	17	19	21	23
◆											

c ▲ × 4 – 3 = ★

▲	22	20	18	16	14	12	10	8	6	4	2
★											

d (● – 4) × 3 = ⬢

●	5	10	15	20	25	30	35	40	45	50	55
⬢											

6 Write the rule for each pattern grid.

a Rule: ★ = ● ÷ __ + __

●	22	20	18	16	14	12
★	15	14	13	12	11	10

b Rule: ▲ = ⬢ × __ + __

⬢	4	8	12	16	20	24
▲	16	28	40	52	64	76

c Rule: ● =  ■ × __ + __

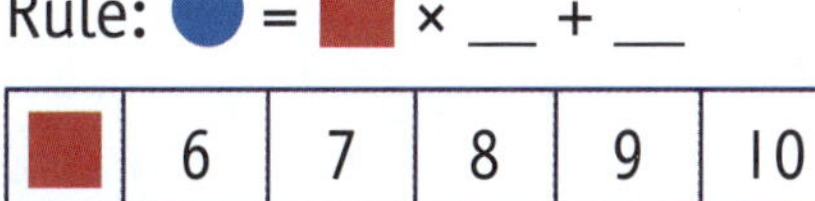

■	6	7	8	9	10	11
●	23	26	29	32	35	38

d Rule: ■ = ▲ ÷ __ – __

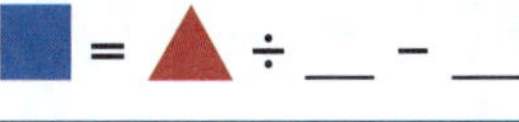

▲	10	32	30	40	20	36
■	2	13	12	17	7	15

REVIEW

7 Complete the grids.

a

Rule	5	6	7	8	9	10
× 2						
+ 3						
– 5						
× 8						
– 3						

b

Rule	10	9	8	7	6	5
× 3						
+ 4						
– 6						
× 9						
+ 5						

8 Complete these equivalent number sentences.

a 24 ÷ 6 = 10 – ☐

b 49 ÷ ☐ = 20 – 13

c ☐ × 9 = 50 – 14

d 56 ÷ 8 = ☐ – 17

e 121 ÷ ☐ = 30 – 19

f ☐ × 5 = 50 – 5

g 9 + ☐ = 3 × 7

h 110 ÷ 10 = 11 × ☐

i ☐ × 4 = 12 × 3

j 27 + ☐ = 9 × 6

9 Write two equivalent number sentences.

a 15 ÷ 3 = __________ = __________

b 5 × 2 = __________ = __________

c 6 × 4 = __________ = __________

d 20 – 13 = __________ = __________

10 Continue these number patterns.

a $\frac{2}{8}, \frac{4}{8}, \frac{6}{8}$, _____, _____, _____, _____

b $\frac{10}{4}, \frac{8}{4}, \frac{6}{4}$, _____, _____, _____, _____

c $\frac{1}{3}, \frac{2}{3}, \frac{3}{3}$, _____, _____, _____, _____

d 0.3, 0.6, 0.9, _____, _____, _____

e 7.2, 7.4, 7.6, _____, _____, _____

f 8.7, 8.2, 7.7, _____, _____, _____

CATCH UP MATHS YEAR 6 BOOK A © PASCAL PRESS ISBN: 9781925726183

11 Continue these sequences that have two patterns.

a 6.4, 3.7, 6.2, 4, 6.0, 4.3, _____, _____, _____, _____, _____, _____

b 8.4, 12.3, 8.8, 12.6, 9.2, 12.9, _____, _____, _____, _____, _____, _____

12 Use the rule to continue the pattern.

a 2.6, ____, ____, ____, ____ Rule: Increase by 0.3

b $1\frac{2}{8}$, ____, ____, ____, ____ Rule: Add $\frac{1}{8}$

c 9.8, ____, ____, ____, ____ Rule: Decrease by 1.4

d 4, ____, ____, ____, ____ Rule: Subtract $\frac{1}{3}$

e 1.7, ____, ____, ____, ____ Rule: Add 0.5

f 6.2, ____, ____, ____, ____ Rule: Decrease by 0.8

13 For each sequence, use red to circle the first term and green to circle the third term.

a 1, 3, 5, 7, 9

b 2, 4, 6, 8, 10

c 25, 20, 15, 10, 5

d 12, 9, 6, 3, 0

e 7, 17, 27, 37, 47

f 24, 26, 28, 30, 32

14 Which term is circled in each number sequence?

a 3, 5, 7, **9**, 11 ______

b 90, **80**, 70, 60, 50 ______

c **4**, 8, 12, 16, 20 ______

d 7, 14, 21, **28**, 35 ______

e 105, **101**, 97, 93, 89 ______

f 29, 36, 43, 50, **57** ______

g 72, 78, **84**, 90, 96 ______

h 198, 196.8, 195.6, **194.4** ______

i 8.2, **8.6**, 9, 9.4, 9.8 ______

j 15, 30, 45, **60**, 75 ______

15 Describe each pattern of terms in Question 14.

a The terms increase / decrease by ___

b The terms increase / decrease by ___

c The terms increase / decrease by ___

d The terms increase / decrease by ___

e The terms increase / decrease by ___

f The terms increase / decrease by ___

g The terms increase / decrease by ___

h The terms increase / decrease by ___

i The terms increase / decrease by ___

j The terms increase / decrease by ___

REVIEW

16 Complete the tables of terms.

a

Order of term	1	2	3	4	5	6
Term	5	10	15			

b

Order of term	1	2	3	4	5	6
Term	6	12	18			

c

Order of term	1	2	3	4	5	6
Term	64	56	48			

d

Order of term	1	2	3	4	5	6
Term	110	100	90			

17

Order of term	1	2	3	4	5	6	7	8	9
Term	7	14	21						

a Complete the table.

b What is the 10th term? ______

c What is the 20th term? ______

d What is the 100th term? ______

18 Continue the number patterns.

a 107, 101, 95, ______, ______, ______, ______, ______, ______

b 95, 102, 109, ______, ______, ______, ______, ______, ______

c 81, 69, 57, ______, ______, ______, ______, ______, ______

d 12.5, 12.25, 12, ______, ______, ______, ______, ______, ______

e 6.3, 5.7, 5.1, ______, ______, ______, ______, ______, ______

f 77, 83, 89, ______, ______, ______, ______, ______, ______

g 61, 57, 53, ______, ______, ______, ______, ______, ______

h 39, 32, 25, ______, ______, ______, ______, ______, ______

i $10\frac{1}{2}$, $10\frac{1}{4}$, 10, ______, ______, ______, ______, ______, ______

j 72.3, 71.9, 71.5, ______, ______, ______, ______, ______, ______

k $6\frac{1}{4}$, 6, $5\frac{3}{4}$, ______, ______, ______, ______, ______, ______

l 1.75, 1.70, 1.65, ______, ______, ______, ______, ______, ______

m 87, 89.5, 92, ______, ______, ______, ______, ______, ______

n 16.24, 16.44, 16.64, ______, ______, ______, ______, ______, ______

o 92.5, 92.0, 91.5, ______, ______, ______, ______, ______, ______

CATCH UP MATHS YEAR 6 BOOK A © PASCAL PRESS ISBN: 9781925726183

19 Plot the points on the Cartesian plane.

A (1,1)
B (−1,4)
C (−3,−5)
D (−2,6)
E (−6,−5)
F (4,−2)
G (2,6)
H (3,2)
I (−4,−4)
J (−5,3)

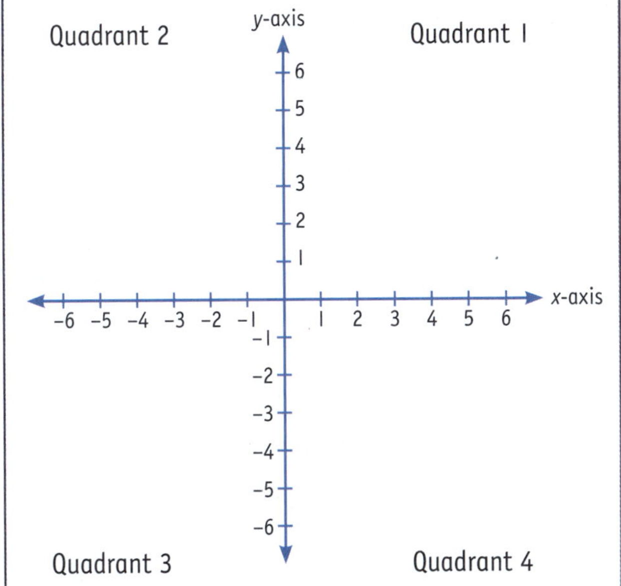

20 Write the ordered pair and quadrant that represents each letter on the Cartesian plane.

Quadrant 2
y-axis
Quadrant 1
J R K S Z T M
x-axis
Y C X D L B A
Quadrant 3
Quadrant 4

	Ordered pair	Quadrant
Z		
X		
A		
S		
C		
K		
J		
Y		
D		
B		
L		
M		
T		
R		

PROBABILITY

SCAN to watch video

Probability is the likelihood of an event happening.
The probability of an event happening is rated from 0 to 1.

0 | 0.1 | 0.2 | 0.3 | 0.4 | 0.5 | 0.6 | 0.7 | 0.8 | 0.9 | 1

- 0 — An impossible event has a probability of 0.
- 0.5 — Even chance
- 1 — A certain event has a probability of 1.

Example 1: Write the likelihoods in the correct boxes.

• even chance • ~~unlikely~~ • impossible • certain • likely

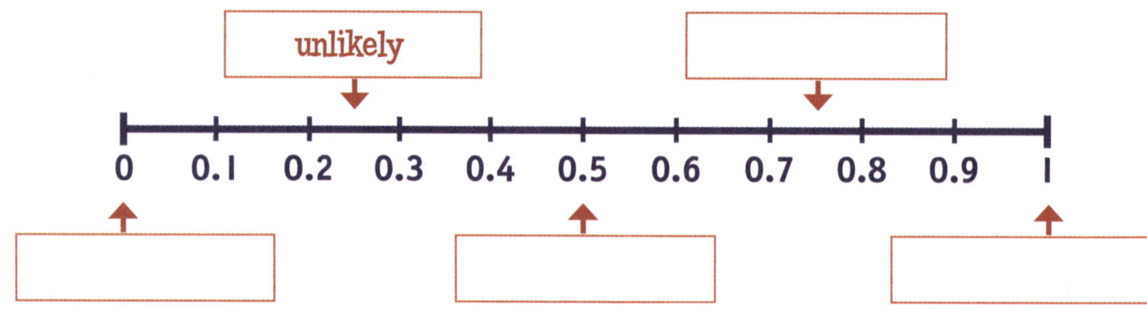

Example 2: Use this spinner to complete the statements.

- The chance of landing on pink is 0.3 because $\frac{3}{10}$ of the parts are pink.
- There is a $\frac{2}{10}$ chance the spinner will land on green.
- There is a __ in 10 chance the spinner will land on blue.
- There is a 0.__ chance the spinner will land on red.

Complete the equivalent fractions and percentages on the probability scale.

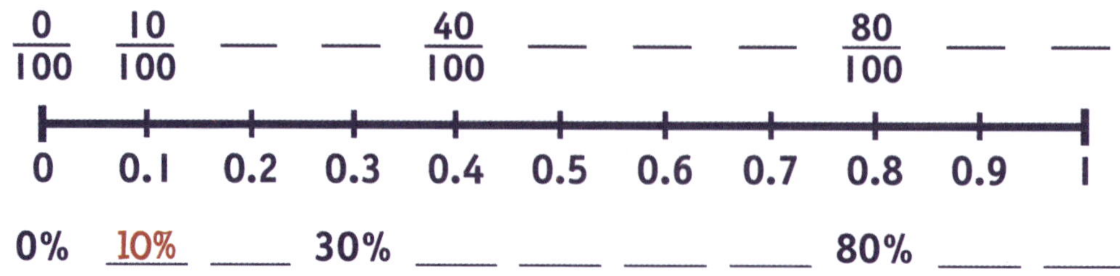

Check your answers
How many did you get correct?

CATCH UP MATHS YEAR 6 BOOK A © PASCAL PRESS ISBN: 9781925726183

PRACTICE

1 Rate the probability of these events on a scale of 0 to 1.

	Rating	Event
●	0.5	A coin tossed will land on tails.
a		When you roll a dice, it will land on ⚄ or ⚅.
b		I will take a white jellybean from a bag of jellybeans that has 7 white jellybeans and 3 red jellybeans.
c		This spinner will land on red:
d		The month after January will be February.
e		You will grow over 5 metres tall.
f		A newborn baby will be a girl.
g		I will take a blue ball from a bag that has 8 blue balls and 2 red balls.

2 Look at the spinner at the right. Draw lines from the descriptions to the decimals on the probability scale.

● The spinner lands on an odd number. (0.5)

a The spinner lands on an even number.

b The spinner lands on the number 5.

c The spinner lands on a multiple of 2.

d The spinner lands on a multiple of 3.

e The spinner lands on a factor of 9.

f The spinner lands on a number below 8.

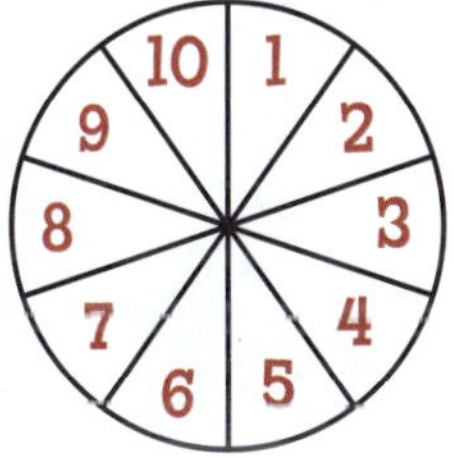

3 Describe an event that matches each probability.

● 0.3 In a bag of 20 lollies, there are 6 red lollies ($\frac{6}{20} = \frac{3}{10} = 0.3$)

a 0 __________

b 0.6 __________

c 0.9 __________

d 0.7 __________

e 1 __________

OUTCOMES

An outcome is any possible result in a chance experiment.

SCAN to watch video

Example 1:

Chris has three coloured blocks to stack.

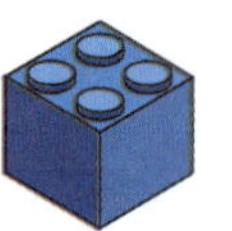  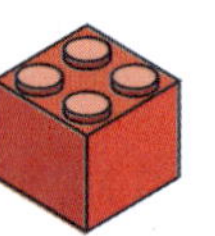

Here are all the possible combinations Chris could make.

  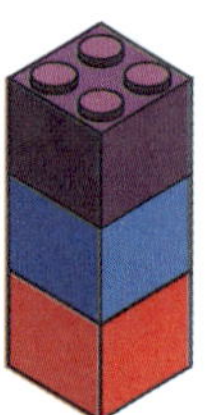 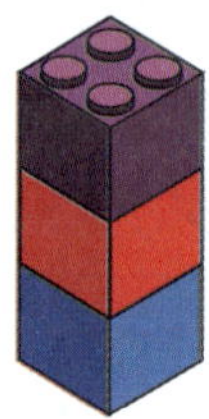 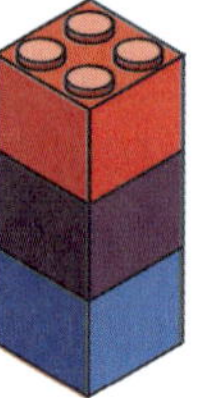 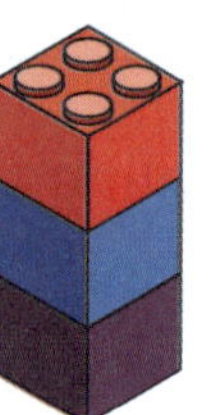

Example 2:

Colour the blocks to show the possible outcomes for stacking these three blocks.

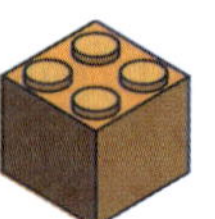 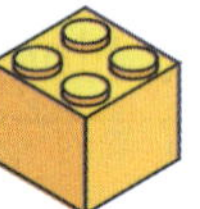 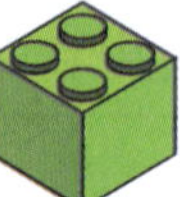

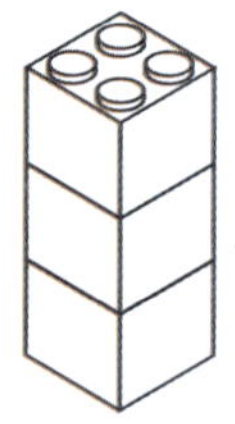 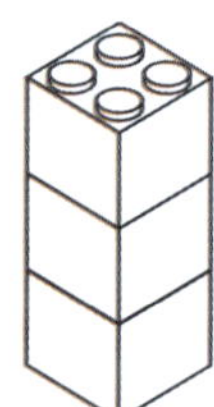 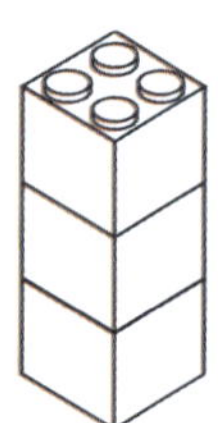 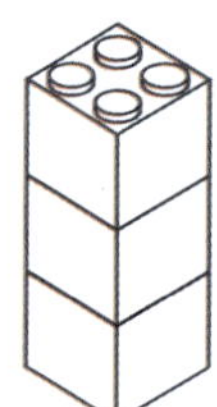 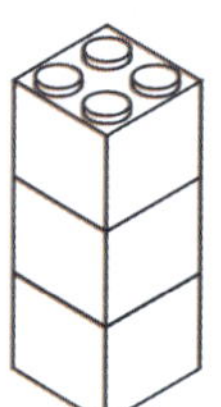 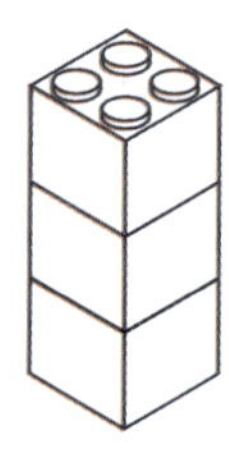

Check your answer on the video!

Draw the possible combinations when you pick three jellybeans from these jars.

●  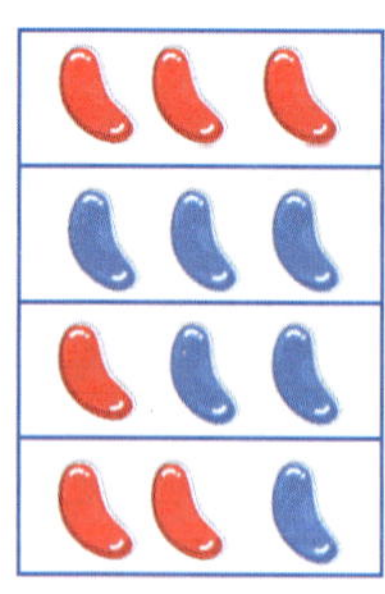

a  

SELF CHECK Tick how you feel		
Got it! ☐	Need help... ☐	I don't get it ☐

Check your answers

How many did you get correct? ☐

CATCH UP MATHS YEAR 6 BOOK A © PASCAL PRESS ISBN: 9781925726183

PRACTICE

1 Bernard is going to buy a new car. His choices are in the box at the right.

Complete the tree diagram for Bernard's new car, and then answer the following questions.

Options for Bernard's New Car	
Colours:	Blue (**B**), White (**W**) or Red (**R**)
Style:	2 door (**2D**) or 4 door (**4D**)
Transmission:	Automatic (**A**) or Manual (**M**)
Wheels:	Plain (**P**) or Mags (**M**)

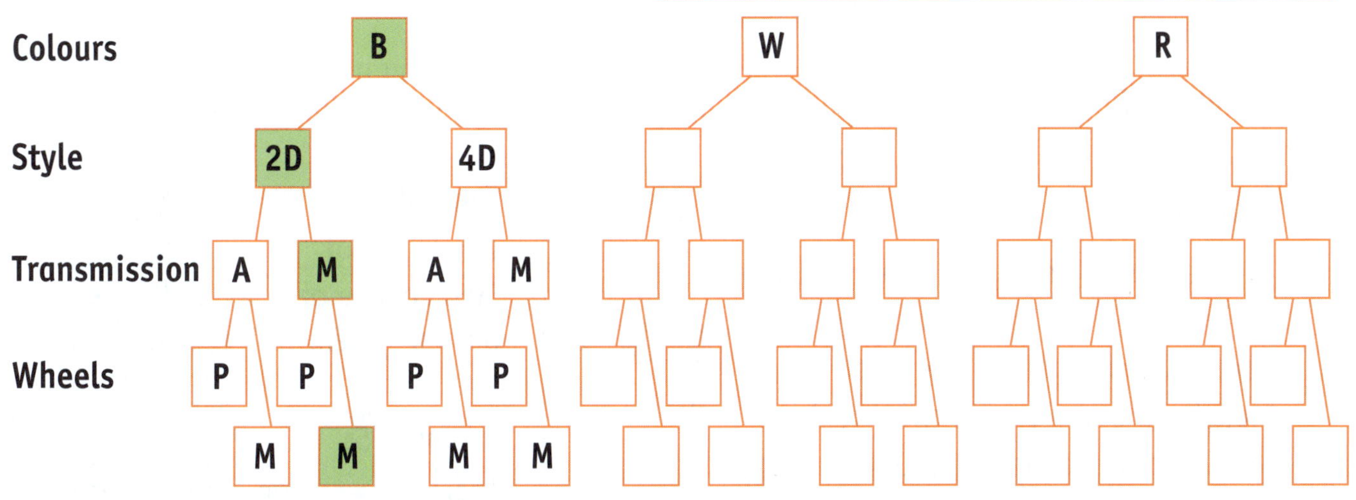

- Use green to colour the path of a blue car with 2 doors, Manual transmission and Mag wheels.

a Use blue to colour the path of a white car with 4 doors, Automatic transmission and Plain wheels.

b Use red to colour the path of a red car with 2 doors, Manual transmission and Mag wheels.

c How many combinations of car did Bernard have to choose from? ____

2 How many different ways can Elise arrange her flowers?

Use the letters R, P, B and Y to write the combinations.

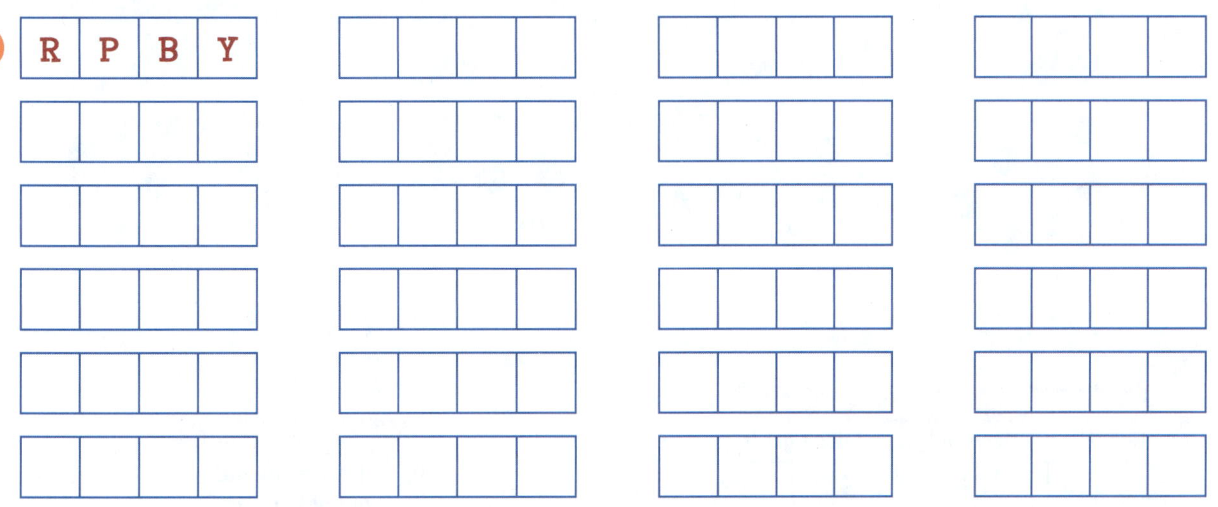

EQUALLY LIKELY OUTCOMES

An even chance is when the possible outcomes are equally likely to happen.

SCAN to watch video

Example 1:

When a coin is tossed, it lands on heads (H) or tails (T).
The chance of either outcome is even.
There are two possible outcomes.

There is a one in two ($\frac{1}{2}$) or 50% chance of landing on heads and a one in two ($\frac{1}{2}$) or 50% chance of landing on tails.

Example 2:

If there are 3 red and 3 yellow jellybeans in a jar, there is a one in two ($\frac{1}{2}$) or 50% chance of picking a red jellybean and a one in two ($\frac{1}{2}$) or 50% chance of picking yellow.

Example 3:

There is an unequal chance of picking a red jellybean from this jar as the numbers of each colour are not equal.

There are more yellow jellybeans than red jellybeans, so picking a yellow jellybean is more likely.

Example 4:

There is not a 50% chance of picking green.

Example 5:

There ________ an even chance of picking blue.

Example 6:

There ________ a one in two chance of picking pink.

Check your answer on the video!

Your turn

Draw extra marbles to show an equal chance ($\frac{1}{2}$) outcome.

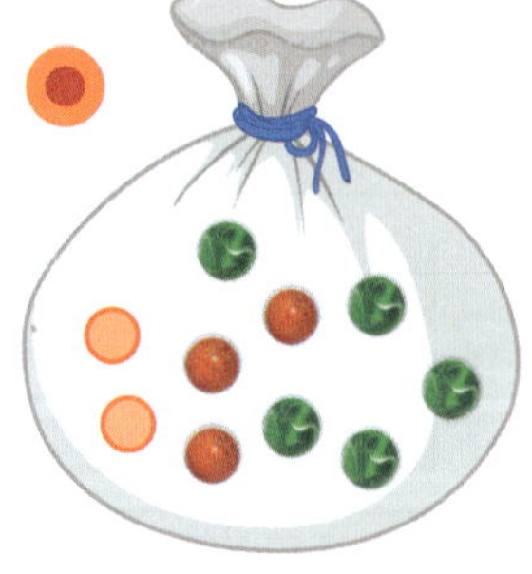

a

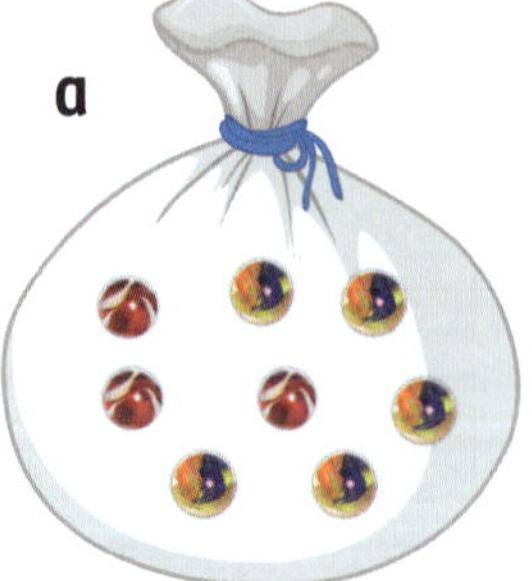

b

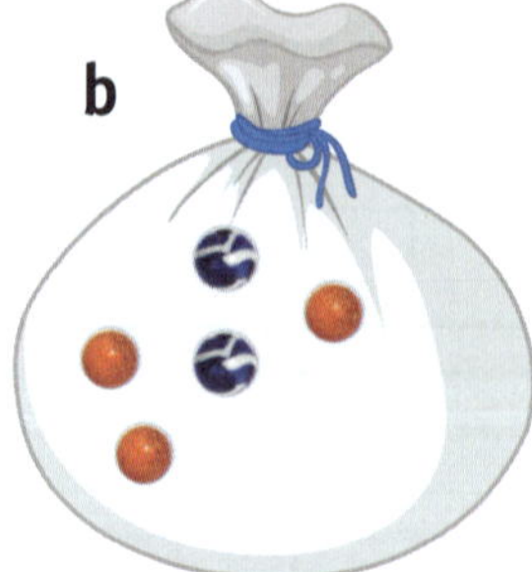

c

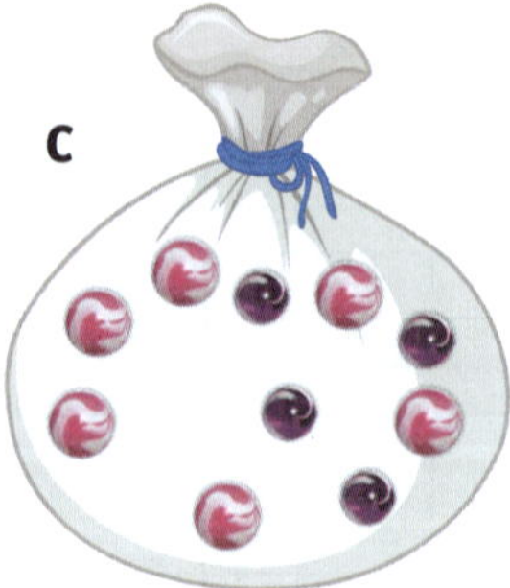

SELF CHECK Tick how you feel

Got it!	Need help...	I don't get it
☐	☐	☐

Check your answers
How many did you get correct? ☐

CATCH UP MATHS YEAR 6 BOOK A © PASCAL PRESS ISBN: 9781925726183

1 Colour the spinners below to show the chance.

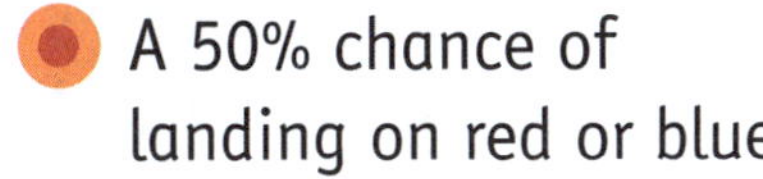
A 50% chance of landing on red or blue

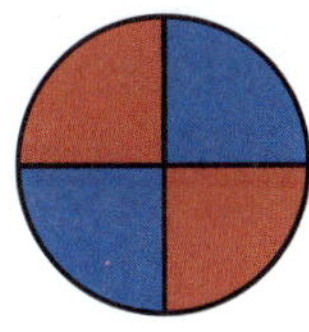

a A $\frac{1}{2}$ chance of landing on pink or purple

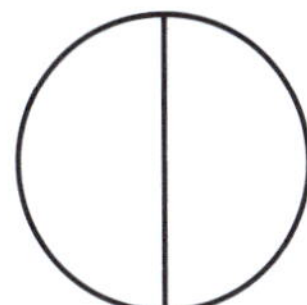

b A one-in-two chance of landing on orange or black.

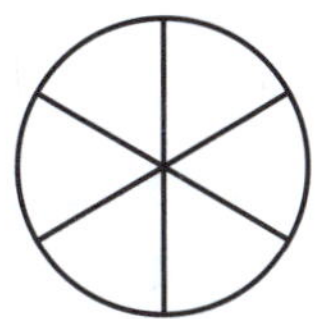

c A 50% chance of landing on yellow or green

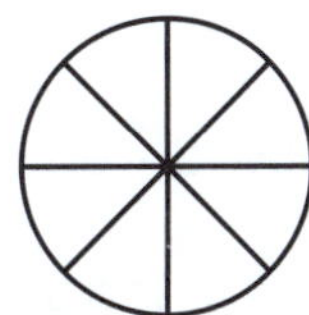

d A greater chance of landing on blue than red

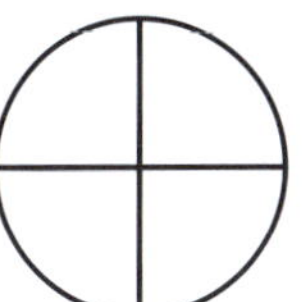

e A greater chance of landing on pink than purple

f A lesser chance of landing on black than orange

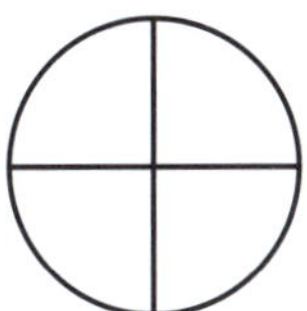

g A lesser chance of landing on yellow than green

2 Draw more marbles in each jar to show 1 in 2 ($\frac{1}{2}$) chance outcomes.

a

b

c

d

3 Tick the even chance events and cross the uneven chance events.

- ☑ A coin is tossed.

a ☐ I have a bag with 3 green and 3 red counters.

b ☐ A jar has 4 black, 4 blue and 4 green jellybeans.

c ☐ A bag has 6 white and 6 blue pegs.

d ☐ A jar has 3 red lollipops and 3 yellow lollipops.

e ☐ I roll a dice.

UNEQUAL CHANCES

When there are two possible outcomes, an unequal chance is any chance outcome that is different from 1 in 2 ($\frac{1}{2}$).

SCAN to watch video

Example 1:

This jar has 10 cars.

I have a 6 out of 10 ($\frac{6}{10}$) or 60% chance of pulling out a green car.

I have a 4 out of 10 ($\frac{4}{10}$) or 40% chance of pulling out a blue car.

It is an unequal chance because there are more green cars than blue cars.

Example 2:

Colour the counters in each jar to show an unequal chance.

Label each jar as showing an equal or unequal chance.

● unequal chance

b ________ chance

d ________ chance

a ________ chance

c ________ chance

e ________ chance

SELF CHECK Tick how you feel		
Got it! ☐	Need help... ☐	I don't get it ☐

Check your answers

How many did you get correct? ☐

CATCH UP MATHS YEAR 6 BOOK A © PASCAL PRESS ISBN: 9781925726183

PRACTICE

1 Max has 10 gummy bears in a jar.
What is the chance of pulling out a gummy bear of each colour?
Write your answer as a fraction.

- Green $\frac{1}{10}$
- a Red ___
- b Blue ___
- c Orange ___
- d Yellow ___
- e Black ___

2 Max has another jar with 100 gummy bears in it.
The fraction for each colour is the same as for the jar with 10 gummy bears.
How many gummy bears of each colour are in the jar?

- Green 10
- a Red ___
- b Blue ___
- c Orange ___
- d Yellow ___
- e Black ___

3 Joy has 20 jellysnakes in a jar.
Write the missing numbers and then describe the chance as equal or unequal.

- There is a 6 in 20 chance ($\frac{6}{20}$) the jellysnake will be orange.
  It is an unequal chance.
- a There is a ___ in ___ chance ($\frac{}{20}$) the jellysnake will be red.
  It is an ___ chance.
- b There is a ___ in ___ chance ($\frac{}{20}$) the jellysnake will be blue.
  It is an ___ chance.
- c There is a ___ in ___ chance ($\frac{}{20}$) the jellysnake will be yellow.
  It is an ___ chance.
- d There is a ___ in ___ chance ($\frac{}{20}$) the jellysnake will be green.
  It is an ___ chance.

4 Use Joy's jar in Question 3 to answer the following questions.

- a The colour of jellysnake with the greatest chance of being picked is ___.
- b The colour of jellysnake with the least chance of being picked is ___.

PROBABILITY AS DECIMALS, FRACTIONS AND PERCENTAGES

Probability can be recorded as decimals, fractions and percentages.

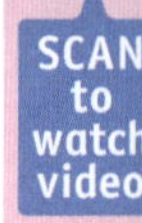

There are 10 gummy bears in this jar.

Gummy bear Colour	Probability of picking out a colour		
	Decimals	Fractions	Percentages
Red	0.40	$\frac{4}{10}$	40%
Orange	0.10	$\frac{1}{10}$	10%
Purple	0.20	$\frac{2}{10}$	20%
Green	0.20	$\frac{2}{10}$	20%
Yellow	0.10	$\frac{1}{10}$	10%

Example 1: There is a __40__% chance of picking a red gummy bear.

Example 2: There is a $\frac{2}{10}$ chance of picking a green gummy bear.

Example 3: There is a 0.__10__ chance of picking an orange gummy bear.

Example 4: There is a ____% chance of picking a yellow gummy bear.

Example 5: There is a 0.____ chance of picking a purple gummy bear.

Check your answer on the video!

Your turn

Use the bag of marbles to complete the table.

	Marble Colour	Probability of picking out a colour		
		Decimals	Fractions	Percentages
	Orange	0.15	$\frac{3}{20}$	15%
a	Purple			
b	Blue			
c	Green			
d	Pink			

SELF CHECK Tick how you feel

Got it!	Need help...	I don't get it
☐	☐	☐

Check your answers
How many did you get correct? ☐

CATCH UP MATHS YEAR 6 BOOK A © PASCAL PRESS ISBN: 9781925726183

PRACTICE

Use different colours to match each decimal with its equivalent fraction and percentage.

	Decimal	Fraction	Percentage
●	0.1	$\frac{10}{100}$	40%
a	0.3	$\frac{100}{100}$	60%
b	0.5	$\frac{80}{100}$	50%
c	0.4	$\frac{75}{100}$	75%
d	0.6	$\frac{30}{100}$	20%
e	0.8	$\frac{25}{100}$	0%
f	1.0	$\frac{0}{100}$	100%
g	0.2	$\frac{50}{100}$	10%
h	0.25	$\frac{40}{100}$	25%
i	0.75	$\frac{20}{100}$	80%
j	0	$\frac{60}{100}$	30%

There are 60 marbles in the box.
Colour them to match the given percentages and then fill in the missing numbers.

● 10% blue = 6 blue marbles

a 20% green = __ green marbles

b 5% yellow = __ yellow marbles

c 30% red = __ red marbles

d 15% orange = __ orange marbles

e 20% black = __ black marbles

f Write the percentage as a decimal for each colour.

Colour	Blue	Green	Yellow	Red	Orange	Black
Decimal	0.10					

CHANCE REVIEW

1 Rate the probability of these events happening on a scale of 0 to 1. Then write the matching fraction and percentage.

	Decimal	Fraction	Percentage	Event
a				Pick a red card out of a deck of cards.
b				Roll a dice and land on an even number.
c				Roll a dice and land on 2 or 4.
d				Roll a dice and land on a factor of 3.
e				Pick a pink ball out of a bag with 4 pink balls and 6 blue balls.
f				Roll a dice and get a 7.
g				Win first prize in a random draw of 10 tickets.
h				Spin red on this spinner:
i				Land on purple or orange on this spinner:
j				Land on green on this spinner:

2 Terry has these four blocks. How many different ways can Terry arrange them? Use the letters B, Y, G and R to write the combinations.

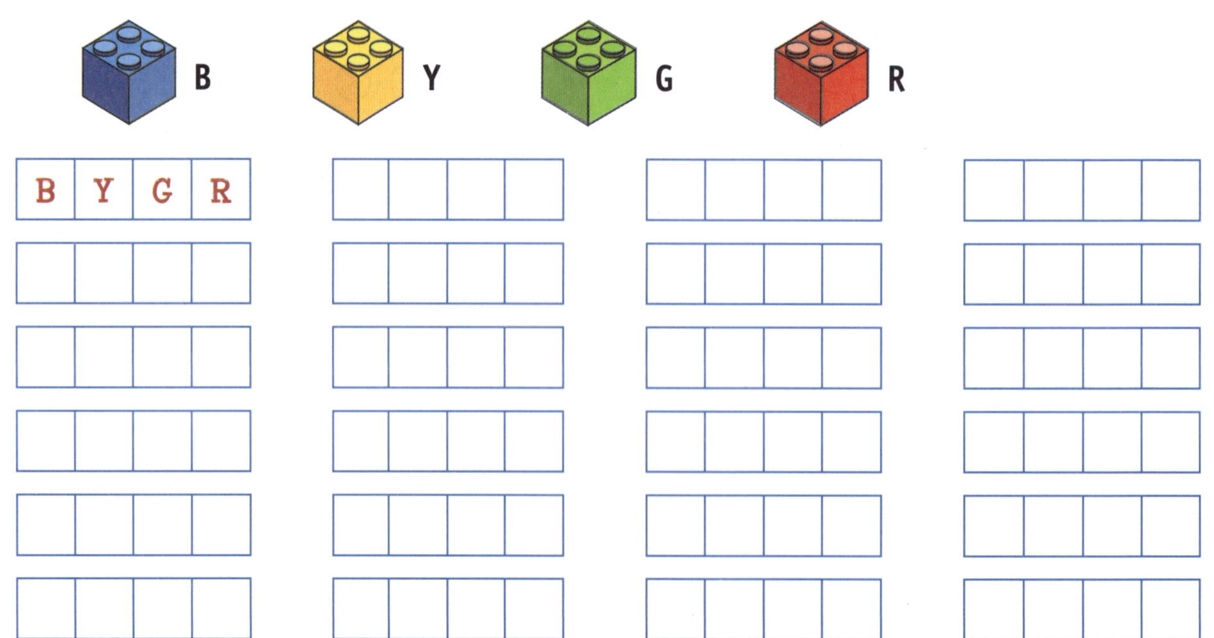

CATCH UP MATHS YEAR 6 BOOK A © PASCAL PRESS ISBN: 9781925726183

Lin is buying a new car. Her choices are in the box at the right.

Complete the tree diagram for Lin's new car, and then answer the following questions.

Options for Lin's New Car	
Colours:	Red (**R**), Black (**B**) or Yellow (**Y**)
Style:	Convertible (**C**) or Hatch (**H**)
Transmission:	Automatic (**A**) or Manual (**M**)
Wheels:	Black (**B**) or Chrome (**C**)

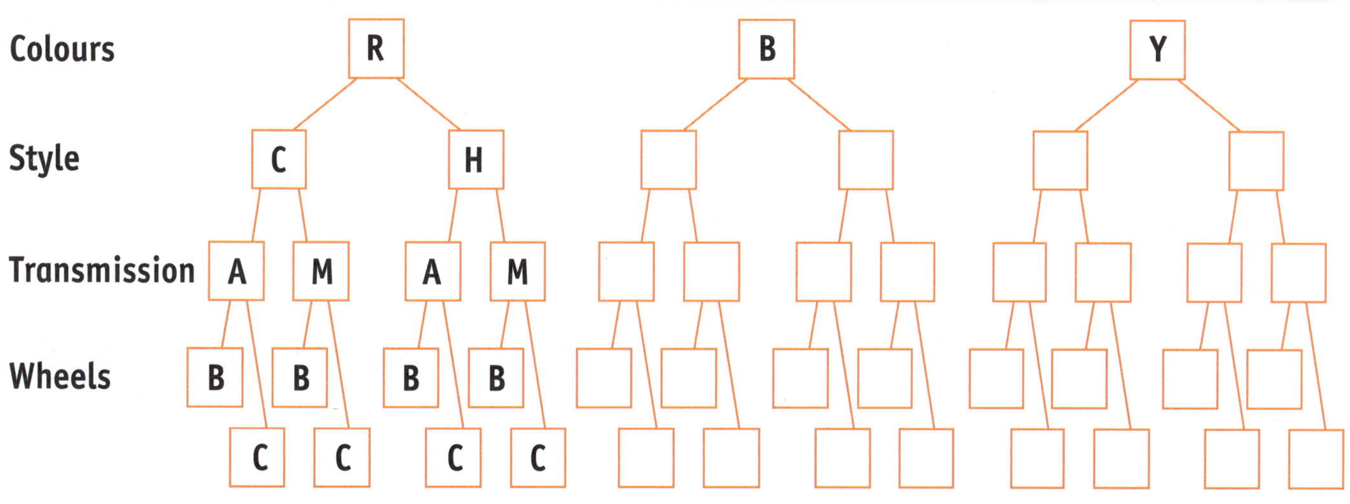

a Use red to colour the path of the combination: Red, Convertible, Automatic, Chrome.

b Use yellow to colour the path of the combination: Yellow, Hatch, Manual, Black.

c Use blue to colour the path of the combination: Black, Convertible, Automatic, Chrome.

Circle the bags of marbles that show a $\frac{1}{2}$ chance.

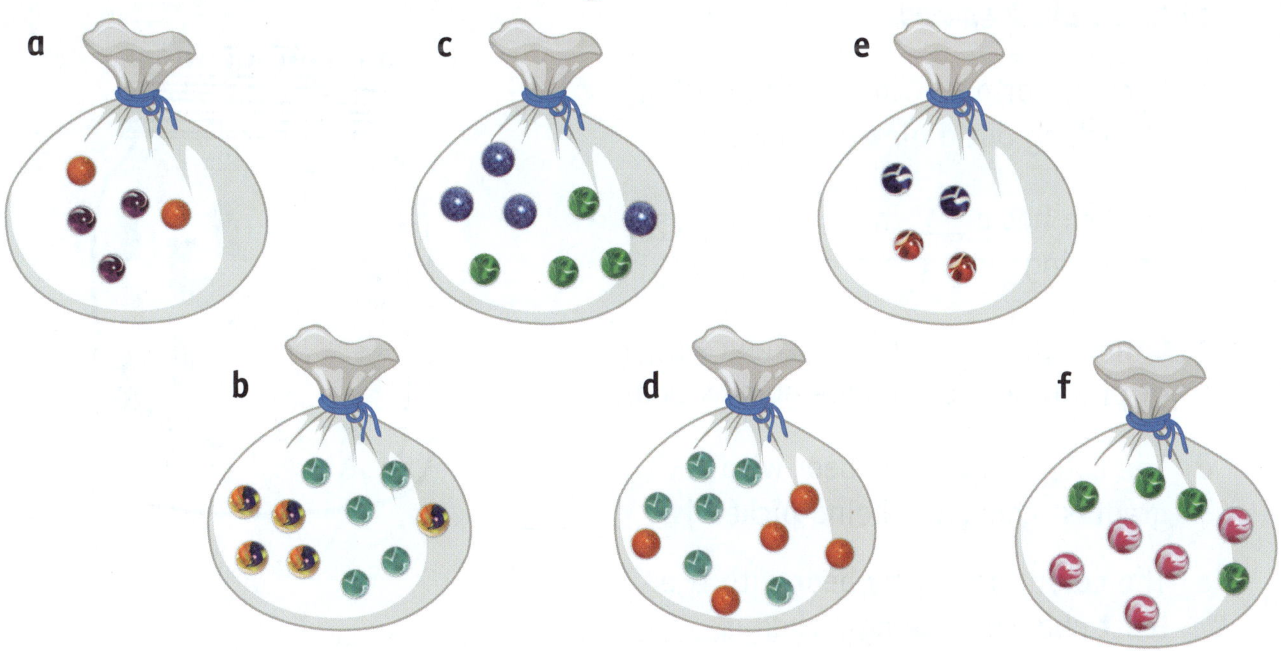

REVIEW

5 Colour the spinners to show the chance.

a A 50% chance of landing on green

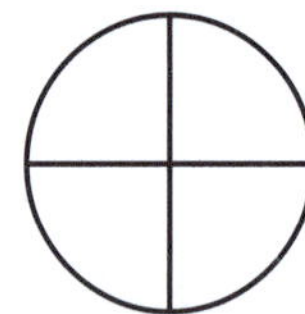

b A 25% chance of landing on black

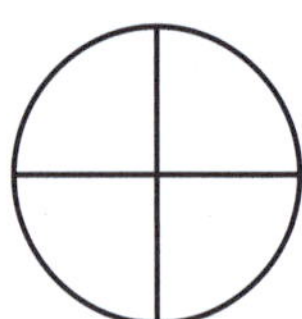

c A $\frac{4}{10}$ chance of landing on yellow

d A $\frac{2}{3}$ chance of landing on red

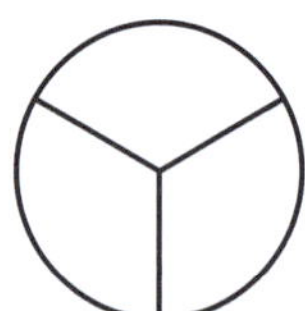

e A greater chance of landing on red than blue

f A lesser chance of landing on blue than green

g A one-in-two chance of landing on pink or black

h A 75% chance of landing on yellow

6 Circle the jars that show an unequal chance.

a

b

c

d

e

f

7 Carmen has 20 gummy bears in a jar. Fill in the blank spaces.

a Carmen has a ___ in ___ chance ($\frac{\quad}{20}$) of picking out a pink gummy bear.

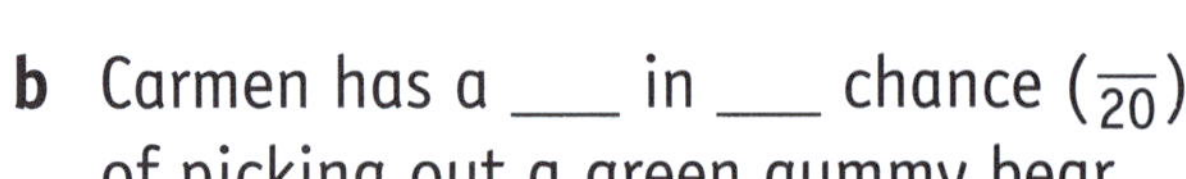

b Carmen has a ___ in ___ chance ($\frac{\quad}{20}$) of picking out a green gummy bear.

c Carmen has a ___ in ___ chance ($\frac{\quad}{20}$) of picking out an orange gummy bear.

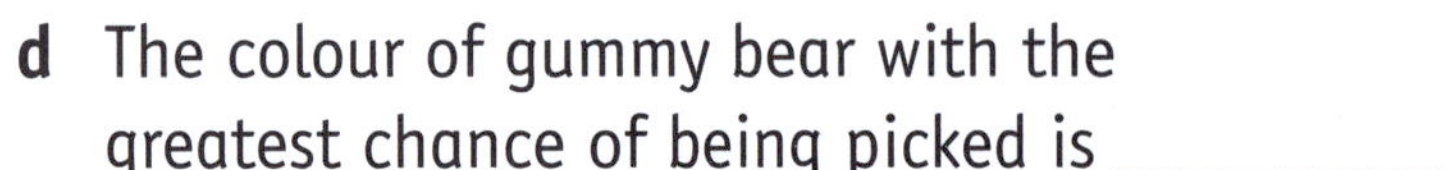

d The colour of gummy bear with the greatest chance of being picked is __________.

e The colour of gummy bear with the least chance of being picked is ___________.

CATCH UP MATHS YEAR 6 BOOK A © PASCAL PRESS ISBN: 9781925726183

8 There are 40 counters in the box.
Colour them to match the given percentages and then fill in the missing numbers.

a 10% red = ___ red counters

b 15% blue = ___ blue counters

c 20% yellow = ___ yellow counters

d 30% green = ___ green counters

e 25% black = ___ black counters

9 Complete the tables.

	Fraction	Decimal	Percentage
a	$\frac{30}{100}$		
b		1.0	
c			70%
d		0.60	
e	$\frac{90}{100}$		
f		0.25	
g			40%
h			90%

	Fraction	Decimal	Percentage
i	$\frac{10}{100}$		
j		0.33	
k			50%
l		0.2	
m	$\frac{80}{100}$		
n			75%
o		0.70	
p	$\frac{67}{100}$		

10 Joe has 20 toy cars in a box.
Use the colours of the cars to fill in the table.

Car Colour	Fraction	Decimal	Percentage
Red			
Blue			
Orange			
Green			

TABLES

Data is information collected after a question is asked.

The question might be: How do you get to school? What pet(s) do you own? What is your favourite music?

Data is often displayed in a table.

Example 1:

The table shows the data collected in answer to the question:

What is the favourite colour of the students in 6F?

6F's Favourite Colours ← This is the information being collected.

Colour	Tally	Total
Red	////	4
Blue	~~////~~ //	7
Green	///	3
Yellow	//	2
Black	~~////~~ /	6
		22

These are the colours that data is collected about.

After counting the tallies, write the total.

The total number of students who provided data (all the totals added together).

Each line is a tally mark. One tally mark means one person voted for that colour. ~~////~~ = 5

You can get a lot of information from this table. For example:

- The students in 6F have 5 different favourite colours.
- Black is the favourite colour of 6 students.
- There are 22 students in 6F.

Use the table to complete the following statements.

a The most popular colour in 6F is __blue__.

b The least popular colour in 6F is __yellow__.

c __4__ students chose red as their favourite colour.

d __3__ students chose green as their favourite colour.

e The difference in votes for black and red is ___.

f ___ is the difference in votes for the most popular colour and the least popular colour.

g ___ students in 6F voted.

CATCH UP MATHS YEAR 6 BOOK A © PASCAL PRESS ISBN: 9781925726183

Example 2: Count the tally marks and write the totals to complete the table. Then answer the following questions.

Native Birds Seen During 6K Excursion

Native Bird	Tally	Total
Kookaburra	卌 卌 |||	
Galah	卌 卌 ||	
Cockatoo	卌 ||||	
Rainbow lorikeet	卌 卌 ||||	
Magpie	||	
Willie wagtail	卌		

a How many galahs were seen? ___

b How many different types of native birds were seen? ___

c The bird seen the most was ________________________.

d The bird seen the least was ________________________.

e There were ___ willie wagtails seen.

f ___ native birds altogether were seen by 6K.

Your turn

Complete the table and then fill in the blank spaces.

● 8 people chose tacos as their favourite food.

a ___ people chose sushi as their favourite food.

b The least popular food in 6T is ______________.

c The most popular food in 6T is ______________.

d The difference in votes for the most popular and least popular foods is ___.

6T's Favourite Food

Food	Tally	Total
Burgers	||||	
Pasta	卌	
Pizza	卌 卌	
Sushi	卌 ||||	
Tacos	卌 |||	

SELF CHECK Tick how you feel

Got it!	Need help...	I don't get it
☐	☐	☐

Check your answers

How many did you get correct? ☐

PRACTICE

1 Amber surveyed her class and asked what their favourite type of music was. She organised the data she collected in this table.

Use Amber's table to answer the questions.

Favourite Music

Music	Tally	Total
Country	|||	3
Rock	~~||||~~	5
Classical	||	2
R'n'B	~~||||~~ |	6
Dance	~~||||~~ ~~||||~~ ||	12
		28

a How many people did Amber survey? ____

b What was the most popular music?

____

c What was the least popular music?

____

d How many people chose R'n'B? ____

e What is the difference in the numbers of votes for Dance and Rock? ____

f What is the difference in the numbers of votes for Country and Classical? ____

2 Organise the information into the table. Cross out the gummy bears as you tally them.

Gummy Bears

Colour	Tally	Total
Black		
Blue		
Yellow		
Green		
Red		
Orange		
Purple	|	
Pink		

3 Use the information in Question 2 to answer the following questions.

a How many different colours of gummy bears were tallied? ____

b There were ____ purple gummy bears.

c ____ green gummy bears were tallied.

d Both ______________ and ______________ had 4 gummy bears tallied.

e The total number of gummy bears is ____.

  ISBN: 9781925726183

TWO-WAY TABLES

A two-way table shows two sets of data.
It shows the relationship between data.

SCAN to watch video

Example 1: This two-way table shows 32 people sorted by hair type (curly or straight) and hair colour (brown, blond, black or red).

Hair colour	Hair type		Total
	Curly	Straight	
Brown	3	6	9
Blond	5	5	10
Black	4	4	8
Red	2	3	5
Total	**14**	**18**	**32**

The table shows that 9 people have brown hair. 3 people have curly brown hair and 6 people have straight brown hair.

Example 2:
Complete the total column and row in this two-way table.

Drink choice	People		Total
	Girls	Boys	
Water	5	3	
Milk	3	5	
Juice	1	2	
Softdrink	4	3	
Total			

Use the two-way table below to answer the questions.

Takeaway choice	People		Total
	Boys	Girls	
Pizza	3		5
Burger		3	8
Hot chips	2	6	
Kebab		3	7
Sushi	6		10
Total			**38**

a Complete the two-way table.

b How many boys was data collected from? ___

c How many people was data collected from? ___

d The most popular takeaway was

_______________.

e The most popular takeaway for girls was _______________.

f The least popular takeaway for boys was _______________.

SELF CHECK Tick how you feel

Got it!	Need help...	I don't get it
☐	☐	☐

Check your answers

How many did you get correct? ☐

ISBN: 9781925726183

PRACTICE

1 Lisa surveyed 100 women and 100 men about the type of electronic communication they used the most.

She presented her results in a two-way table and in a side-by-side column graph.

Main Electronic Communication Methods

Communication method	People		Total
	Women	Men	
Social media	18	5	23
Text message	24	40	64
Phonecall	55	50	105
Email	3	5	8
Total	**100**	**100**	**200**

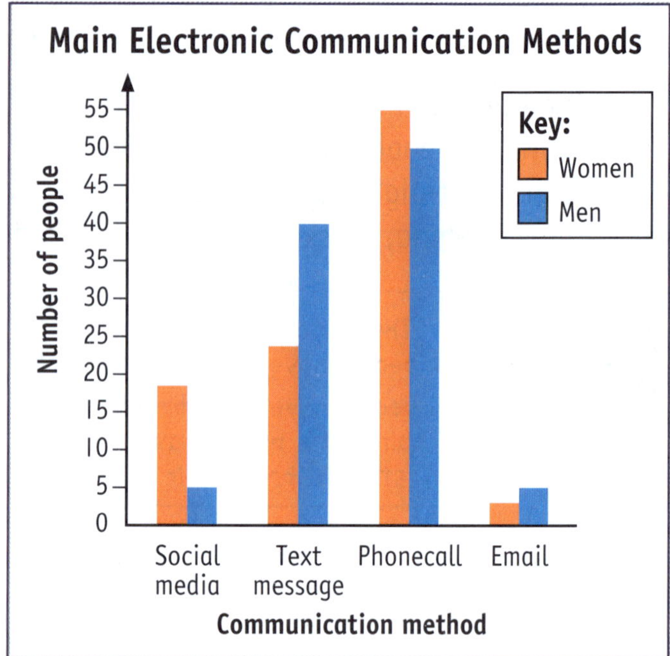

Use Lisa's data to answer the following questions.

a What is the least popular method of communication for women? ____________

b What is the most popular method of communication for men? ____________

c Which group had a higher use of social media – women or men? __________

d Which group had a higher use of text messaging? __________

e Which group had a higher use of phonecalls? __________

f Which group had a higher use of email? __________

g How many people were surveyed? _____

2 Some Year 6 students were asked if they were left handed or right handed. The results are in the table.

Dominant hand

Group	Handed		Total
	Right	Left	
Female	48	4	52
Male	57	7	64
Total	**105**	**11**	**116**

a How many females were asked? ____

b How many people were asked? ____

c Were there more or less right-handed females than right-handed males? ________

d The difference between left-handed males and left-handed females is ____.

e ____ more males than females were surveyed.

CATCH UP MATHS YEAR 6 BOOK A © PASCAL PRESS ISBN: 9781925726183

200 people were asked about their favourite sport to watch on TV. The data is in this two-way table.

Favourite Sport to Watch on TV

Sport	People		Total
	Male	Female	
Basketball	31	41	72
Soccer	15	20	35
Ice hockey	35	19	54
Tennis	20	15	35
Golf	4	0	4
Total	**105**	**95**	**200**

a Were more females or males surveyed?

____

b What was the most popular sport for males?

____

c What was the most popular sport for females?

____

d What was the least popular sport for females? __________________

e What was the least popular sport for males? __________________

f What is the difference between the number of females voting for soccer and the number of females voting for basketball?

_____

g What is the difference between the number of males voting for ice hockey and the number of females voting for tennis?

_____

h What sport was not voted for by any females? __________________

Nemo saw 300 fish at the Great Barrier Reef. The data is in this two-way table.

Fish Seen at the Great Barrier Reef

Fish	Time of Day		Total
	Morning	Afternoon	
Angelfish	34	19	53
Cardinalfish	33	62	95
Clownfish	16	18	34
Parrotfish	23	24	47
Gobies	17	29	46
Grouper	10	15	25
Total	**133**	**167**	**300**

a Were more fish seen in the morning or in the afternoon?

____

b The fish seen least was the

_______________.

c The fish seen most was the

_______________.

d ___ clownfish were seen altogether.

e ___ fish were seen in the afternoon.

f 47 _____________ were seen altogether.

PICTURE GRAPHS

A picture graph uses pictures to represent data.
The key tells you what the pictures mean.

Example 1: Use the picture graph to answer the following questions.

It is easy to see that 6T raised the most money.

Money raised for Year 6 Farewell

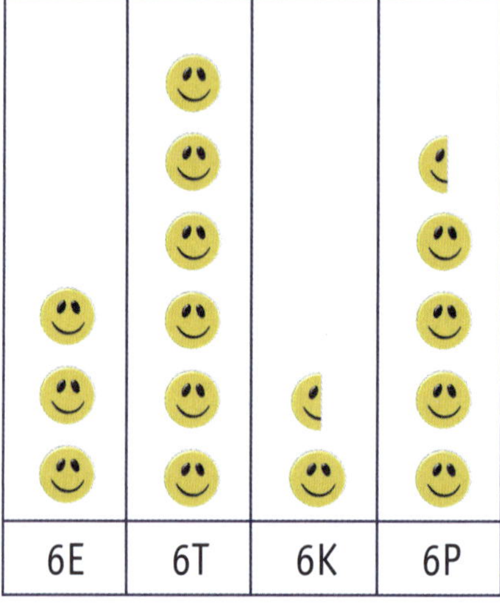

a Class 6K raised the least amount of money.

b $450 was raised by 6P.

c $300 was raised by 6E.

d The difference in money raised by 6T and 6K is $450.

e The total amount of money raised for the Year 6 farewell is $1500.

Example 2: Use the picture graph to complete the table.

Sausages Sold at Sausage Sizzle

April	▲▲▲▲▲▲◢
May	▲▲▲▲▲▲▲▲▲
June	▲▲▲▲▲▲▲▲◢
July	▲▲▲▲▲▲▲▲◢

Key: ▲ = 20 sausages

Sausages Sold at Sausage Sizzle

Month	Total
April	130
May	
June	
July	

CATCH UP MATHS YEAR 6 BOOK A © PASCAL PRESS ISBN: 9781925726183

Example 3: Use the information in the box to complete the picture graph below.

- 75 students chose Pizza Day.
- Mufti Day got 10 more votes than Pizza Day.
- 50 students chose Crazy Hair Day.
- Both School Disco and Crazy Sock Day got 15 less votes than Crazy Hair Day.

SRC Fundraising Activities

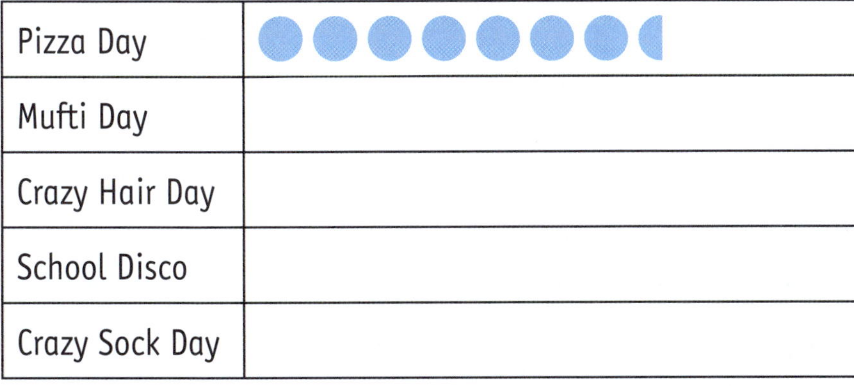

Pizza Day	
Mufti Day	
Crazy Hair Day	
School Disco	
Crazy Sock Day	

Key: = 10 students

Use the picture graph to complete the statements.

- February was the month that most cakes were sold.

a ________ was the month the least number of cakes were sold.

b ___ cakes were sold in January.

c 70 cakes were sold in ________.

d 50 cakes were sold in ________.

e ____ cakes were sold during autumn.

f ____ cakes altogether were sold in the months that have 31 days.

Cakes Sold at Stall

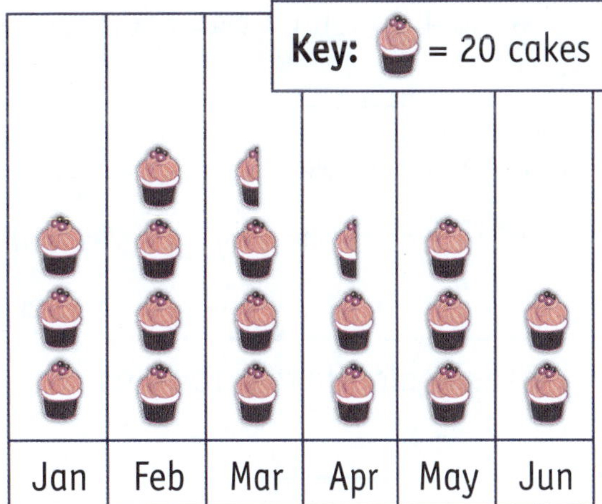

Check your answers
How many did you get correct?

PRACTICE

1 **Use this graph to answer the following questions.**

Cars Built by Ford in June

2018	
2019	
2020	
2021	

Key: = 1000 cars

- How many cars were built in June 2018? 5000

a How many cars were built in June 2019? ______

b In what year were the most cars built? ______

c In what year were the least cars built? ______

d What is the difference between the numbers of cars built in June 2019 and June 2021? ______

e How many cars were built from June 2018 to 2021? ________

2 **6P made this graph of their favourite ice-cream flavours. Use it to answer the questions.**

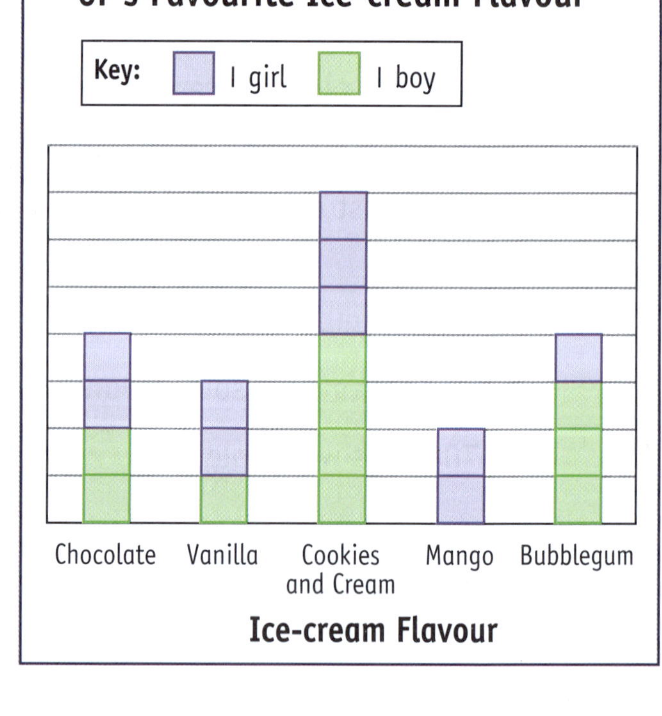

- 2 girls chose chocolate as their favourite ice-cream flavour.

a The most popular flavour for boys is ____________________.

b Mango was chosen by ___ girls.

c Vanilla was chosen by ___ students.

d ____________________ is the most popular flavour for girls.

e ____________________ is the most popular flavour.

f Cookies and cream was chosen by ___ boys.

g ___ students are in 6P.

h Write two more questions you could ask about this graph.

____

____

CATCH UP MATHS YEAR 6 BOOK A © PASCAL PRESS ISBN: 9781925726183

COLUMN GRAPHS

A column graph uses columns to represent data.
The columns can be vertical or horizontal.

Example 1: This is a vertical column graph.

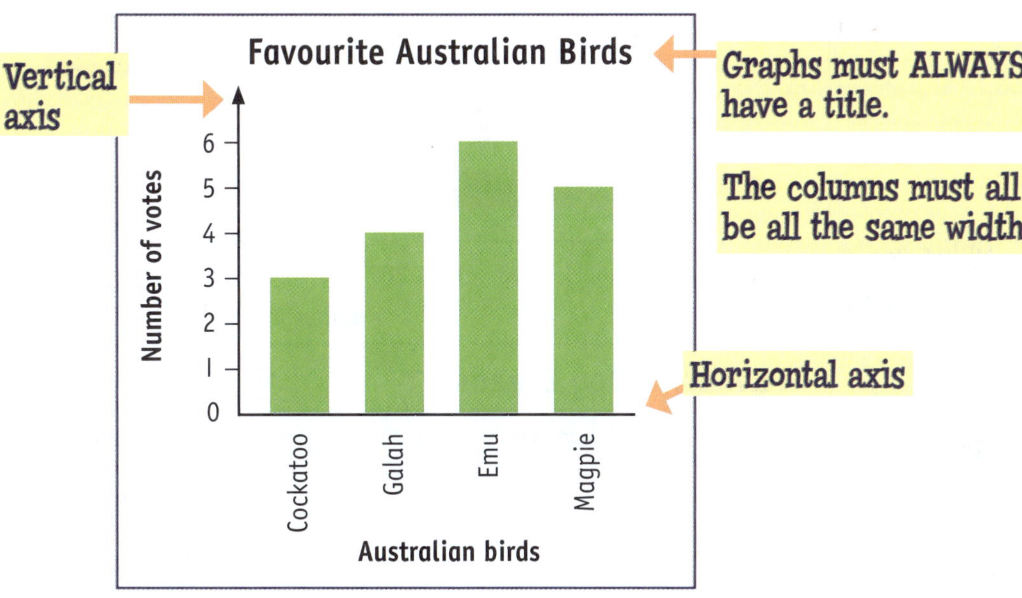

Example 2:
This horizontal column graph shows the same data as Example 1.

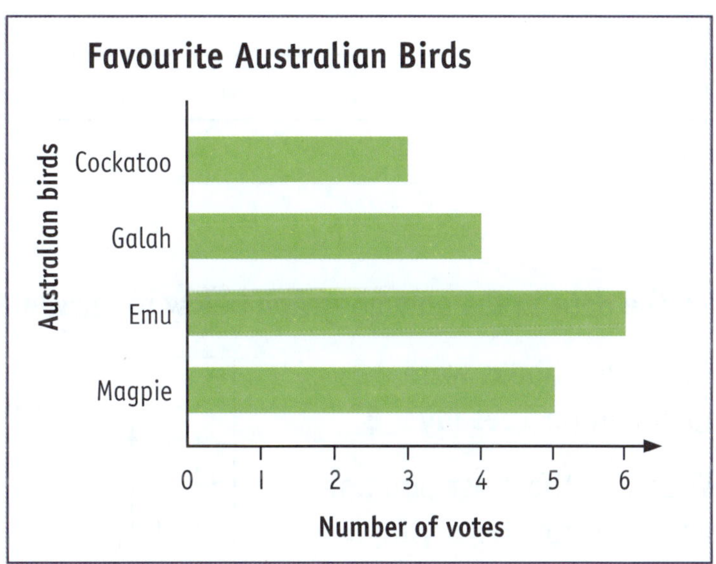

Example 3:
Use the information in the graph above to complete the table.

Favourite Australian Birds

Bird	Tally	Total
Cockatoo	|||	3
Galah	||||	4
Emu	~~||||~~ |	6
Magpie	~~||||~~	5
		18

Example 4:

Use the data in the column graph to complete the table.

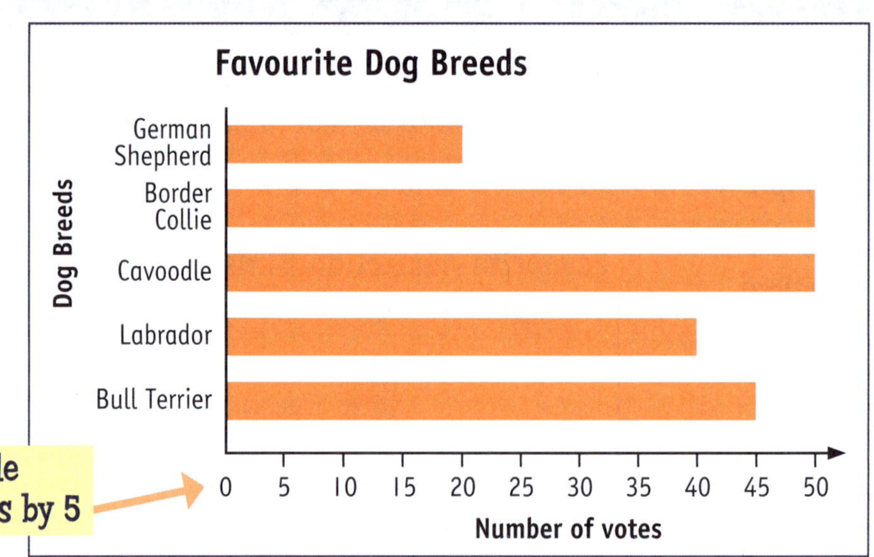

Favourite Dog Breeds

Dog breed	Tally	Total
German Shepherd		
Border Collie		
Cavoodle		
Labrador		
Bull Terrier		

Check your answer on the video!

Your turn

Use the data in the column graph below to answer the questions.

- How many books were borrowed on Tuesday? 8

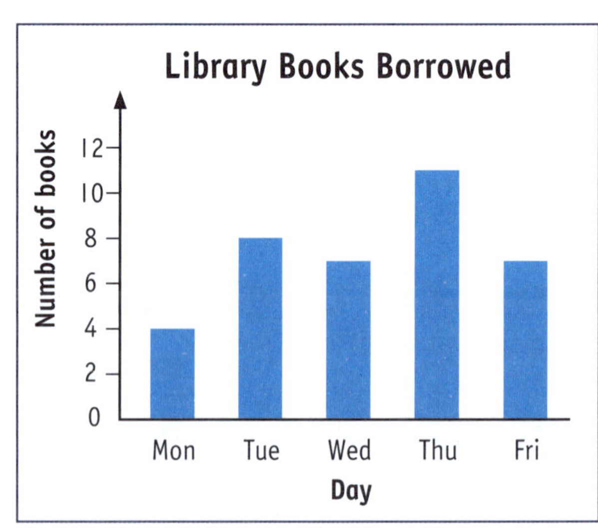

a What was the most popular day to borrow library books? ____________

b How many library books were borrowed in total? ____

c What is the difference in the numbers of books borrowed on Thursday and Friday? ____

SELF CHECK Tick how you feel

Got it!	Need help...	I don't get it

Check your answers

How many did you get correct?

CATCH UP MATHS YEAR 6 BOOK A © PASCAL PRESS ISBN: 9781925726183

1 Use the graph to answer the questions.

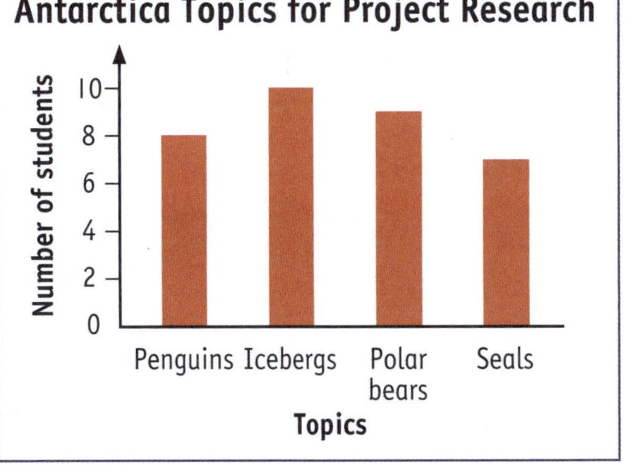

- What was the most popular research topic? Icebergs
- **a** What was the least popular research topic? ______
- **b** What is the difference in the number of students studying polar bears and the number studying penguins? ____
- **c** How many students are doing research on Antarctica? ____
- **d** What is the total number of students researching seals and penguins? ____

2 Use the graph at the right to complete the following.

Students at Annie's Dance School

Dance style	Group		Total
	Boys	Girls	
Jazz	6	2	
Hip Hop			
Tap			
Ballet			
Country			
Total			

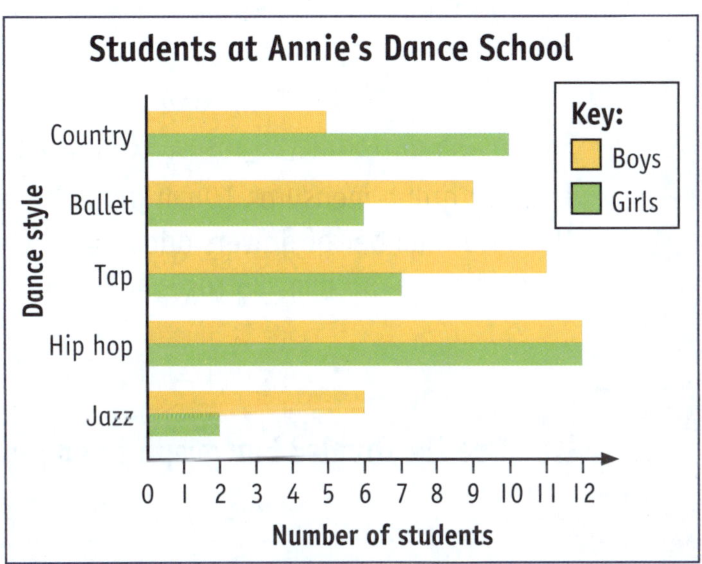

- **a** Fill in the two-way table using the data in the column graph.
- **b** Do more girls or boys attend Annie's Dance School? ______
- **c** What is the most popular dance class for boys? ______
- **d** What is the least popular dance class for girls? ______
- **e** What is the difference between the number of boys doing tap and the number of girls doing ballet? ______
- **f** What is the difference between the number of girls doing hip hop and the number of boys doing ballet? ______
- **g** How many boys attend Annie's Dance School? ____
- **h** How many girls attend Annie's Dance School? ____

DIVIDED BAR GRAPHS

In a divided bar graph, the bar is divided into different sections to show different values of data. The complete bar represents the whole group.

Example 1:

Sports Played By 6T

Soccer	Football	Tennis	Golf

It is easy to see that soccer is the most popular sport.

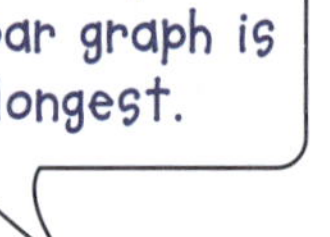

Example 2:

Use the following information to complete the divided bar graph.

Sid's Pot Plants

Daffodils	Roses	

Sid has 35 daffodils, 25 roses, 10 daisies, 15 tulips and 15 violets.
This bar graph measures 10 cm across.
The total number of flowers adds up to 100, so each flower takes 1 mm.
35 daffodils = 35 mm, 25 roses = 25 mm.

Use the divided bar graph to answer the questions below.

Favourite School Subject

Maths	Art	English	Science

- The most popular subject is maths.

a The least popular subject is ________.

b Is English more or less popular than science? ________

c Is maths more or less popular than English? ________

d Is art more or less popular than science? ________

SELF CHECK Tick how you feel

Got it!	Need help...	I don't get it
☐	☐	☐

Check your answers
How many did you get correct? ☐

CATCH UP MATHS YEAR 6 BOOK A © PASCAL PRESS ISBN: 9781925726183

PRACTICE

1 **Elias got $200 for his birthday. He plans to spend 50% of it on clothes, 25% on shoes and the rest on a baseball cap.**

a Finish dividing the bar graph to show these amounts.

Clothes	

b How much money did Elias spend on clothes? ________

c How much money did Elias spend on shoes? ________

d How much was the baseball cap? ________

2 **Use this divided bar graph to answer the questions.**

Fish	Reptiles	Birds	Cats	Dogs

a What title could be given to the divided bar graph?

____

b What types of pets were the most popular? ________________

c Are there more or less birds than fish? ________

d Are there more or less fish than reptiles? ________

e How many types of pets are shown on the divided bar graph? ___

f If the graph is for a total of 300 animals, there are about ____ dogs, ____ birds and ____ fish.

3 **There are 64 students in Year 6. Use this information to write the number of students for each section of the graph below.**

Famous Australians on Australian Money Researched by Year 6 Students

Banjo Paterson	Mary Reibey	David Unaipon	Sir John Monash	Rev. John Flynn	Queen Elizabeth II	Edith Cowan
16						

DOT PLOTS

A dot plot uses dots to represent data.
You can use a dot plot to show one-to-one data and many-to-one data.

SCAN to watch video

Many-to-one data means that one dot can represent 2 students, or 10 students. You don't need a dot for each student.

Example 1: Use the dot plot to complete the table.

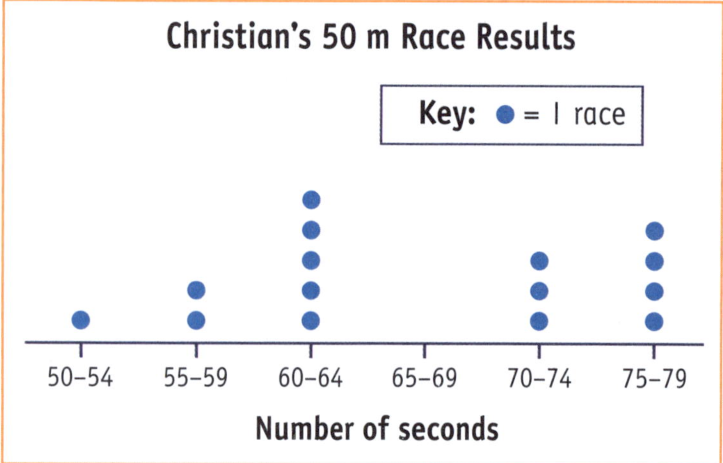

Christian's 50 m Race Results

Seconds	Tally	Total
50–54	/	1
55–59		
60–64		
65–69		
70–74		
75–79		

Example 2:

30 students were asked how many texts they sent in one day.
Their answers are in the box below.

Finish graphing the data on the dot plot and then answer the questions on the next page.

10, 5, 2, 10, 5, 5, 2,
2, 10, 5, 8, 6, 8, 6,
10, 2, 1, 3, 4, 4, 1, 6,
7, 7, 9, 9, 8, 2, 3, 1

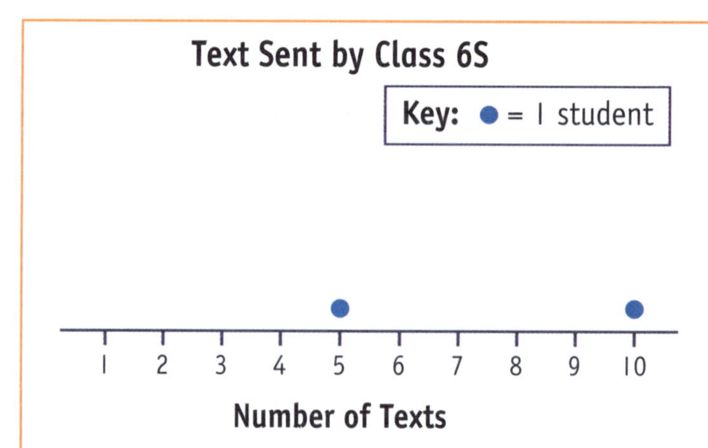

CATCH UP MATHS YEAR 6 BOOK A © PASCAL PRESS ISBN: 9781925726183

a What was the highest number of texts sent?

_____

b How many students sent 3 texts? _____

c How many more students sent 10 texts than 6 texts?

_____

d How many texts were sent altogether?

_____

e Complete the table at the right to show the information in the dot plot.

Texts Sent by Class 6S

Number of Texts	Tally	Total
1		
2		
3		
4		
5		
6		
7		
8		
9		
10		

Your turn

Use the dot plot below to answer the following questions.

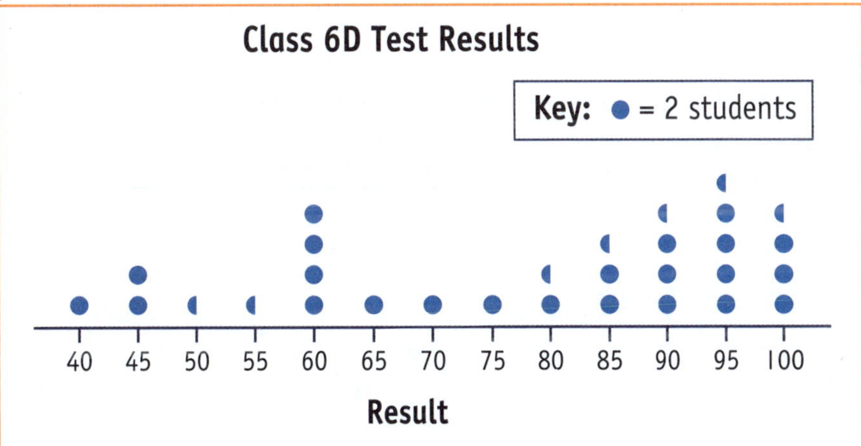

● How many students scored 40? 2

a How many students are in 6D? _____

b How many students scored below 60? _____

c How many students scored above 80? _____

d If 50 was the pass mark, how many students did not pass the test? _____

SELF CHECK Tick how you feel

Got it!	Need help...	I don't get it
☐	☐	☐

Check your answers
How many did you get correct? ☐

  ISBN: 9781925726183

PRACTICE

Use the dot plot to answer the questions.

● 21 students were 140 cm tall.

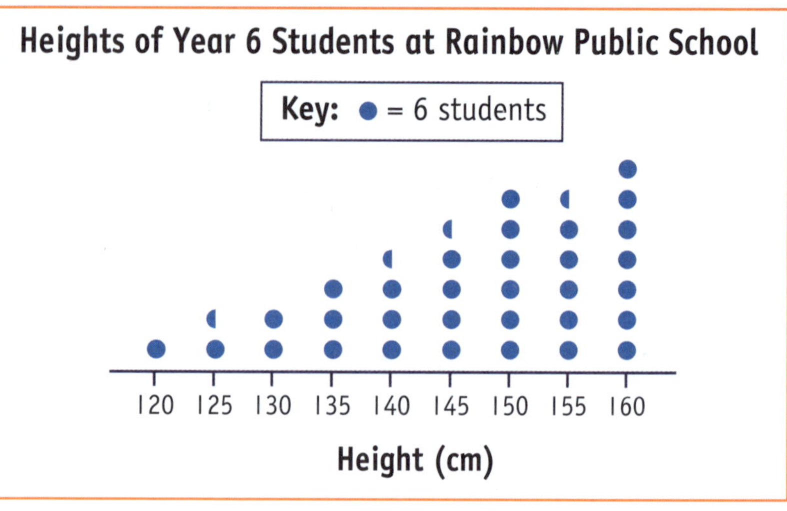

a ___ students were more than 150 cm tall.

b ___ students were 140 cm or less.

c The shortest students were _____ cm tall.

d The tallest students were _____ cm tall.

e There are _____ students in Year 6 at Rainbow Public School.

40 students were asked how many text messages they sent on Saturday. Their answers are in the box at the right.

Graph this data on the dot plot. The first one has been done. Then answer the following questions.

0, 1, 3, 4, 2, 6, 7, 2, 2, 1,
3, 4, 4, 2, 6, 6, 1, 1, 2, 3,
3, 0, 2, 5, 5, 3, 2, 1, 1, 0,
4, 2, 4, 5, 5, 6, 3, 2, 1, 3

Title: ______________________________

Key: ● = 1 text message

0 1 2 3 4 5 6 7

Number of Text Messages

a What was the highest number of texts sent? ___

b How many students sent no texts? ___

c How many students sent 6 text messages? ___

d What was the most common number of text messages sent? ___

CATCH UP MATHS YEAR 6 BOOK A © PASCAL PRESS ISBN: 9781925726183

LINE GRAPHS

A line graph is when data points on a set of axes are joined by a line. The set of axes has a horizontal axis called the x-axis and a vertical axis called the y-axis.
Line graphs are often used when small changes occur over time.

SCAN to watch video

Example 1: Use the graph to complete the data in the table below.
Then complete the following questions.

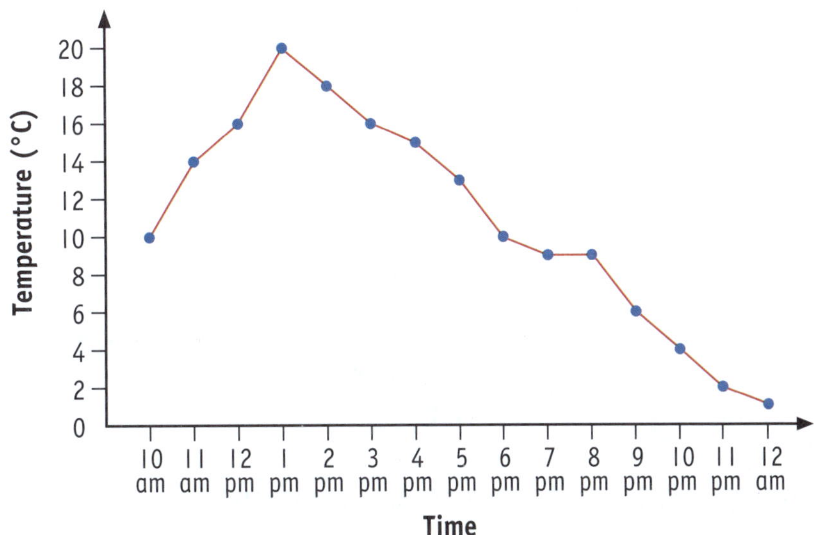

Temperature on Saturday 12 June

Time	10 am	11 am	12 noon	1 pm	2 pm	3 pm	4 pm	5 pm	6 pm	7 pm	8 pm	9 pm	10 pm	11 pm	12 am
Temp. (°C)	10			20						9					1

a What was the temperature at 3 pm? 16 °C

b The highest temperature recorded was 20 °C at 1 pm.

c The lowest temperature recorded was 1 °C at 12 midnight.

d What time was it when these temperatures were recorded?

2 °C 11 pm   9 °C 7 pm, 8 pm   6 °C 9 pm

e Estimate the temperature at these times:

9:30 pm 5 °C   12:30 pm 18 °C   5:30 pm 11.5 °C

Example 2: This line graph shows the numbers of visitors to Thredbo Mountain Park in 2022.

There were 150 visitors to the park in November, and 250 visitors in December. Add this information to the graph.

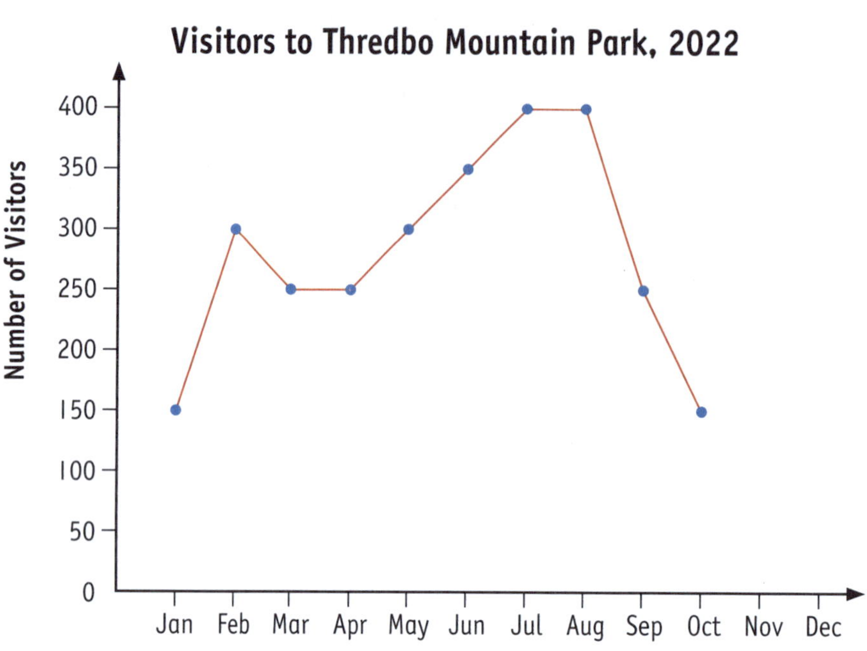

Check your answer on the video!

Your turn

Use the line graph above to answer the questions.

a Use the graph to complete the table.

Visitors to Thredbo Mountain Park, 2022

Month	Jan	Feb	Mar	Apr	May	Jun	Jul	Aug	Sep	Oct	Nov	Dec
No. of Visitors												

b Which month had the same number of visitors as February? ________

c How many people visited in June? _______

d How many people visited in winter? _______

e How many people visited in summer? _______

f What is the difference in the numbers of visitors during autumn and during spring? _______

g How many people visited the park during 2022? ________

SELF CHECK Tick how you feel

Got it!	Need help...	I don't get it
☐	☐	☐

Check your answers

How many did you get correct? ☐

CATCH UP MATHS YEAR 6 BOOK A © PASCAL PRESS ISBN: 9781925726183

PRACTICE

Jaban drives to Bathurst. He left home at 9:30 am.

Add the following data to Jaban's driving graph and join the points with a line. Then use the graph to answer the following questions.

- 12 pm: 175 km
- 1 pm: 175 km
- 2 pm: 300 km
- 3 pm: 450 km

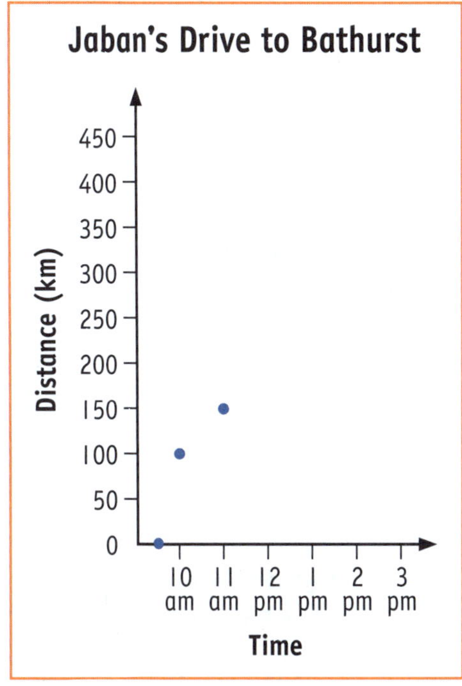

a When did Jaban take a rest?

____

b Between which times did Jaban drive 150 km? ______________

c How far was Jaban's drive to Bathurst?

____

Adriano recorded the temperature on Sunday between 10 am and 3 pm.

- 10 am: 15 °C
- 11 am: 20 °C
- 12 pm: 25 °C
- 1 pm: 30 °C
- 2 pm: 35 °C
- 3 pm: 35 °C

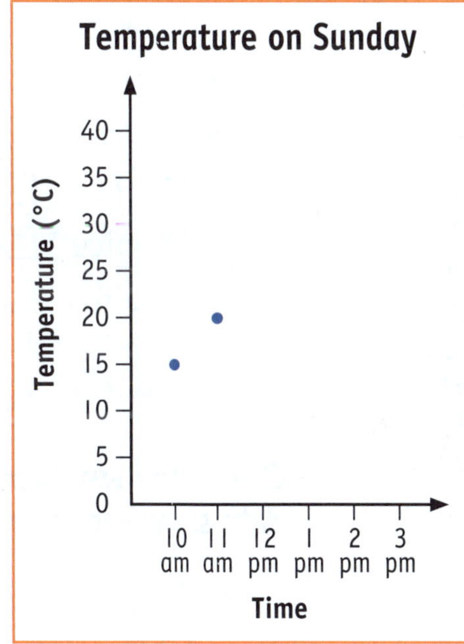

a Finish plotting the temperatures on the graph and join the points with a line.

b What is the highest temperature recorded by Adriano?

______

c What is the lowest temperature recorded by Adriano?

______

d What was the temperature at 1 pm? _____

e What was the temperature at 2 pm? _____

f What was the temperature at 3 pm? _____

g What is the difference between the highest and lowest temperatures? _______

PIE GRAPHS

A pie graph is drawn in a circle, like a pie. Different sectors, or pieces, represent data.

SCAN to watch video

Example 1:

Distance from School to Home for Students in 4T

There are 4 sectors in this pie graph.

Pie graphs make it easy to see the smallest and largest data groups.

Key:
- < 5 km
- 5–10 km
- 10–15 km
- > 20 km

Pie graphs are also called pie charts.

The red sector is clearly the largest. Match it to the key to find out that it represents < 5 km.

Example 2:

Colour the pie graph to show this data:

- 50% dogs ☐
- 25% cats ☐
- 13% fish ☐
- 8% mice ☐
- 4% birds ☐

Pets Sold At Pet Palace

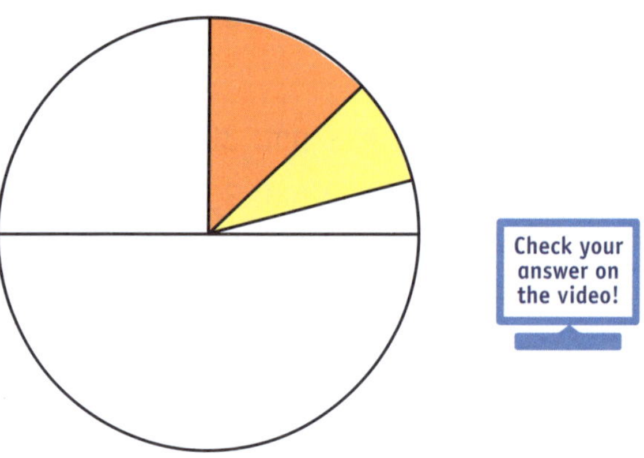

Check your answer on the video!

Colour the key to match the data shown in the pie graph.

- ☐ Jelly 50%
- ☐ Ice cream 25%
- ☐ Chocolate cake 10%
- ☐ Cheese cake 10%
- ☐ Caramel slice 5%

Favourite Desserts

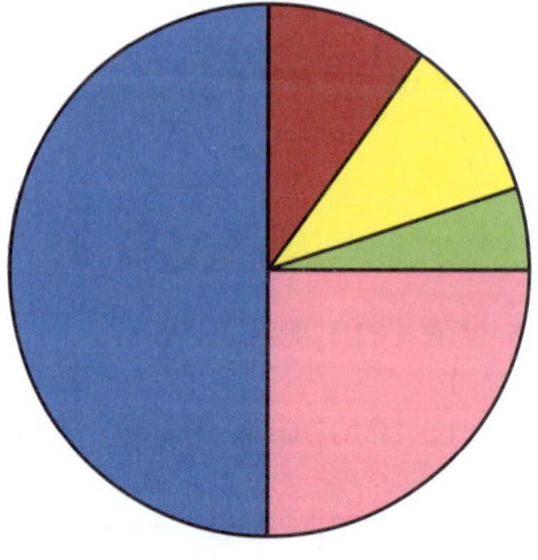

SELF CHECK Tick how you feel		
Got it! ☐	Need help... ☐	I don't get it ☐

Check your answers
How many did you get correct? ☐

CATCH UP MATHS YEAR 6 BOOK A © PASCAL PRESS ISBN: 9781925726183

PRACTICE

These pie graphs show how much of a cheese pizza and a meatlovers pizza were eaten by four people.

Use the graphs to answer the questions.

Cheese Pizza

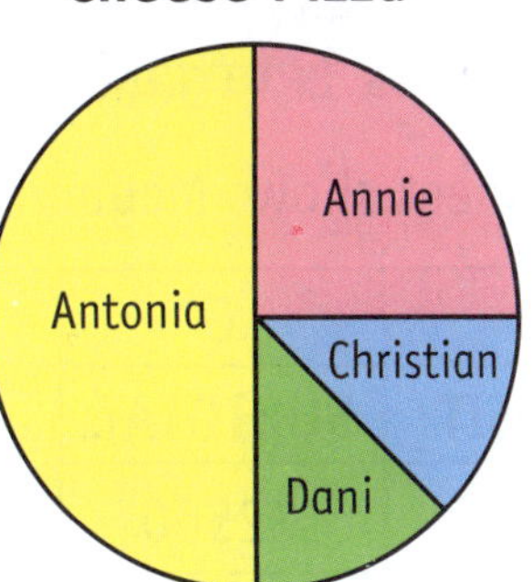

Meatlovers Pizza

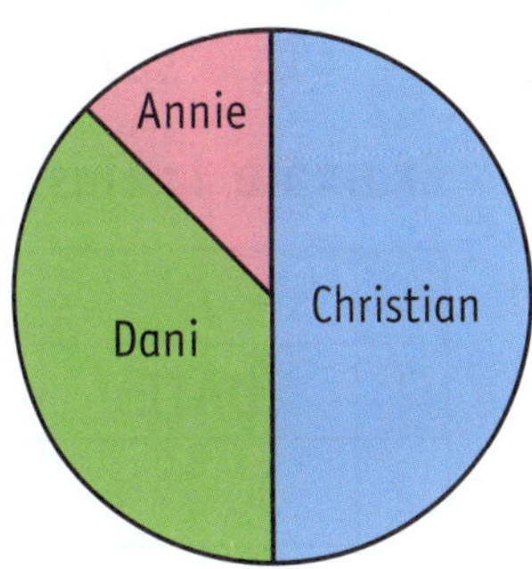

a How many different types of pizza were eaten? ___

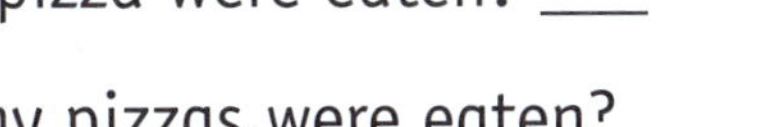

b How many pizzas were eaten? ___

c Who ate the most cheese pizza? ___________

d Who ate the least meatlovers pizza? ___________

e Who ate no meatlovers pizza? ___________

f Who ate the same amount of cheese pizza as Christian? ___________

g Who ate as much meatlovers pizza as Dani ate cheese pizza? ___________

h Who ate the most pizza? ___________

i Who ate the least pizza? ___________

These pie charts have the labels a to h instead of values.

Use the pie charts to answer the questions.

Pie Chart A

a
b
c

Pie Chart B

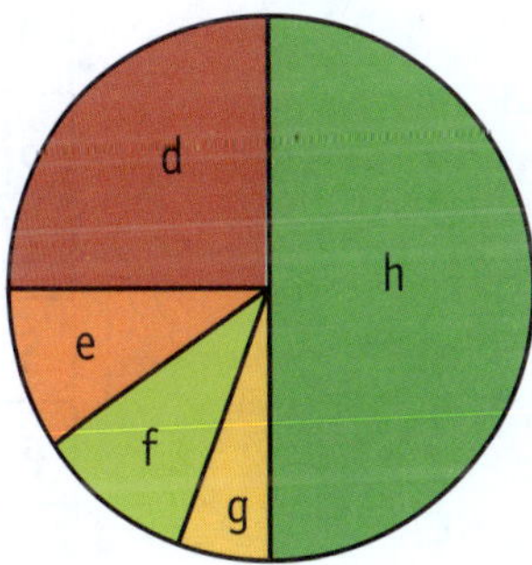

a How many sections make up the total of Pie Chart A? ___

b How many sections make up the total of Pie Chart B? ___

c Which value is 75% of the total? ____

d Which value is 25% of the total? ____

e Which value is 50% of the total? ____

f If a is 10, what is the total value of pie chart A? _____

g If d is 40, what is the total value of pie chart B? _____

h If a is 10 and d is 40, which label shows the largest value? ____

SPREADSHEETS

In a spreadsheet, data is organised into rows and columns.

Example 1: This spreadsheet shows Megumi's savings.

	A	B	C	
1	**DATE**	**DEPOSIT**	**SUBTOTAL**	
2	30 JAN	$2560	$2560	= B2
3	28 FEB	$1590	$4150	= C2 + B3
4	30 MAR	$3250	$7400	= C3 + B4
5	30 APR	$5210	$12 610	= C4 + B5

Work out the subtotal by adding the C column value to the last B column value.
For example, on 28 February, $1590 was deposited.
Add the C column (C2) to the deposit (B3).
$2560 + $1590 = $4150 (The value of C3).

Example 2: This spreadsheet shows the deposits from Medowie Public School's Canteen. Use it to answer the following questions.

	A	B	C	
1	**DATE**	**DEPOSIT**	**SUBTOTAL**	
2	1 FEB	$525.00	$525.00	= B2
3	8 FEB	$600.00	$1125.00	= C2 + B3
4	9 FEB	$430.00	$1555.00	= C3 + B4
5	15 FEB	$715.00	$2270.00	= C4 + B5
6	20 FEB	$49.50		= C5 + B6
7	23 FEB	$125.00		= C6 + B7
8	3 MAR	$370.50		= C7 + B8
9	10 MAR	$249.00		= C8 + B9

a Fill in the missing information on the spreadsheet.

b What was the largest deposit made? $715.00

c What date was the smallest deposit made? __________

d How much money was deposited on 23 February? __________

e What was the subtotal on 10 March? __________

CATCH UP MATHS YEAR 6 BOOK A © PASCAL PRESS ISBN: 9781925726183

Example 3: Add these items to the Sixers Basketball Team spreadsheet and then answer the following questions.

- 25 July: bought balls for $275.50
- 30 July: bought awards for $129.75
- 31 July: bought trophies for $796.85

Sixers Basketball Club Expenses					
	A	B	C	D	
1	DATE	ITEM	COST	BALANCE	
2	3 JUL	Opening	–	$30 250.00	
3	7 JUL	Registrations	$1295.00	$28 955.00	= D2 – C3
4	12 JUL	Jerseys bought	$1349.50	$27 605.50	= D3 – C4
5	15 JUL	Bags bought	$1375.95	$26 229.55	= D4 – C5
6	20 JUL	Shorts bought	$865.45	$25 364.10	= D5 – C6
7	25 JUL				= D6 – C7
8	30 JUL				= D7 – C8
9	31 JUL				= D8 – C9

a Complete the balance column.

b How much were the jerseys bought on 12 July? __________

c What was the largest expense? ________________

d What was the balance (subtotal) on 31 July? ____________

Complete the subtotal column in the spreadsheet and then answer the questions.

	A	B	C	
1	DATE	DEPOSIT	SUBTOTAL	
2	1 MAY	$1000	$1000	= B2
3	1 JUN	$1000		= C2 + B3
4	1 JUL	$1200		= C3 + B4
5	1 AUG	$500		= C4 + B5
6	1 SEP	$1000		= C5 + B6

● How much money was deposited on 1 May?

$1000

a What was the total deposited? _______

b How many deposits were made? ______

c What was the balance (subtotal) on 1 August? ________

SELF CHECK Tick how you feel

Got it!	Need help...	I don't get it
☐	☐	☐

Check your answers

How many did you get correct? ☐

PRACTICE

1 Use this spreadsheet to answer the following questions.

Bundeenal Baseball Club					
	A	B	C	D	
1	DATE	ITEM	DEPOSIT	SUBTOTAL	
2	1 JUN	Opening	--	$1500	
3	3 JUN	Fundraiser	$1200	$2700	= D2 + C3
4	10 JUN	Donations	$550		= D3 + C4
5	21 JUN	Fundraiser	$1550		= D4 + C5
6	28 JUN	Fees paid to club	$650		= D5 + C6
7	30 JUN	Fundraiser	$1650		= D6 + C7

a Complete the missing amounts in the subtotal column.

b What date was the smallest deposit made? __________

c How much was the largest deposit made? __________

d How much money was deposited on 21 June? __________

e What was the balance (subtotal) on 30 June? __________

2 Add the following items to the Lilly Pilly spreadsheet.

- Item B6: the club bought trophies for $1550
- Item B7: the club bought socks for $290
- Item B8: the club bought food for a fundraiser for $575
- Item B9: the club bought goals for $1200

Lilly Pilly Football Club Expenses					
	A	B	C	D	
1	DATE	ITEM	COST	BALANCE	
2	1 JULY	Opening	–	$10 000	
3	3 JULY	Registrations	$1250	$8750	= D2 – C3
4	15 JULY	Soccer shirts bought	$1500	$7250	= D3 – C4
5	18 JULY	Bags bought	$1000	$6250	= D4 – C5
6	19 JULY				= D5 – C6
7	23 JULY				= D6 – C7
8	27 JULY				= D7 – C8
9	29 JULY				= D8 – C9

CATCH UP MATHS YEAR 6 BOOK A © PASCAL PRESS ISBN: 9781925726183

DATA REVIEW

Complete the table and then answer the questions.

a ___ people were asked for their favourite food.

b ___ people chose pasta.

c ___ people chose steak.

d ________________ was the most popular.

e ________________ was the least popular.

f The difference between the numbers of people choosing pasta and pizza was ___.

6K's Favourite Food

Food	Tally	Total
Fish		4
Pasta	卌 I	
Sausages	II	
Chicken	卌 IIII	
Pizza		7
Steak		3

Complete the two-way table and use the data to construct a side-by-side column graph. Then answer the following questions.

Hair colour	Hair type		Total
	Curly	Straight	
Brown	3		8
Blond		9	16
Black	9		11
Red		4	8
Total			

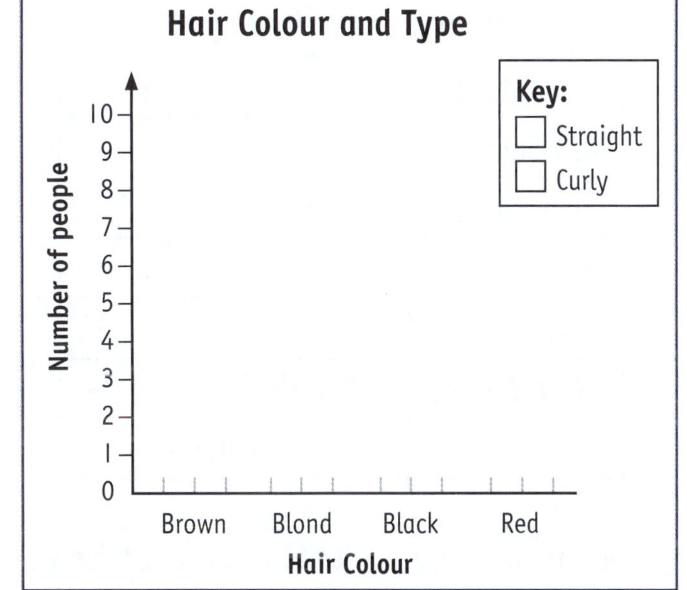

a How many people have curly hair? _____

b How many people have straight hair? _____

c What was the most popular colour of curly hair? ____________

d What was the less popular hair type? ____________

e Which two hair colours had the same total? ________________________

f What hair colour was the most popular? ____________

g What was the more popular type of black hair? ____________

h ___ people have black hair.

i Did more or less people have straight blond hair than curly blond hair? ________

REVIEW

Use the picture graph to answer the questions.

a Class ___ raised the most money.

b ___ raised $350.

c ___ raised the least money.

d The difference in money raised by 6T and 6K is _________.

e 6E and 6K raised ______ altogether.

f __________ in total was raised by Year 6.

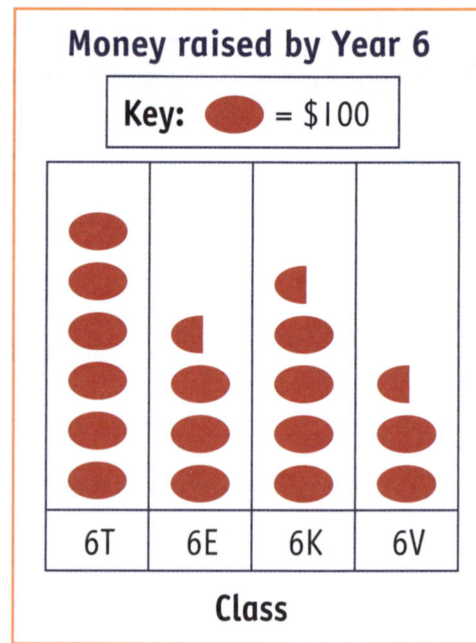

Use the data in the table to construct a column graph.

States Researched by 6T

State	Tally	Total
New South Wales	𝍸	5
Queensland	||||	4
Victoria	𝍸 |||	8
Western Australia	𝍸 |||	8
South Australia	𝍸 |	6
Tasmania	𝍸 𝍸	10

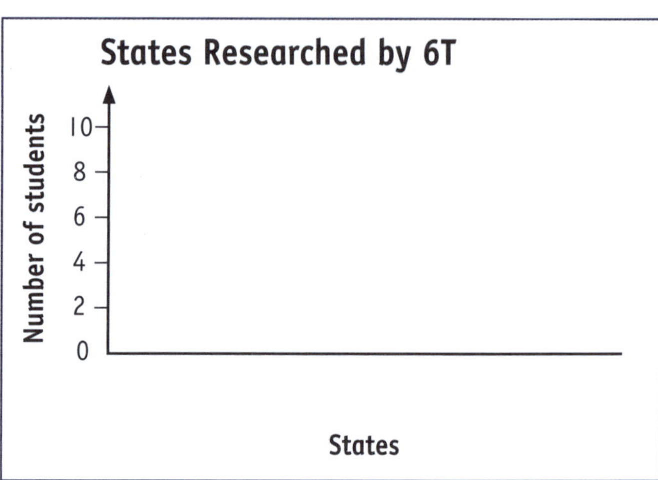

a How many students researched New South Wales? ___

b ___ students researched South Australia.

c Tasmania was researched by ___ students.

d The total number of students who researched Queensland and Western Australia is _____.

e The difference in students who researched Tasmania and Victoria is ___.

f The states researched by the same numbers of students are ________________ and ________________.

g There are ____ students in 6T.

CATCH UP MATHS YEAR 6 BOOK A © PASCAL PRESS ISBN: 9781925726183

5 Divide the bar graph to show the favourite chocolate flavours of 100 people.

Favourite Chocolate Flavour	
Plain	50
Peanut	25
Caramel	10
Honeycomb	10
Peppermint	5
Total	**100**

Favourite Chocolate Flavour

a What chocolate flavour was the most popular? ________________

b How many flavours of chocolate were included in the data? ____

c Do more or less people like caramel, compared to peanut? ________

d ____ people prefer peppermint chocolate.

e The total number of people who chose honeycomb, peanut or plain chocolate is ____.

f How many people were surveyed? ____

6 Use the dot plot to answer the following questions.

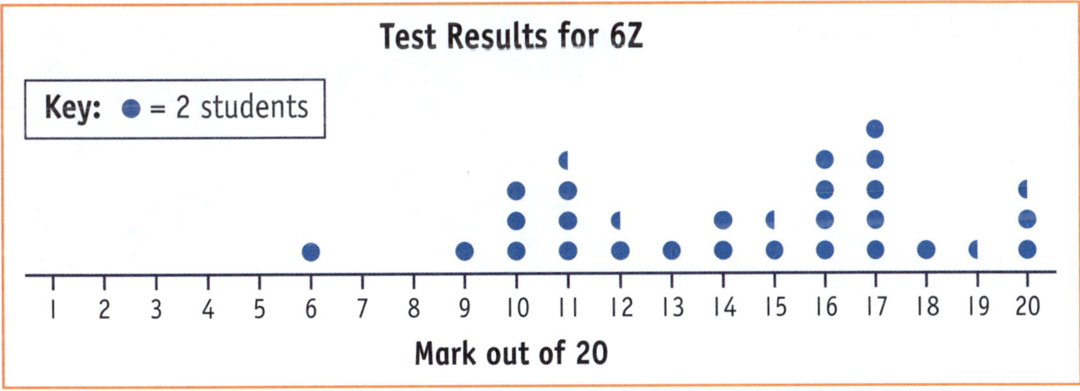

a ___ students are in 6Z.

b ___ students scored less than 11.

c ___ students scored 14 or more.

d ___ students scored full marks.

e How many students scored 15 or less? ___

f How many students scored from 15 to 18? ___

REVIEW

7 Use the data in the table to construct a line graph. Then answer the questions.

Temperature on 1 January

Time	8 am	10 am	12 pm	2 pm	4 pm	6 pm	8 pm	10 pm	12 am
Temperature (° C)	10	15	25	29	30	28	25	21	15

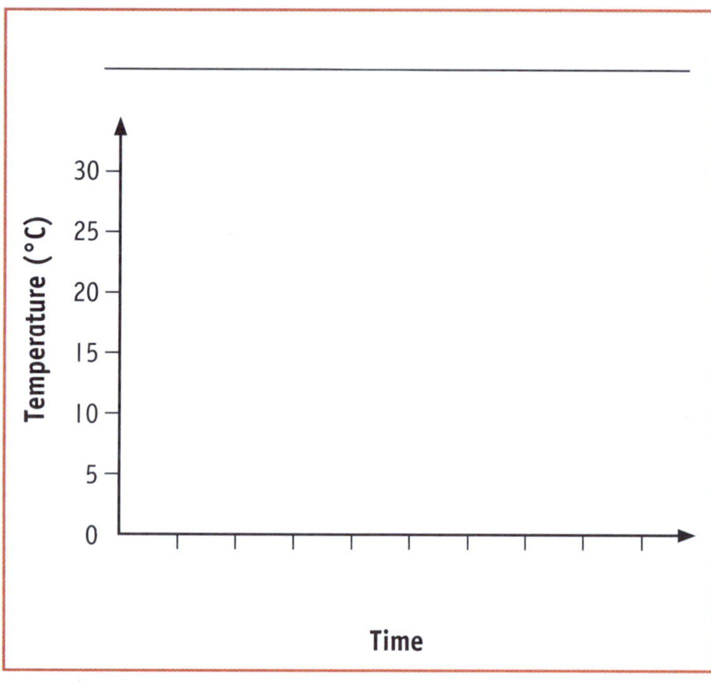

a What was the temperature at 10 am? _____

b The highest temperature recorded was _____ at ___ pm.

c The lowest temperature was _____ at _______.

d What time was it 21 °C? _____

e What time was it 29 °C? _____

f What time was it 30 °C? _____

8 These pie graphs have the labels a to h instead of values.

Use the pie charts to answer the questions.

Pie Chart X

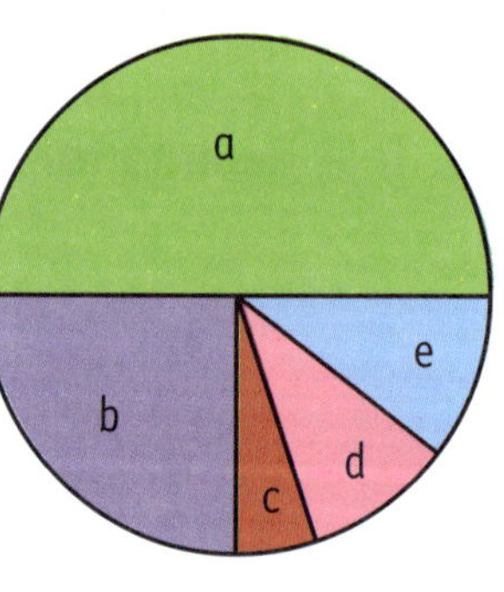

Pie Chart Y

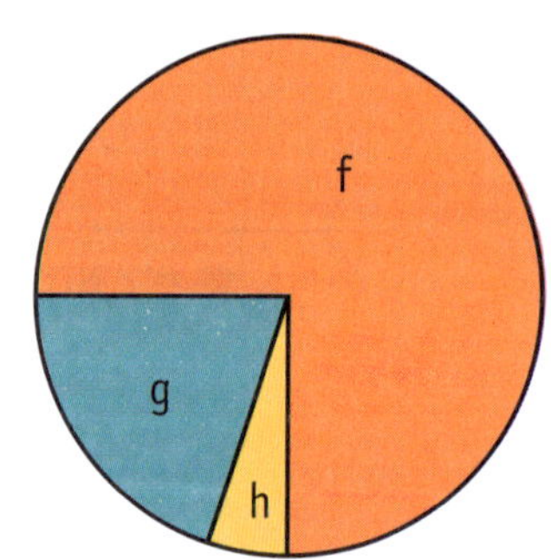

a Which value is 25% of the total? ___

b What percentage of the total is f? _____

c What percentage of the total is b? _____

d Which value is 50% of the total? _________

e If a is worth 50, what is the value of b? _________

f If g + h is worth 100, what is f worth? _____

g Which sector has the least value in pie graph X? ___

h Which sector has the least value in pie graph Y? ___

CATCH UP MATHS YEAR 6 BOOK A © PASCAL PRESS ISBN: 9781925726183

Complete the spreadsheet and answer the questions.

Power Flower			
	A	B	C
1	DATE	DEPOSIT	SUBTOTAL
2	15 FEB	$1250	= B2
3	15 MAR	$675	= C2 + B3
4	15 APR	$249	= C3 + B4
5	15 MAY	$1220	= C4 + B5
6	15 JUN	$1463	= C5 + B6
7	15 JUL	$2425	= C6 + B7
8	15 AUG	$745	= C7 + B8

a How much money was deposited on 15 March? ___________

b What was the balance on 15 May? ___________

c How many deposits were made? ____

d How much money was deposited altogether? ___________

Complete the balance column and then answer the following questions.

St James Tennis Club Expenses				
	A	B	C	D
1	DATE	ITEM	COST	BALANCE
2	2 AUG	Opening	–	$14 235.80
3	5 AUG	Buy tennis balls	$250.00	= D2 – C3
4	10 AUG	Registrations	$1454.50	= D3 – C4
5	14 AUG	Insurance	$2853.80	= D4 – C5
6	20 AUG	Buy tennis net	$563.20	= D5 – C6
7	20 AUG	Buy uniforms	$602.50	= D6 – C7
8	22 AUG	Buy tennis shoes	$150.00	= D7 – C8
9	26 AUG	Buy food for fundraiser	$352.75	= D8 – C9
10	30 AUG	Buy tennis racquets	$1523.63	= D9 – C10

a The club's biggest expense was ______________________.

b Uniforms were bought on ______________.

c Registrations cost ______________.

d The balance on 22 August was ______________.

e The balance after registrations were made was ______________.

f The club's smallest expense was ______________________.

g How many expenses did St James Tennis Club have in August? _____

1 WHOLE NUMBERS

Three-digit, Four-digit and Five-digit Numbers

Page 1 – Your Turn

3-digit: 103, 326, 453, 954
4-digit: 1359, 2496, 9420, 1002
5-digit: 58 320, 94 371, 57 195, 42 858

Page 2 – Practice

1 a three hundred and forty-nine
b four hundred and twenty-eight
c five hundred and three
d one thousand, nine hundred and thirty-seven
e five thousand, six hundred and forty-two
f seven thousand and thirty
g twenty-four thousand, three hundred and forty-six
h forty-one thousand, nine hundred and thirty
i fifty-seven thousand and five

2

	Number	Ten thousands	Thousands	Hundreds	Tens	Ones
a	749	0	0	7	4	9
b	903	0	0	9	0	3
c	6245	0	6	2	4	5
d	7142	0	7	1	4	2
e	29 307	2	9	3	0	7
f	99 390	9	9	3	9	0

3 3 digits: 140, 359, 752, 203, 157
4 digits: 2413, 3008, 2893, 1439, 9981
5 digits: 23 205, 50 000, 82 493, 41 580, 71 345

Place Value THREE-DIGIT, FOUR-DIGIT AND FIVE-DIGIT NUMBERS

Page 3 – Your Turn

1 a 352
b 93526
c 4150
d 41384
e 580
f 5836
g 1429
h 409
i 177
j 5389
k 83715

Page 4 – Practice

1 a hundreds
b hundreds
c thousands
d ones
e hundreds
f thousands
g tens
h tens
i ten thousands
j thousands
k hundreds

2 a 65923
b 514
c 71490
d 4830
e 95243
f 781
g 40854
h 5821

3 a hundreds
b ten thousands
c ones
d tens
e ones
f tens
g hundreds

4 a 3263
b 948
c 70 046
d 94 351
e 50 003
f 6437

Value THREE-DIGIT, FOUR-DIGIT AND FIVE-DIGIT NUMBERS

Page 5 – Your Turn

Use red: 526, 8500, 73 581, 72 531, 549, 1532, 7539
Use blue: 1453, 852, 753, 258
Use green: 105, 2985, 645

Page 6 – Practice

1 a 60
b 6000
c 6
d 60 000
e 600
f 6000
g 600
h 60
i 6
j 6
k 60

2 Adult to check

3 a 439
b 1384
c 593
d 879
e 41589
f 2491
g 34285
h 1509
i 16073
j 97329
k 42444
l 555
m 9009
n 24823
o 60389

4 a 562, 6593, 34 531
b 57 111, 7000, 47 328
c 93 246, 90 331, 98 363
d 61 592, 59 472, 1582
e 4165, 3265, 37 469

5 Adult to check

Expanded Numbers

THREE-DIGIT, FOUR-DIGIT AND FIVE-DIGIT NUMBERS

Page 7 – Your Turn

1 a 7000 + 300 + 50 + 1
b 500 + 1
c 50 000 + 6000 + 200 + 4
d 200 + 90 + 3
e 1000 + 400 + 80 + 2

Page 8 – Practice

1 a 70 000 + 6000 + 900 + 40 + 1
b 1000 + 500 + 30 + 6
c 30 000 + 200 + 40 + 5
d 300 + 70
e 500 + 30 + 9
f 7000 + 60 + 5
g 90 000 + 3000 + 200 + 50
h 4000 + 300 + 50 + 8
i 100 + 60

2 a 20 000 + 4000 + 600 + 20 + 3
b 10 000 + 5000 + 90 + 4
c 3000 + 300 + 40 + 3
d 1000 + 700 + 50
e 200 + 40 + 9
f 500 + 3

3

	Number	Place Value	Value
a	249	tens	40
b	64 731	thousands	4000
c	49 387	ten thousands	40 000
d	62 354	ones	4
e	104	ones	4
f	4321	thousands	4000
g	43 877	ten thousands	40 000
h	473	hundreds	400
i	34 920	thousands	4000

Ordering Numbers

THREE-DIGIT, FOUR-DIGIT AND FIVE-DIGIT NUMBERS

Page 9 – Your Turn

1 a ascending
b descending
c ascending
d descending
e descending
f ascending
g ascending

Page 10 – Practice

1 a 126, 162, 261, 612, 621
b 347, 374, 437, 734, 743
c 1526, 2651, 5261, 6251, 6512
d 4395, 4539, 5349, 5934, 9453
e 13 426, 31 462, 41 623, 62 431
f 30 856, 80 653, 83 560, 85 630
g 26 783, 32 867, 62 378, 73 862

2 a 909, 800, 646, 472, 384
b 890, 803, 642, 380, 246
c 6132, 3162, 2361, 1623, 1236
d 6841, 4382, 3981, 2357, 1459
e 83 621, 62 813, 32 183, 26 138, 13 826
f 72 415, 71 524, 45 147, 24 715, 15 472

3 f d c b a e

Rounding to the nearest 100, 1000 and 10 000

Page 11 – Your Turn

1 a 7200 b 400 c 49 600 d 500 e 63 400

CATCH UP MATHS YEAR 6 BOOK A © PASCAL PRESS ISBN: 9781925726183

ANSWERS

Page 12 – Practice

1 a 1500 b 8700 c 400 d 15 400 e 152 500 f 1 573 200 g 7600

2 a 2000 b 93 000 c 17 000 d 124 000 e 75 000 f 3000 g 6 373 000

3 a 40 000 b 270 000 c 20 000 d 3 260 000 e 20 000 f 7460 000 g 80 000

4

	Number	Round to nearest 100	Round to nearest 1000	Round to nearest 10 000
a	34 568	34 600	35 000	30 000
b	59 731	59 700	60 000	60 000
c	54 836	54 800	55 000	50 000
d	97 425	97 400	97 000	100 000
g	580 263	580 300	580 000	580 000
h	742 589	742 600	743 000	740 000
i	1 429 632	1 429 600	1 430 000	1 430 000
j	5 643 859	5 643 900	5 644 000	5 640 000

Place Value SIX-DIGIT AND SEVEN-DIGIT NUMBERS

Page 13 – Your Turn

6-digit numbers: 120 036, 950 133, 236 158, 415 273, 847 311, 641 209
7-digit numbers: 8 542 187, 1 248 970, 5 341 286, 3 436 219, 7 436 281

Page 14 – Practice

1 a 694 387 b 9 377 425

2 a three hundred and seven thousand, two hundred and eighty-one
b eight million, seven hundred and two thousand, nine hundred

3 a thousands b hundred thousands c hundred thousands d ones e ten thousands

4 a hundred thousands b hundreds c hundred thousands d ten thousands e tens

5 a 6 489 305 b 4 208 000 c 704 900

Value SIX-DIGIT AND SEVEN-DIGIT NUMBERS

Page 15 – Your Turn

Circled yellow: 736 489, 4 759 328, 5 768 430, 9 759 015, 6 700 000
Circled black: 4 873 581, 4 759 328, 4 873 581, 4 573 000

Page 16 – Practice

1 a 800 000 b 800 000 c 8 000 000 d 800 e 80 000 f 80 g 800 000 h 8 i 80 000 j 800 k 80

2 a 732 495, 1 b 73 249, 5, 1 c 7324, 9, 5, 1 d 732, 4, 9, 5, 1 e 73, 2, 4, 9, 5, 1 f 7, 3, 2, 4, 9, 5, 1

3 a 58 328, 4 b 5832, 8, 4 c 583, 2, 8, 4 d 58, 3, 2, 8, 4 e 5, 8, 3, 2, 8, 4

4 Adult to check

Expanded Numbers SIX-DIGIT AND SEVEN-DIGIT NUMBERS

Page 17 – Your Turn

1 a 4 000 000 + 300 000 + 70 000 + 8000 + 200 + 30 + 1
$4 \times 10^6 + 3 \times 10^5 + 7 \times 10^4 + 8 \times 10^3 + 2 \times 10^2 + 3 \times 10 + 1 \times 1$
b 500 000 + 30 000 + 6000 + 900 + 20 + 8
$5 \times 10^5 + 3 \times 10^4 + 6 \times 10^3 + 9 \times 10^2 + 2 \times 10 + 8 \times 1$

Page 18 – Practice

1 a 1 000 000 + 300 000 + 70 000 + 2000 + 400 + 80 + 9
b 800 000 + 30 000 + 9000 + 200 + 4
c 5 000 000 + 600 000 + 80 000 + 2000 + 400 + 70 + 3
d 500 000 + 60 000 + 9000 + 300 + 20

2 a 3 424 219 b 537 637 c 4 200 352 d 807 020

3 a $2 \times 10^5 + 4 \times 10^4 + 3 \times 10^3 + 6 \times 10^2 + 5 \times 10 + 9 \times 1$
b $1 \times 10^6 + 4 \times 10^5 + 7 \times 10^4 + 8 \times 10^3 + 8 \times 10^2 + 3 \times 10 + 1 \times 1$
c $4 \times 10^6 + 3 \times 10^4 + 6 \times 10^3 + 5 \times 10^2 + 2 \times 10 + 8 \times 1$
d $7 \times 10^5 + 5 \times 10^4 + 8 \times 10^3 + 3 \times 10^2 + 6 \times 10 + 1 \times 1$

4 a 567 201 b 8 479 154 c 940 308 d 332 133 e 6 004 659

Ordering Numbers SIX-DIGIT AND SEVEN-DIGIT NUMBERS

Page 19 – Your Turn

a descending b ascending c descending d ascending e ascending

Page 20 – Practice

1 a 233 846, 632 483, 843 262, 1 384 236
b 2 469 137, 3 716 924, 4 269 713, 6 719 423
c 102 483, 410 283, 438 120, 832 014
d 104 731, 109 413, 471 013, 901 341
e 3 296 345, 4 932 523, 5 436 923, 9 234 352

2 a 2, 5, 1, 3, 4 b 5, 1, 3, 4, 2 c 4, 3, 1, 2, 5 d 5, 4, 1, 3, 2

3 a 1, 5, 4, 3, 2 b 5, 2, 3, 1, 4 c 1, 4, 5, 3, 2 d 1, 3, 2, 5, 4

Odd and Even Numbers

Page 21 – Your Turn

Odd: 989 113, 14 243, 5945, 1115, 654 321, 274 447, 1 824 739, 937 937, 7
Even: 42, 143 744, 4 836 420, 54 982, 2 459 998, 73 248, 1 473 846, 1 824 739, 876 534, 120

Page 22 – Practice

1 Odd: 24 893, 1591, 24 983, 3579, 287 369, 44 467, 131
Even: 2 456 702, 1460, 434 362, 708, 1 475 672, 13 504, 85 342, 1 493 246

2 a 149 355, 149 357, 149 359 b 3 762 453, 3 762 455, 3 762 457 c 289 404, 289 406, 289 408 d 142, 144, 146 e 6850, 6852, 6854 f 37 241, 37 243, 37 245 g 4 037 503, 4 037 505, 4 037 507

3 a 1 783 143 b 584 253 c 6 831 421 d 673 541 e 3 871 435 f 764 155 g 8 437 281

Greater Than, Less Than, Equal To

Page 23 – Your Turn

1 a T b T c F d T e T f F g F h T i T j T k F l T m T

Page 24 – Practice

1 a No b Yes c No d Yes e Yes f Yes g No h Yes i Yes j No k Yes l Yes m Yes n No o No

2 a > b < c > d = e < f < g < h > i < j = k < l < m < n > o =

3 Correct words:
a is equal to b is less than c is equal to d is less than e is less than f is greater than

1 WHOLE NUMBERS CONTINUED

Largest and Smallest Numbers

Page 25 – Your Turn

a 30 059, 95 300
b 102 356, 653 210
c 2 344 689, 9 864 432

Page 26 – Practice

1

	Smallest number	Largest number
a	3049	9 430
b	245 678	876 542
c	30 059	95 300
d	1 146	6 411
e	388	883
f	1 377 899	9 987 731
g	2 456 889	9 886 542
h	134 589	985 431
i	346 789	987643
j	2 368 999	9 998 632

2 a correct
b incorrect
c correct
d correct
e incorrect
f incorrect
g correct
h correct
i correct
j correct

Rounding To 100 000 and 1 000 000

Page 27 – Your Turn

1 a 3 700 000
b 500 000
2 a 5 000 000
b 8 000 000

Page 28 – Practice

1 a 6 600 000
b 500 000
c 300 000
d 6 000 000
e 9 900 000
f 700 000
g 500 000
h 900 000
i 400 000
j 600 000
k 8 900 000

2 a 10 000 000
b 8 000 000
c 7 000 000
d 5 000 000
e 1 000 000
f 6 000 000
g 7 000 000
h 4 000 000
i 3 000 000
j 2 000 000
k 2 000 000

3 a 5 200 000, 5 000 000
b 9 500 000, 10 000 000
c 8 500 000, 9 000 000
d 6 300 000, 6 000 000
e 3 100 000, 3 000 000
f 5 300 000, 5 000 000
g 8 700 000, 9 000 000
h 9 900 000, 10 000 000

The Role of Zero

Page 29 – Your Turn

a ten thousands
b hundreds
c thousands
d thousands

Page 30 – Practice

1 a tens
b thousands
c hundred thousands
d tens
e ones
f thousands
g thousands
h thousands
i ones
j hundreds
k hundred thousands
l ones
m tens
n ones
o tens
p tens
q hundreds

2 Adult to check

Abbreviations of Large Numbers

Page 31 – Your Turn

1 a 615K
b 14K
c 824K

2 a 716 thous.
b 179 thous.
c 623 thous.

Page 32 – Practice

1 a 197K
b 27K
c 503K
d 6K
e 90K
f 810K
g 403K
h 20K
i 38K
j 524K
k 100K

2 a 23 thous.
b 17 thous.
c 840 thous.
d 99 thous.
e 103 thous.
f 711 thous.
g 616 thous.
h 8 thous.
i 400 thous.
j 430 thous.
k 595 thous.

3 a 162 K
b 89 thous.
c 511 thous.
d 670 K
e 403 thous.
f 820 K
g 4 thous.
h 309K
i 5K
j 248 thous.
k 425 thous.

Factors

Page 33 – Your Turn

a 3
b 2
c 4
d 6
e 2

Page 34 – Practice

1 a 3
b 3
c 4
d 3
e 5
f 6
g 4

2 a 1, 3, 9, 27
b 1, 2, 4, 5, 8, 10, 20, 40
c 1, 2, 4, 5, 10, 20, 25, 50, 100
d 1, 2, 3, 6, 9, 18
e 1, 2, 3, 4, 6, 8, 12, 24
f 1, 5, 25
g 1, 2, 4, 7, 14, 28
h 1, 19
i 1, 2, 3, 4, 6, 8, 9, 12, 16, 18, 24, 36, 48, 72, 144
j 1, 2, 4, 8, 16, 32, 64
k 1, 7
l 1, 2, 5, 10, 25, 50
m 1, 2, 4, 7, 8, 14, 28, 56
n 1, 2, 13, 26

Highest Common Factor (HCF)

Page 35 – Your Turn

a 3
b 30

Page 36 – Practice

1 a 16: 1, 2, 4, 8, 16
20: 1, 2, 4, 5, 10, 20
HCF = 4
b 5: 1, 5
10: 1, 2, 5, 10
HCF = 5
c 21: 1, 3, 7, 21
24: 1, 2, 3, 4, 6, 8, 12, 24
HCF = 3
d 10: 1, 2, 5, 10
24: 1, 2, 3, 6, 8, 12, 24
HCF = 2
e 20: 1, 2, 4, 5, 10, 20
36: 1, 2, 3, 4, 6, 9, 12, 18, 36
HCF = 4
f 18: 1, 2, 3, 6, 9, 18
36: 1, 2, 3, 4, 6, 9, 12, 18, 36
HCF = 18
g 33: 1, 3, 11, 33
55: 1, 5, 11, 55
HCF = 11
h 60: 1, 2, 3, 4, 5, 6, 10, 12, 15, 20, 30, 60
42: 1, 2, 3, 6, 7, 14, 21, 42
HCF = 6
i 50: 1, 2, 5, 10, 25, 50
110: 1, 2, 5, 10, 11, 22, 55, 110
HCF = 10
j 91: 1, 7, 13, 91
65: 1, 5, 13, 65
HCF = 13

Multiples

Page 37 – Your Turn

a 3, 6, 9, 12
b 5, 10, 15, 20
c 10, 20, 30. 40

Page 38 – Practice

1 a 12, 24, 36, 48, 60, 72, 84, 96
b 6, 12, 18, 24, 30, 36, 42, 48
c 1, 2, 3, 4, 5, 6, 7, 8
d 3, 6, 9, 12, 15, 18, 21, 24
e 5, 10, 15, 20, 25, 30, 35, 40
f 8, 16, 24, 32, 40, 48, 56, 64
g 10, 20, 30, 40, 50, 60, 70, 80
h 2, 4, 6, 8, 10, 12, 14, 16
i 7, 14, 21, 28, 35, 42, 49, 56

2 a 35, 42, 49, 56
b 40, 45, 50, 55
c 72, 80, 88, 96
d 72, 81, 90, 99

3 a 37
b 12
c 29
d 78
e 20
f 33
g 58

Lowest Common Multiple (LCM)

Page 39 – Your Turn

a 12
b 9

Page 40 – Practice

1 a 4: 4, 8, 12, 16, 20
5: 5, 10, 15, 20
LCM = 20
b 4: 4, 8, 12, 16, 20
10: 10, 20, 30, 40
LCM = 20
c 2: 2, 4, 6, 8
6: 6, 12, 18, 24
LCM = 6
d 1: 1, 2, 3, 4, 5
5: 5, 10, 15, 20
LCM = 5
e 5: 5, 10, 15, 20
10: 10, 20, 30, 40
LCM = 10
f 4: 4, 8, 12, 16, 20
8: 8, 16, 24, 32, 40
LCM = 8
g 2: 2, 4, 6, 8, 10
8: 8, 16, 24, 32
LCM = 8
h 2: 2, 4, 6, 8, 10
10: 10, 20, 30, 40
LCM = 10
i 3: 3, 6, 9, 12, 15
4: 4, 8, 12, 16, 20
LCM = 12
j 7: 7, 14, 21, 28, 35
5: 5, 10, 15, 20, 25, 30, 35
LCM = 35

Integers

Page 41 – Your Turn

1 a −4 b −3 c −2

2 −4, 126, 72, −3, −24, −123, −5, 243, 84

Page 42 – Practice

1 4, −4; 5, −5; 2, −2; 1, −1; 7, −7; 6, −6; 8, −8

2 a −8, −3, 1, 5, 9
b −32, −19, −6, 16, 32
c −100, −24, 12, 15, 29
d −140, −32, 1, 27, 53

3 a 15 °C
b 40 °C
c 10 °C
d 5 °C
e 20 °C
f 25 °C
g 15 °C
h 15 °C
i 40 °C

4 a −1 b −1 c 0

Prime and Composite Numbers

Page 43 – Your Turn

a 1, 2, 3, 4, 6, 12, composite
b 1, 3, 7, 21, composite
c 1, 13: prime

Page 44 – Practice

1 prime: 17, 11, 101, 7
composite: 16, 39, 9, 14, 56, 30, 82, 121, 24, 36, 51

2 a 4, 6, 8, 9, 10, 12, 14, 15, 16, 18,
b 50, 51, 52, 54, 55, 56, 57, 58, 60, 62, 63, 64, 65, 66, 68, 69, 70, 72, 74, 75
c 126, 128, 129, 130, 132, 133, 134, 135, 136, 138, 140, 141, 142, 143, 144, 145, 146, 147, 148

3 a 2, 3, 5, 7, 11, 13, 17, 19
b 53, 59, 61, 67, 71, 73, 79, 83, 89, 97
c 113, 127, 131, 137, 139, 149

4 a 67 b 47 c 139
5 a 4 b 112 c 87
6 a 17 b 29 c 167
7 a 30 b 39 c 150

Whole Numbers Review Page 45

1 248, 585, 215, 385
2 6632, 3541, 6422, 7410, 9043
3 40 052, 52 437, 13 959, 82 634
4 a two hundred and fifty-three
b eight hundred and twenty
c seven hundred and five
d five hundred and nine
e one thousand, five hundred and thirty-six
f eight thousand, two hundred and ninety
g four thousand and fifty-three
h two thousand and two
i eighty-nine thousand, two hundred and sixteen
j seventy thousand and fourteen
k sixty thousand and fifty
l eighty thousand, nine hundred and twenty-seven
m fifty thousand and five
n twenty-two thousand, four hundred and sixty-five
o nineteen thousand, two hundred and forty
p ninety-two thousand, three hundred

5

	Number	Ten thousands	Thousands	Hundreds	Tens	Ones
a	503	0	0	5	0	3
b	857	0	0	8	5	7
c	960	0	0	9	6	0
d	1795	0	1	7	9	5
e	2803	0	2	8	0	3
f	5036	0	5	0	3	6
g	6100	0	6	1	0	0
h	37 389	3	7	3	8	9
i	34 090	3	4	0	9	0
j	26 000	2	6	0	0	0

6 a ones
b tens
c hundreds
d tens
e ones
f thousands

7 a tens
b thousands
c ones
d thousands
e ten thousands
f hundreds

8 a 6258
b 974
c 50 009
d 16 534
e 3763
f 80 045

9 a 2
b 200
c 2000
d 200
e 200
f 2
g 2
h 20 000
i 20
j 2000
k 200
l 20

10 Adult to check
11 Adult to check
12 Adult to check

13 a 300 + 20 + 5
b 100 + 30
c 500 + 80 + 9
d 1000 + 500 + 40 + 2
e 6000 + 900 + 70
f 9000 + 50 + 3
g 20 000 + 4000 + 900 + 5
h 10 000 + 5000 + 400 + 10 + 9
i 60 000 + 300 + 20 + 6

14 a ones, 7
b hundreds, 700
c tens, 70
d ten thousands, 70 000
e hundreds, 700
f thousands, 7000
g hundreds, 700
h ones, 7
i hundreds, 700
j tens, 70
k hundreds, 700
l ten thousands, 70 000

15 a 105, 501, 510, 515, 551
b 1536, 1653, 3165, 5136, 6315
c 34,790, 40,973, 74 390, 79 430, 94 307

16 a 743, 734, 437, 374, 347
b 9201, 9102, 2190, 1209, 1029
c 64 313, 63 431, 36 413, 33 461, 13 346

17 a five hundred and sixty-two thousand, four hundred and thirty-one
b seven hundred and three thousand, two hundred and thirty-five

1 WHOLE NUMBERS CONTINUED

c four hundred and ninety-eight thousand, one hundred and seven
d five million and ninety thousand, five hundred
e seven million and six thousand, two hundred and fifty
f eight million, four hundred and twenty nine thousand, six hundred and twenty-one

18 285 362, 624 938, 402 581

19 2 846 443, 5 738 246, 4 254 124

20

	Number	Millions	Hund. Thous.	Ten Thous.	Thous.	Hundreds	Tens	Ones
a	583 296	0	5	8	3	2	9	6
b	145 370	0	1	4	5	3	7	0
c	209 460	0	2	0	9	4	6	0
d	602 000	0	6	0	2	0	0	0
e	1 672 438	1	6	7	2	4	3	8
f	4 053 203	4	0	5	3	2	0	3
g	5 374 002	5	3	7	4	0	0	2
h	3 006 060	3	0	0	6	0	6	0

21 a hundred thousands b ones c hundreds d millions e tens f hundred thousands g tens h millions

22 a hundreds b millions c hundred thousands d ten thousands e hundreds f hundred thousands g ten thousands h ones

23 a 4 370 215 b 3 060 047 c 8 050 068

24 a 654 276 b 895 753 c 530 020

25 a 1000 b 1 000 000 c 1000 d 100 000 e 10 000 f 1 g 10 000 h 1 000 000

26 Adult to check

27 a 200 000 + 30 000 + 8000 + 200 + 9
b 100 000 + 40 000 + 7000 + 200 + 80 + 3
c 500 000 + 90 000 + 400 + 50 + 1
d 2 000 000 + 400 000 + 80 000 + 6000 + 200 + 80 + 1
e 3 000 000 + 10 000 + 5000 + 200 + 10 + 5
f 7 000 000 + 9
g 9 000 000 + 300 000 + 20 000 + 500 + 50 + 3

28 a $6 \times 10^5 + 3 \times 10^4 + 2 \times 10^3 + 4 \times 10^2 + 9 \times 10 + 7 \times 1$
b $2 \times 10^5 + 4 \times 10^4 + 9 \times 10^3 + 3 \times 10^2 + 5 \times 10 + 2 \times 1$
c $4 \times 10^5 + 5 \times 10^4 + 1 \times 10^3 + 6 \times 10^2 + 4 \times 10 + 3 \times 1$
d $5 \times 10^6 + 3 \times 10^5 + 4 \times 10^4 + 7 \times 10^3 + 2 \times 10^2 + 4 \times 10 + 3 \times 1$
e $6 \times 10^6 + 8 \times 10^5 + 4 \times 10^4 + 2 \times 10^3 + 1 \times 10^2 + 5 \times 10$
f $2 \times 10^6 + 9 \times 10^5 + 5 \times 10^4 + 2 \times 10^2 + 4 \times 10 + 3 \times 1$
g $3 \times 10^6 + 2 \times 10^5 + 2 \times 10^3 + 5 \times 10^2 + 3 \times 10 + 6 \times 1$

29 a 3 425 325 b 7 158 216 c 357 370 d 405 602 e 6 205 417 f 513 153 g 1 684 725 h 5 242 738 i 4 136 857

30

	Millions	Hund. Thous.	Ten Thous.	Thous.	Hundreds	Tens	Ones
a	2	4	1	6	8	0	2
b	0	9	4	2	4	6	3
c	0	8	1	1	7	5	0
d	1	2	1	2	4	3	7
e	0	5	5	6	8	0	3
f	1	0	0	0	0	0	2

31 a three million, four hundred and twenty-seven thousand and ninety-three
b two million, five hundred and eighty-one thousand, three hundred and twenty
c eight hundred and forty-two thousand, five hundred and one
d ninety-two thousand, six hundred and eighty-nine

32 a (5) 973 181 b (7) 349 413 c (5) 282 112 d (6) 493 651 e (4) 382 400 f (4) 059 368 g (1) 937 397

33 Adult to check

34 Adult to check

35 a 234 785, 243 875, 524 378, 735 842, 874 532
b 98 643, 489 326, 629 438, 698 432, 943 286
c 2 389 546, 3 596 849, 4 682 359, 5 469 283, 6 459 382

36 Odd: 1527, 947, 23 465, 4 374 125, 8009, 78 475
Even: 324, 43 824, 324 190, 7342, 641 382, 943 515

37 a 247 130, 247 132, 247 134 b 502 143, 502 145, 502 147 c 3 553 421, 3 553 423, 3 553 425 d 8 409 520, 8 409 522, 8 409 524

38 a F b T c F d F e T f F g T h F

39 a 234 489, 984 432 b 101 235, 532 110 c 1 045 567, 7 655 410 d 1 345 689, 9 865 431

40 a 300 b 500 c 3600 d 5900 e 591 300 f 1 471 300

41 a 4000 b 3000 c 7000 d 11 000 e 482 000 f 2 494 000

42 a 40 000 b 50 000 c 70 000 d 490 000 e 680 000 f 850 000 g 6 850 000 h 8 780 000

43 a 600 000 b 800 000 c 3 400 000 d 1 500 000 e 8 800 000 f 400 000 g 4 300 000 h 7 100 000

44 a 6 000 000 b 3 000 000 c 1 000 000 d 4 000 000 e 5 000 000 f 7 000 000 g 10 000 000 h 1 000 000

45

	Number	Nearest 100	Nearest 1000	Nearest 10 000	Nearest 100 000	Nearest 1 000 000
a	2 836 129	2 836 100	2 836 000	2 840 000	2 800 000	3 000 000
b	3 439 312	3 439 300	3 439 000	3 440 000	3 400 000	3 000 000
c	7 632 790	7 632 800	7 633 000	7 630 000	7 600 000	8 000 000
d	5 153 251	5 153 300	5 153 000	5 150 000	5 200 000	5 000 000
e	6 758 469	6 758 500	6 758 000	6 760 000	6 800 000	7 000 000
f	9 529 612	9 529 600	9 530 000	9 530 000	9 500 000	10 000 000
g	8 135 293	8 135 300	8 135 000	8 140 000	8 100 000	8 000 000
h	4 689 424	4 689 400	4 689 000	4 690 000	4 700 000	5 000 000

46 a ten thousands b ones c thousands d ten thousands e thousands f thousands g tens h ten thousands i thousands j ones k tens l hundreds

47 Adult to check

48 a 2K b 51K c 27K d 243K e 49K f 103K

49 a 16 thous. b 215 thous. c 84 thos. d 929 thous. e 102 thous. f 92 thous.

50 a 1, 2, 3, 4, 6, 12 b 1, 3, 9, 27 c 1, 2, 4, 8, 16 d 1, 2, 3, 6, 9, 18 e 1, 2, 3, 6, 8, 12, 24 f 1, 2, 5, 10, 25, 50

51 a 10: 1, 2, 5, 10
12: 1, 2, 3, 4, 6, 12
HCF = 2
b 2: 1, 2
3: 1, 3
HCF = 1
c 20: 1, 2, 4, 5, 10, 20
24: 1, 2, 3, 4, 6, 8, 12, 24
HCF = 4
d 15: 1, 3, 5, 15
20: 1, 2, 4, 5, 10, 20
HCF = 5
e 12: 1, 2, 3, 4, 6, 12
15: 1, 3, 5, 15
HCF = 3

52 a 5, 10, 15, 20, 25 b 11, 22, 33, 44, 55 c 3, 6, 9, 12, 15 d 10, 20, 30, 40, 50

CATCH UP MATHS YEAR 6 BOOK A © PASCAL PRESS ISBN: 9781925726183

53 a 40, 44, 48 b 55, 60, 65 c 24, 27, 30 d 48, 54, 60

54 a 3: 3, 6, 9, 12, 15
5: 5, 10, 15
LCM = 15
b 4: 4, 8, 12, 16
8: 8, 16, 24
LCM = 8
c 3: 3, 6, 9, 12, 15, 18, 21, 24
8: 8, 16, 24
LCM = 24
d 3: 3, 6, 9, 12, 15
4: 4, 8, 12, 16
LCM = 12
e 7: 7, 14, 21, 28, 35
4: 4, 8, 12, 16, 20, 24, 28
LCM = 28

55 a −3 b +7 c +4 d −5 e +10 f +8 g +1 h −2 i +6 j −9

56 a −8, −5, −1, 0, 2, 6
b −3, −2, 1, +4, +6, 14
c −32, −24, −11, 2, 18, 30
d −26, −10, −3, 2, 5, 7
e −18, −8, −7, 0, 2, 3
f −50, −36, −24, 3, 16, 20

57 Numbers to cross out:
a $\frac{1}{2}$, 4.3, $-1\frac{1}{2}$, $3\frac{1}{3}$
b $-3\frac{1}{4}$, 3.18
c −20.8, $15\frac{1}{2}$, 4.9
d $14\frac{1}{2}$, 8.4, $6\frac{1}{4}$
e $14\frac{1}{4}$, 9.4
f 2.384, $2\frac{1}{2}$, $+3\frac{1}{4}$, $-2\frac{1}{2}$

58 a 10°C b 10°C c 20°C d 20°C e 25°C f 25°C g 15°C h 20°C i 10°C j 5°C

59 a −2 b −2 c −5 d 0

60 a 1, 2, 3, 6, composite
b 1, 2, 7, 14, composite
c 1, 23, prime

61 prime: 101, 3, 67, 153, 151, 23, 7
composite: 32, 27, 78, 14, 150, 24, 168, 15, 100, 30, 54, 28

2 ADDITION

Addition Without Trading

TWO-DIGIT AND THREE-DIGIT NUMBERS

Page 60 – Your Turn

1 a 686 b 398 c // d 789 e 79

Page 61 – Practice

1 a 93 b 69 c 88 d 96
2 a 47 b 97 c 79 d 69
3 a 437 b 785 c 579 d 298
4 a 748 b 487 c 877 d 199
5 a 697 b 589 c 969 d 879
6 a 579 b 799 c 907 d 888

Addition With Trading

TWO-DIGIT AND THREE-DIGIT NUMBERS

Page 62 – Your Turn

1 a 100 b 756 c 740

Page 63 – Practice

1 a 101 b 115 c 85
2 a 102 b 81 c 105
3 a 504 b 828 c 951
4 a 714 b 351 c 519
5 a 918 b 906 c 567
6 a 1390 b 482 c 877

Addition Without Trading

FOUR-DIGIT AND FIVE-DIGIT NUMBERS

Page 64 – Your Turn

a 5696 b 88 676 c 7970

Page 65 – Practice

1 a 5998 b 6798 c 7128
2 a 7438 b 3999 c 5999
3 a 8497 b 6939 c 75 688 d 84891 e 67 775
4 a 97 559 b 75 863 c 86 323 d 77977 e 48 375

Addition With Trading

FOUR-DIGIT AND FIVE-DIGIT NUMBERS

Page 66 – Your Turn

a 9080 b 63 654

Page 67 – Practice

1 a 9467 b 12 725 c 75 027 d 52 900 e 103 817

2 a 2520 b 6370 c 8527 d 10 978 e 8788 f 9492 g 78 972 h 67 857 i 40 015 j 24 709 k 98 035 l 78 427 m 80 231 n 98 817

Rounding To Estimate Answers

Page 68 – Your Turn

a 490, 180, 670 b 800, 200, 1000

Page 69 – Practice

1 a 240, 190, 430 b 740, 160, 900 c 450, 220, 670 d 190, 530, 720 e 620, 740, 1360 f 180, 280, 460 g 120, 530, 650 h 240, 540, 780 i 330, 410, 740

2 a 300, 200, 500 b 500, 300, 800 c 400, 400, 800 d 700, 300, 1000 e 400, 200, 600 f 800, 100, 900 g 300, 300, 600

Using The Jump Strategy To Add

TWO-DIGIT AND THREE-DIGIT NUMBERS

Page 70 – Your Turn

1 a 58 + 24 = 82
b 526 + 72 = 598

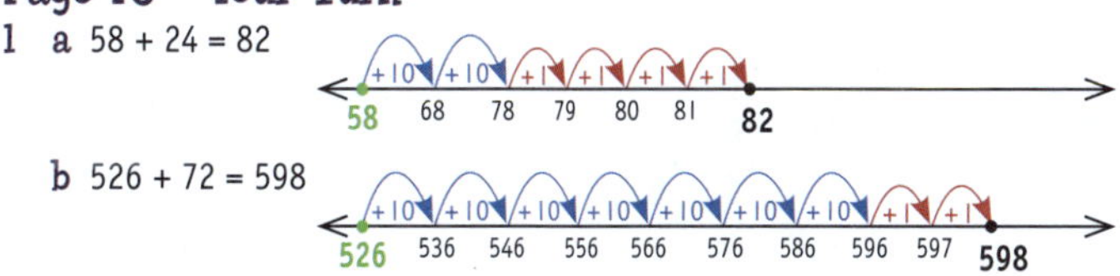

Page 71 – Practice

1 a 95 b 36 c 65 d 677 e 448 f 283 g 861 h 749

Using The Jump Strategy To Add

THREE-DIGIT AND FOUR-DIGIT NUMBERS

Page 72 – Your Turn

1 a 351 + 504 = 855

+100 +100 +100 +10 +10 +10 +10 +10 +1
504 604 704 804 814 824 834 844 854 855

b 8295 + 1430 = 9725

+1000 +100 +100 +100 +100 +10 +10 +10
8295 9295 9395 9495 9595 9695 9705 9715 9725

2 ADDITION CONTINUED

Page 73 – Practice

a 1125 b 1647 c 732 d 2344 e 8981 f 5947 g 12 873 h 8120

Using the Split Strategy to Add

TWO-DIGIT AND THREE-DIGIT NUMBERS

Page 74 – Your Turn

a 41 + 13
40 + 10 = 50
1 + 3 = 4
50 + 4 = 54

b 133 + 56
100
30 + 50 = 80
3 + 6 = 9
100 + 80 + 9 = 189

Page 75 – Practice

1 a 30 + 20 = 50
7 + 1 = 8
50 + 8 = 58

b 40 + 30 = 70
8 + 0 = 8
70 + 8 = 78

c 80 + 10 = 90
5 + 4 = 9
90 + 9 = 99

d 200 = 200
50 + 30 = 80
3 + 6 = 9
200 + 80 + 9 = 289

e 400 = 400
40 + 20 = 60
1 + 3 = 4
400 + 60 + 4 = 464

f 800 = 800
0 + 40 = 40
2 + 7 = 9
800 + 40 + 9 = 849

g 100 + 500 = 600
50 + 30 = 80
4 + 2 = 6
600 + 80 + 6 = 686

h 800 + 200 = 1000
30 + 40 = 70
7 + 0 = 7
1000 + 70 + 7 = 1077

i 500 + 200 = 700
50 + 40 = 90
6 + 3 = 9
700 + 90 + 9 = 799

j 900 + 600 = 1500
0 + 40 = 40
5 + 2 = 7
1500 + 40 + 7 = 1547

Using The Split Strategy To Add

THREE-DIGIT AND FOUR-DIGIT NUMBERS

Page 76 – Your Turn

a 7271 + 1516
7000 + 1000 = 8000
200 + 500 = 700
70 + 10 = 80
1 + 6 = 7
8000 + 700 + 80 + 7 = 8787

Page 77 – Practice

1 a 200 + 300 = 500
60 + 10 = 70
1 + 7 = 8
500 + 70 + 8 = 578

b 100 + 300 = 400
40 + 50 = 90
3 + 2 = 5
400 + 90 + 5 = 495

c 700 + 200 = 900
40 + 50 = 90
1 + 8 = 9
900 + 90 + 9 = 999

d 1000 = 1000
400 + 300 = 700
30 + 60 = 90
5 + 2 = 7
1000 + 700 + 90 + 7 = 1797

e 2000 = 2000
500 + 400 = 900
30 + 20 = 50
8 + 0 = 8
2000 + 900 + 50 + 8 = 2958

f 6000 + 2000 = 8000
500 + 400 = 900
40 + 50 = 90
3 + 4 = 7
8000 + 900 + 90 + 7 = 8997

g 7000 + 2000 = 9000
100 + 600 = 700
50 + 30 = 80
3 + 3 = 6
9000 + 700 + 80 + 6 = 9786

h 4000 + 2000 = 6000
900 + 700 = 1600
30 + 10 = 40
8 + 0 = 8
6000 + 1600 + 40 + 8 = 7648

i 6000 + 0 = 6000
500 + 100 = 600
90 + 0 = 90
2 + 3 = 5
6000 + 600 + 90 + 5 = 6695

Rounding to the Nearest 5 Cents

Page 78 – Your Turn

a $2.45 b $8.60 c $10.65 d $20.55 e $140.20

Page 79 – Practice

1 a $6.45 b $17.50 c $42.65 d $21.50 e $89.45 f $1.00 g $66.65 h $15.60 i $0.95 j $121.10 k $0.15 l $19.40 m $12.05 n $38.55 o $19.95

2 a $0.15 b $65.75 c $39.40 d $8.30 e $66.15 f $5.55 g $149.00 h $3.55 i $24.00 j $100.00 k $57.60 l $39.80 m $104.05 n $209.10 o $49.30

Working Out Change

Page 80 – Your Turn

a $2.25 b $8.45

Page 81 – Practice

1 a $3.15 b $2.80 c $2.35 d $18.75 e $5.05 f $22.60 g $3.55 h $27.85 i $15.05 j $83.60 k $41.40

Spending And Saving Money

Page 82 – Your Turn

a $77.50

Page 83 – Practice

1 a Maria $160.20, Carmen $245.00
b $129.25
c $127.50
d Carmen
e Maria
f 6 weeks

2 a

Deposit	Withdraw	Balance
		$ 0.00
$ 129.25		$ 129.25
	$ 20.00	$ 109.25

b

Deposit	Withdraw	Balance
		$ 0.00
$ 127.50		$ 127.50
	$ 20.00	$ 107.50

3 a 5 b $1960 c 3 d $803.70 e $100 f $760 g $981.30 h 30/5 i 5/5 j $1524.50 k $543.20 l $975.00

4

Date	Deposit	Withdrawal	Balance
5/12	$50.00		$50.00
8/12	$35.00		$85.00
10/12	$435.00		$520.00
13/12		$75.00	$445.00
15/12		$62.55	$382.45
18/12	$189.20		$571.65
21/12		$325.00	$246.65
27/12	$520.00		$766.65
29/12		$415.75	$350.90
30/12	$120.20		$471.10

a $520.50 e $520 i $1349.40 m $766.65
b $246.65 f 4 j December n $50.00
c $766.65 g 6 k 8/12
d $471.10 h $878.30 l 13/12

5

Earnings	
Item	**Amount**
Pay	$987.45
Pay for chores	$150.00
Pay for walking dogs	$80.00
Total	$1217.45

Expenses	
Item	**Amount**
Food	$240.00
Movies	$30.00
Karate fees	$140.00
Clothes	$135.50
New drum	$50.00
Total	$595.50

a $1217.45 c Food e $595.50 g $1042.25
b $621.95 d Walking dogs f $230

6 Answers may vary according to how students classify items.
Necessities: Clothes, Milk, Bread, Medicine, Meat, Fruit and vegetables. Total: $224.20
Luxuries: Take away, Chocolate, Toy car, Movies, Action figure, Lollies. Total: $154
a Answers will vary c $177.20 e $70.20
b $55 d Necessities f Answers will vary

Word Problems Using Addition

Page 89 – Your Turn

a 114 b 147 c 192 d 372 minutes

Page 90 – Practice

1 a 377 people e 276 km i $157 469
b 453 punnets f 7106 homes j $3 019 160
c 347 square metres g 9693 sheets k 135 308 km
d 638 cupcakes h 127 page l 221 750 m²

Addition Review Page 92

1 a 69 e 89 i 168 m 999
b 67 f 464 j 295 n 989
c 69 g 359 k 396 o 899
d 75 h 268 l 546

2 a 104 e 90 i 871 m 960
b 101 f 212 j 805 n 935
c 112 g 792 k 1285 o 1195
d 101 h 561 l 1001

3 a 8356 d 5689 g 6889 j 9787
b 9269 e 7455 h 4764 k 9585
c 3297 f 8825 i 8657 l 8584

4 a 14 387 d 56 668 g 48 348 j 89 594
b 24 778 e 27 938 h 59 839 k 88 976
c 71 193 f 85 486 i 78 899 l 56 999

5 a 1376 d 4432 g 3562 j 8424
b 5311 e 2918 h 9569 k 6534
c 7421 f 6712 i 10 340 l 12 330

6 a 12 413 d 74 487 g 77 283 j 127 413
b 24 743 e 85 759 h 97 621 k 91 326
c 13 661 f 67 428 i 59 657 l 107 878

7 a 630, 40, 670 c 570, 150, 720
b 180, 430, 610 d 590, 130, 720

8 a 200, 300, 500 c 100, 200, 300
b 700, 400, 1100 d 600, 200, 800

9 a 115 c 161 e 1510 g 10 196
b 200 d 869 f 4363 h 11 613

10 a 52 + 41
50 + 40 = 90
2 + 1 = 3
90 + 3 = 93

b 24 + 35
20 + 30 = 50
4 + 5 = 9
50 + 9 = 59

c 52 + 25
50 + 20 = 70
2 + 5 = 7
70 + 7 = 77

d 325 + 44
300 = 300
20 + 40 = 60
5 + 4 = 9
300 + 60 + 9 = 369

e 712 + 63
700 = 700
10 + 60 = 70
2 + 3 = 5
700 + 70 + 5 = 775

f 875 + 22
800 = 800
70 + 20 = 90
5 + 2 = 7
800 + 90 + 7 = 897

g 428 + 131
400 + 100 = 500
20 + 30 = 50
8 + 1 = 9
500 + 50 + 9 = 559

h 239 + 250
200 + 200 = 400
30 + 50 = 80
9 + 0 = 9
400 + 80 + 9 = 489

i 523 + 654
500 + 600 = 1100
20 + 50 = 70
3 + 4 = 7
1100 + 70 + 7 = 1177

j 5243 + 712
5000
200 + 700 = 900
40 + 10 = 50
3 + 2 = 5
5000 + 900 + 50 + 5 = 5955

k 2762 + 235
2000
700 + 200 = 900
60 + 30 = 90
2 + 5 = 7
2000 + 900 + 90 + 7 = 2997

l 3824 + 143
3000
800 + 100 = 900
20 + 40 = 60
4 + 3 = 7
3000 + 900 + 60 + 7 = 3967

m 4735 + 1244
4000 + 1000 = 5000
700 + 200 = 900
30 + 40 = 70
5 + 4 = 9
5000 + 900 + 70 + 9 = 5979

n 6145 + 2743
6000 + 2000 = 8000
100 + 700 = 800
40 + 40 = 80
5 + 3 = 8
8000 + 800 + 80 + 8 = 8888

o 5215 + 2463
5000 + 2000 = 7000
200 + 400 = 600
10 + 60 = 70
5 + 3 – 8
7000 + 600 + 70 + 8 = 7678

11 a $2.15 f $4.30 k $15.55 p $0.00 u $29.65
b $7.50 g $1.50 l $32.60 q $30.25 v $59.70
c $8.10 h $27.65 m $0.20 r $25.75 w $26.25
d $16.50 i $12.65 n $0.55 s $30.50 x $20.05
e $10.10 j $20.55 o $0.35 t $8.45

12 a $5.50 b $6.45 c $9.75 d $5.80

13 a $235 b $15.75 c Food d $140.00

14 a Antonia: $210, Annie: $325 c $469 f Annie
b $545 d Annie g 3 weeks
e Antonia

15 a June e $1055.50 i $1055.50 m 15/6
b 3 f $130 j $1156 n 10/6
c $1476 g $490 k $582
d $420.00 h $1156 l $300, 10/6

2 ADDITION CONTINUED

16

Date	Deposit	Withdraw	Balance
4/11	$40.00		$40.00
5/11	$50.00		$90.00
7/11	$175.00		$265.00
7/11	$120.00		$385.00
9/11		$70.00	$315.00
11/11		$115.50	$199.50
13/11	$65.00		$264.50
24/11	$72.00		$336.50
28/11		$179.00	$157.50
30/11	$526.50		$684.00

a $90
b $336.50
c $199.50
d $157.50
e $526.50
f 7
g 3
h November
g $1048.50
h $364.50
i 13/11
j 9/11
k $684
l $40

17 Earnings: Pay $1943, Chores $150, Piano lessons $60
Expenses: Food $325, Movies $35, Golf $25, Medicine $40, Take away $30
a $2153.00 b $1698.00 c $455 d $2213.20

18 Necessities: Food, Clothes, Medicine
Luxuries: Take away, Toys, Bike

19 a 97 mistakes b 84 animals c 177 d 660 minutes

3 SUBTRACTION

Using The Jump Strategy
TWO-DIGIT AND THREE-DIGIT NUMBERS

Page 102 – Your Turn

a 76 – 35 = 41

b 424 – 236 = 188

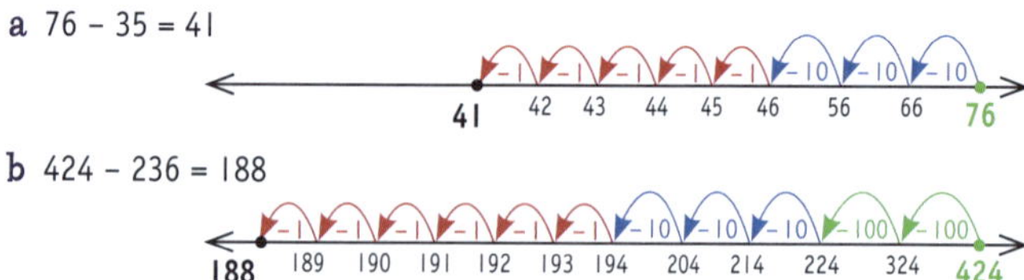

Page 103 – Practice

1 a 61 b 72 c 61 d 827 e 766 f 269 g 394 h 255

Using The Jump Strategy
THREE-DIGIT AND FOUR-DIGIT NUMBERS

Page 104 – Your Turn

a 2998 – 1225 = 1773

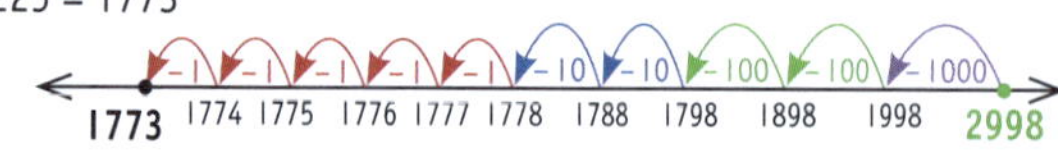

b 9774 – 530 = 9244

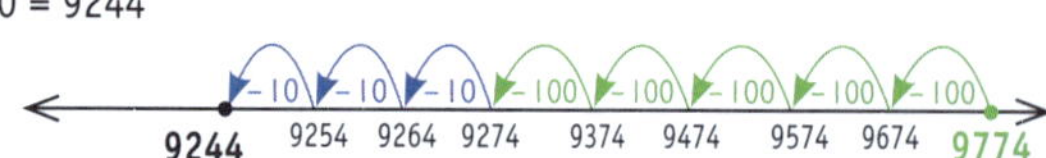

Page 105 – Practice

1 a 8601 b 2482 c 2892 d 7755 e 2826 f 4825 g 3824 h 1541

Using The Split Strategy
TWO-DIGIT AND THREE-DIGIT NUMBERS

Page 106 – Your Turn

a 547 – 23
500 – 0 = 500
40 – 20 = 20
7 – 3 = 4
500 + 20 + 4 = 524

b 562 – 51
500 – 0 = 500
60 – 50 = 10
2 – 1 = 1
500 + 10 + 1 = 511

c 859 – 257
800 – 200 = 600
50 – 50 = 0
9 – 7 = 2
600 + 0 + 2 = 602

Page 107 – Practice

1 a 83 – 21
80 – 20 = 60
3 – 1 = 2
60 + 2 = 62

b 96 – 85
90 – 80 = 10
6 – 5 = 1
10 + 1 = 11

c 76 – 44
70 – 40 = 30
6 – 4 = 2
30 + 2 = 32

d 68 – 47
60 – 40 = 20
8 – 7 = 1
20 + 1 = 21

e 793 – 82
700 – 0 = 700
90 – 80 = 10
3 – 2 = 1
700 + 10 + 1 = 711

f 497 – 73
400 – 0 = 400
90 – 70 = 20
7 – 3 = 4
400 + 20 + 4 = 424

g 542 – 22
500 – 0 = 500
40 – 20 = 20
2 – 2 = 0
500 + 20 = 520

h 689 – 74
600 – 0 = 600
80 – 70 = 10
9 – 4 = 5
600 + 10 + 5 = 615

i 856 – 245
800 – 200 = 600
50 – 40 = 10
6 – 5 = 1
600 + 10 + 1 = 611

j 763 – 452
700 – 400 = 300
60 – 50 = 10
3 – 2 = 1
300 + 10 + 1 = 311

Using The Split Strategy
THREE-DIGIT AND FOUR-DIGIT NUMBERS

Page 108 – Your Turn

a 8794 – 762
8000 = 8000
700 – 700 = 0
90 – 60 = 30
4 – 2 = 2
8000 + 0 + 30 + 2 = 8032

Page 109 – Practice

1 a 8237 – 126
8000 – 0 = 8000
200 – 100 = 100
30 – 20 = 10
7 – 6 = 1
8000 + 100 + 10 + 1
= 8111

b 9983 – 742
9000 – 0 = 9000
900 – 700 = 200
80 – 40 = 40
3 – 2 = 1
9000 + 200 + 40 + 1
= 9241

  ISBN: 9781925726183

c 6556 – 242
6000 – 0 = 6000
500 – 200 = 300
50 – 40 = 10
6 – 4 = 4
6000 + 300 + 10 + 4
= 6314

d 7295 – 185
7000 – 0 = 7000
200 – 100 = 100
90 – 80 = 10
5 – 5 = 0
7000 + 100 + 10
= 7110

e 9873 – 5562
9000 – 5000 = 4000
800 – 500 = 300
70 – 60 = 50
3 – 2 = 1
4000 + 300 + 10 + 1
= 4311

f 8839 – 6428
8000 – 6000 = 4000
800 – 400 = 400
30 – 20 = 10
9 – 8 = 1
2000 + 400 + 10 + 1
= 2411

g 7864 – 3521
7000 – 3000 = 4000
800 – 500 = 300
60 – 20 = 40
4 – 1 = 3
4000 + 300 + 40 + 3
= 4343

h 4987 – 2843
4000 – 2000 = 2000
900 – 800 = 100
80 – 40 = 40
7 – 3 = 4
2000 + 100 + 40 + 4
= 2144

i 6593 – 4212
6000 – 4000 = 2000
500 – 200 = 300
90 – 10 = 80
3 – 2 = 1
2000 + 300 + 80 + 1
= 2381

Subtraction Without Trading
TWO-DIGIT AND THREE-DIGIT NUMBERS

Page 110 – Your Turn
a 422 b 951 c 24 d 413 e 38

Page 111 – Practice
1 a 36 b 20 c 442 d 651 e 380 f 833 g 302

2 a 600 b 63 c 141 d 340 e 302 f 314 g 23 h 551 i 100 j 811 k 713

3 a 78 – 24 = 54
b 489 – 224 = 265
c 994 – 471 = 523
d 852 – 530 = 320
e 99 – 17 = 82
f 534 – 121 = 413
g 652 – 341 = 311

Subtraction Without Trading
THREE-DIGIT AND FOUR-DIGIT NUMBERS

Page 112 – Your Turn
a 9312 b 4413 c 3151 d 7117 e 1050

Page 113 – Practice
1 a 5712 b 6230 c 8115 d 1025 e 1451 f 1441 g 512

2 a 4240 b 8362 c 5416 d 9162 e 6125 f 5315 g 14 h 2213 i 9341

Subtraction Without Trading
FOUR-DIGIT AND FIVE-DIGIT NUMBERS

Page 114 – Your Turn
a 55 343
b 71 237

Page 115 – Practice
1 a 71 315 b 54 100 c 91 130 d 16 104 e 31 031 f 31 420 g 10 061 h 42 451

2 a 68 111 b 71 012 c 50 002 d 91 002 e 32 112

Subtraction With Trading
TWO-DIGIT AND THREE-DIGIT NUMBERS

Page 116 – Your Turn
a 89 b 641 c 190 d 170 e 437

Page 117 – Practice
1 a 58 b 29 c 19 d 19 e 468 f 666 g 866 h 886 i 789 j 178 k 169

2 a 836 – 529 = 307
b 903 – 362 = 541
c 623 – 457 = 166
d 714 – 486 = 228
e 589 – 467 = 122
f 894 – 526 = 368
g 657 – 188 = 469

3 a 307 + 529 = 836
b 541 + 362 = 903
c 166 + 457 = 623
d 228 + 486 = 714
e 122 + 467 = 589
f 368 + 526 = 894
g 469 + 188 = 657

Subtraction With Trading
THREE-DIGIT AND FOUR-DIGIT NUMBERS

Page 118 – Your Turn
a 4067 b 2789 c 4892 d 479 e 8396

Page 119 – Practice
1 a 9170 b 8185 c 2848 d 1954 e 3990 f 1161 g 893 h 5661 i 3071 j 3571 k 3958

2 a 4253 b 5065 c 1642 d 5632 e 833 f 6282 g 3760

Subtraction With Trading
FIVE-DIGIT NUMBERS

Page 120 – Your Turn
a 19 077 b 65 274 c 53 232

Page 121 – Practice
1 a 35 346 b 72 785 c 15 838 d 15 284 e 74 909 f 48 509 g 10 449 h 35 826

2 a 35 346 + 3083 = 38 429
b 72 785 + 399 = 73 184
c 15 838 + 68 285 = 84 123
d 15 284 + 88 = 15 372
e 74 909 + 29 = 74 938
f 48 509 + 7893 = 56 402
g 10 449 + 15 932 = 26 381
h 35 826 + 47 198 = 83 024

3 a 43 719 b 52 953

Trading From Higher Place Values

Page 122 – Your Turn
a 737 b 65 672 c 56 355

Page 123 – Practice
1 a 60 127 b 5614 c 237 d 1719 e 314 396 f 285 640 g 9796 h 18 625 i 14 617 j 5164 k 265 437

2 a $142.55 b $106.53 c $3.25 d $66.35 e $363.50 f $452.47 g $3.25 h $126.50

3 SUBTRACTION CONTINUED

Difference

Page 124 – Your Turn

a 6103
b 4288

Page 125 – Practice

1 a $102 736 b $244 658 c $272 013
2 a $156 711 b $272 014 c $401 369 d $102 737 e $1

Rounding To Estimate Answers

Page 126 – Your Turn

1 a 330 – 200 = 130 b 540 – 460 = 180
2 a 600 – 400 = 200 b 800 – 400 = 400

Page 127 – Practice

1 a 6430, 1360, 5070
b 7490, 5640, 1850
c 140, 130, 10
d 340, 200, 140
e 5820, 3220, 2600
2 a 9300, 3600, 5700
b 4900, 2700, 2200
c 300, 200, 100
d 800, 400, 400
e 900, 200, 700
3 a 15 400, 11 000, 4400
b 84 200, 38 900, 45 300
c 73 300, 44 200, 29 100
d 19 200, 11 400, 7800

Word Problems Using Subtraction

Page 128 – Your Turn

1 a 37 – 21 = 16 goals
b 2293 – 129 = 2164 punnets
c $59 624 – $39 515 = $20 109
d $71 425 – $21 325 = $50 100

Page 130 – Practice

1 a 877 – 352 = 525 km
b 3848 – 2812 = 1036 km
c 16 019 – 2943 = 13 076 km
d 754 218 – 314 020 = $440 198
e 73 – 54 = 19 kg
f 37 – 24 = 13 km
g 236 – 112 = 124 km
h 73 268 – 68 249 = $5019
i 86 000 – 84 132 = 1868 litres
j 16 425 – 12 236 = $4189
k 6 536 259 – 2 436 582 = $4 099 677
l 13413 – 2599 = 10 814 m

Subtraction Review Page 132

1 a 58 b 247 c 768 d 531 e 330 f 5000 g 6969 h 5147 i 5803

2 a 84 – 21
80 – 20 = 60
4 – 1 = 3
60 + 3 = 63
b 78 – 34
70 – 30 = 40
8 – 4 = 4
40 + 4 = 44
c 499 – 38
400 – 0 = 400
90 – 30 = 60
9 – 8 = 1
400 + 60 + 1 = 461
d 697 – 56
600 – 0 = 600
90 – 50 = 40
7 – 6 = 1
600 + 40 + 1 = 641
e 784 – 153
700 – 100 = 600
80 – 50 = 30
4 – 3 = 1
600 + 30 + 1 = 631
f 882 – 481
800 – 400 = 400
80 – 80 = 0
2 – 1 = 1
400 + 0 + 1 = 401
g 7598 – 463
7000 – 0 = 7000
500 – 400 = 100
90 – 60 = 30
8 – 3 = 5
7000 + 100 + 30 + 5 = 7135
h 9897 – 544
9000 – 0 = 9000
800 – 500 = 300
90 – 40 = 50
7 – 4 = 3
9000 + 300 + 50 + 3 = 9353
i 6984 – 1273
6000 – 1000 = 5000
900 – 200 = 700
80 – 70 = 10
4 – 3 = 1
5000 + 700 + 10 + 1 = 5711
j 8168 – 4034
8000 – 4000 = 4000
100 – 0 = 100
60 – 30 = 30
8 – 4 = 4
4000 + 100 + 30 + 4 = 4134
k 9864 – 8731
9000 – 8000 = 1000
800 – 700 = 100
60 – 30 = 30
4 – 1 = 3
1000 + 100 + 30 + 3 = 1133

3 a 42 b 52 c 73 d 35 e 502 f 602 g 941 h 642 i 3731 j 1430 k 9100 l 1016
4 a 83 361 b 51 201 c 51 302 d 54 114 e 30 210 f 58 112
5 a 79 b 59 c 39 d 3 e 7 f 883 g 779 h 381 i 484 j 175 k 7188 l 7908 m 1089 n 6119 o 1056 p 3997
6 a 57 953 b 68 636 c 52 867
7 a 751 b 542 c 323 d 264 e 3672 f 8921 g 568 h 529
8 a 5207 b 29 676 c 59 592 d 356 550 e 47 691 f 257 485
9 a 36 b 14 c 60 d 686 e 821 f 604 g 3332 h 2805 i 7392
10 a 410, 330, 80
b 910, 480, 430
c 750, 40, 710
d 530, 80, 450
e 5420, 470, 4950
f 6560, 430, 6130
g 7980, 1330, 6650
h 4880, 2840, 2040
11 a 500, 300, 200
b 800, 200, 600
c 700, 100, 600
d 700, 400, 300
e 4400, 100, 4300
f 7300, 400, 6900
g 7100, 500, 6600
h 9900, 3600, 6300
12 a 59 – 24 = 35 dolls
b 816 – 634 = 182 km
c 5638 – 643 = 4995 km
d 79 348 – 25 291 = $54 057
e 895 268 – 624 254 = $271 014
f 759 982 – 657 998 = $101 984
13 a 75 – 24 = 51
b 92 – 45 = 47
c 83 – 54 = 29
d 64 – 37 = 27
e 89 – 47 = 42
f 145 – 32 = 113
g 514 – 36 = 478
h 843 – 145 = 698
i 907 – 256 = 651
j 616 – 294 = 322
k 5923 – 147 = 5776
l 6097 – 4214 = 1883
m 7623 – 497 = 7126
n 4314 – 1795 = 2519
o 8448 – 4352 = 4096
p 5826 – 2537 = 3289
14 a 64 073 – 4136 = 59 937
b 75 923 – 14 654 = 61 269
c 91 110 – 73 427 = 17 683
d 82 483 – 60 736 = 21 747
e 45 329 – 3548 = 41 781
f 25 963 – 1374 = 24 589
g 49 230 – 5624 = 43 606
h 97 426 – 47 327 = 50 099
i 72 495 – 14 198 = 58 297
15 Christian scored 0.
Correct answers:
1 711
2 461
3 5364
4 3414
5 43 627
6 $67.42
7 $184.57
8 577 788

CATCH UP MATHS YEAR 6 BOOK A © PASCAL PRESS ISBN: 9781925726183

4 MULTIPLICATION

Product, Factors & Multiples

Page 140 – Your Turn

1 a 9 b 3 c 2

2 a 21 b 45 c 27

Page 141 – Practice

1 a 11 b 6 c 2 d 8 e 7 f 9 g 10 h 6 i 8 j 7 k 11 l 10 m 10 n 6 o 9 p 4 q 7 r 1 s 6 t 6

2 a 24 b 27 c 20 d 88 e 40 f 36 g 8 h 30 i 64 j 49 k 0

3 a 6, 12, 18, 24, 30, 36
b 9, 18, 27, 36, 45, 54
c 10, 20, 30, 40, 50, 60
d 12, 24, 36, 48, 60, 72
e 7, 14, 21, 28, 35, 42
f 1, 2, 3, 4, 5, 6
g 8, 16, 24, 32, 40, 48
h 11, 22, 33, 44, 55, 66
i 5, 10, 15, 20, 25, 30

4 a 1, 2, 7, 14
b 1, 2, 4, 8
c 1, 2, 5, 10
d 1, 3, 5, 15
e 1, 2, 3, 4, 6, 8, 12, 24
f 1, 2, 4, 5, 10, 20
g 1, 2, 3, 4, 6, 9, 12, 18, 36
h 1, 2, 5, 10, 25, 50
i 1, 2, 4, 5, 10, 20, 25, 50, 100
j 1, 2, 3, 4, 5, 6, 10, 12, 15, 20, 30, 60

Multiplying 2-Digit Numbers By 1-Digit Numbers

Page 142 – Your Turn

a Known facts
30 × 4 = 120
120 + 4 + 4 = 128
Multiply by place value
3 tens ×43 + 4 twos
= 120 + 8 = 128
Area model

	30	2
4	120	8

= 120 + 8 = 128

b Known facts
40 × 3 = 120
120 + 8 + 8 + 8 = 144
Multiply by place value
4 tens × 3 + 3 eights
= 120 + 24 = 144
Area model

	40	8
3	120	24

= 120 + 24 = 144

Page 143 – Practice

1 a 128 b 102 c 426 d 270 e 156

Formal Algorithms

Page 144 – Your Turn

1 a

	H	T	O
		7	1
×			4
	2	8	4

b

	H	T	O
		+3 3	7
×			5
	1	8	5

c

	H	T	O
		+2 4	9
×			3
	1	4	7

d

	H	T	O
		+2 8	5
×			5
	4	2	5

e

	H	T	O
		+4 1	6
×			7
	1	1	2

Page 145 – Practice

1 a 96 b 369 c 219 d 126 e 455 f 249 g 156 h 216 i 184 j 477 k 222 l 207 m 444 n 425 o 372

2 a 438 b 116 c 570 d 296 e 217 f 246 g 396 h 378

Multiply 3-Digit & 4-Digit Numbers By 1-Digit Numbers

Page 147 – Your Turn

a 5932 × 4
(5000 × 4) + (900 × 4) + (30 × 4) + (2 × 4)
= 20 000 + 3600 + 120 + 8
= 23 728

	5000	900	30	2
4	20 000	3600	120	8

= 20 000 + 3600 + 120 + 8
= 23 728

b 3842 × 5
(3000 × 5) + (800 × 5) + (40 × 5) + (2 × 5)
= 15 000 + 4000 + 200 + 10
= 19 210

	3000	800	40	2
5	15 000	4000	200	10

= 15 000 + 4000 + 200 + 10
= 19 210

Page 148 – Practice

1 a 22 068 b 3924 c 1716 d 71 560

2 a 1371 b 13 182 c 3572 d 36 025 e 1707 f 55 280 g 6786 h 9214

3 a 24 870 b 2776 c 63 704 d 1620 e 58 410 f 4340 g 9594

Multiply 2-Digit And 3-Digit Numbers By 2-Digit Numbers

Page 151 – Your Turn

a

			+1 +3 6	4
×			3	8
	1	5	1	2
+	1	9	2	0
	2	4	3	2

1920 + 512 = 2432

	60	4	
30	1800	120	1920
8	480	32	512
			2432

b

			+1 7	+1 +2 4	9
×				2	3
		1 2	1 2	4	7
+	1	4	9	8	0
	1	7	2	2	7

14 980 + 2247 = 17 227

	700	40	9	
20	14 000	800	180	14 980
3	2100	120	27	2 247
				17 227

c

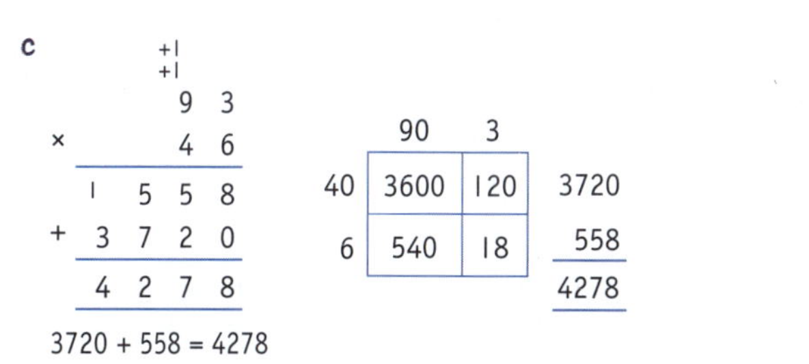

4 MULTIPLICATION CONTINUED

1 a 11 653 c 4158 e 3894 g 29 778
b 57 521 d 2997 f 49 914 h 1536

2 a 1634 c 2418 e 33 912 g 41 944 i 18 914
b 5152 d 1998 f 38 023 h 35 360

Multiply 4-Digit Numbers By 1-Digit And 2-Digit Numbers

Page 154 – Your Turn

a
```
     +3 +1 +2
     +2 +1 +2
      2  4  2  3
×              8  7
   1  1 6 1 9  6  1
+  1  9  3  8  4  0
   2  1  0  8  0  1
```

b
```
        +2    +2
        +5    +4
         5  6  0  5
×              4  9
   1   1 5 1 0 1 4  4  5
+  2  2  4  2  0  0
   2  7  4  6  4  5
```

Page 155 – Practice

1 a 21 354 c 19 485 e 8030 g 26 136
b 17 280 d 41 293 f 4206 h 84 357

2 a 259 840 c 190 971 e 850 122 g 79 050
b 158 160 d 371 304 f 247 920 h 393 635

Multiply By 10, 100 And 1000

Page 156 – Your Turn

1 a 490 b 6470 c 1820

2 a 74 100 b 17 600 c 59 300

3 a 623 000 b 7000 c 42 000

Page 157 – Practice

1 a 630 d 50 g 160 j 45 710
b 4710 e 7110 h 5060 k 86 230
c 1090 f 1000 i 21 930

2 a 200 d 4000 g 63 000 j 9800
b 27 500 e 11 900 h 10 000 k 521 500
c 87 700 f 17 500 i 290 300

3 a 7000 d 103 000 g 2 711 000 j 574 000
b 24 000 e 17 000 h 208 000 k 6 258 000
c 520 000 f 1 582 000 i 5 387 000

4 a 6150, 61 500, 615 000
b 260, 2600, 26 000
c 4160, 41 600, 416 000
d 80, 800, 8000
e 3920, 39 200, 392 000
f 8150, 81 500, 815 000
g 520, 5200, 52 000
h 6070, 60 700, 607 000

Order Of Operations

Page 158 – Your Turn

a 25 ÷ 5 = 5 b 100 − 26 = 74 c 67 × 4 + 3 = 271

Page 159 – Practice

1 a 6 c 5 e 6 g 4 i 84
b 19 d 8 f 50 h 37

2 a 2 d + g × j − m −
b − e + h − k ÷ n +
c 6 f × i × l × o ×

Multiplication Review Page 160

1 a 3 c 3 e 5 g 3 i 6 k 6
b 7 d 9 f 7 h 12 j 5 l 0

2 a 27 c 32 e 0 g 3 i 11 k 49
b 4 d 18 f 30 h 64 j 48 l 60

3 a 14, 21, 28, 35, 42
b 6, 9, 12, 15, 18
c 10, 15, 20, 25, 30
d 4, 6, 8, 10, 12
e 16, 24, 32, 40, 48
f 8, 12, 16, 20, 24
g 2, 3, 4, 5, 6
h 12, 18, 24, 30, 36

4 a 1 × 30 = 30
2 × 15 = 30
3 × 10 = 30
5 × 6 = 30
b 1 × 10 = 10
2 × 5 = 10
c 1 × 32 = 32
2 × 16 = 32
4 × 8 = 32
d 1 × 16 = 16
2 × 8 = 16
4 × 4 = 16
e 1 × 20 = 20
2 × 10 = 20
4 × 5 = 20
f 1 × 28 = 28
2 × 14 = 28
4 × 7 = 28

5 a 24 × 5 = 120 b 63 × 8 = 504 c 54 × 7 = 378

6 a 168 d 568 g 558 j 261
b 63 e 368 h 343 k 434
c 84 f 136 i 288 l 801

7 a 249 × 4 = 996 c 789 × 5 = 3945
b 374 × 6 = 2244 d 436 × 7 = 3052

8 a 2565 d 2300 g 24 030
b 3066 e 3053 h 13 581
c 6624 f 14 421 i 29 992

9 a 1645 c 4015 e 29 406
b 2046 d 23 464 f 82 838

10 a 11 924 c 68 128 e 32 977
b 19 212 d 22 158 f 15 414

11 a 375 396 c 234 104 e 463 166
b 533 566 d 184 452 f 463 608

12

	Number	× 10	× 100	× 1000
a	6	60	600	6000
b	49	490	4900	49 000
c	317	3170	31 700	317 000
d	482	4820	48 200	482 000
e	5896	58 960	589 600	5 896 000
f	4970	49 700	497 000	4 970 000
g	689	6890	68 900	689 000
h	435	4350	43 500	435 000
i	28	280	2800	28 000
j	7	70	700	7000

13 a 25 d 10 g 47 j 73 m 100 p 70
b 38 e 6 h 32 k 36 n 23 q 30
c 31 f 40 i 33 l 80 o 20 r 16

  ISBN: 9781925726183

5 DIVISION

Quotient, Divisor And Dividend

Page 166 – Your Turn

1 a 6 b 5 c 8 d 10 e 2
2 a 7 b 11 c 4 d 9 e 9
3 a 64 b 80 c 14 d 132 e 81

Page 167 – Practice

1 a 4 b 10 c 3 d 7 e 8 f 1 g 2 h 20 i 20 j 4 k 30
2 20 ÷ 4, 45 ÷ 9, 15 ÷ 3, 40 ÷ 8, 5 ÷ 1, 30 ÷ 6
3 a 4 b 9 c 10 d 3 e 7 f 10 g 6 h 40 i 7 j 3 k 5
4 a 4 b 16 c 12 d 9 e 7 f 3 g 7 h 3 i 3 j 2 k 1 l 7 m 10 n 4 o 2 p 6 q 9

5

	Dividend	Divisor	Quotient
a	28	7	4
b	63	7	9
c	100	10	10
d	169	13	13
e	54	6	9
f	108	12	9
g	72	12	6

Formal Division

Page 168 – Your Turn

a 10 b 3 c 10 d 10 e 8 f 2 g 9

Page 169 – Practice

1 a 4, 4 × 6 = 24
b 7, 7 × 8 = 56
c 11, 11 × 11 = 121
d 9, 9 × 12 = 108
e 11, 11 × 12 = 132
f 9, 9 × 7 = 63
g 4, 4 × 2 = 8
h 1, 1 × 6 = 6
i 9, 9 × 10 = 90
j 11, 11 × 7 = 77
k 10, 10 × 5 = 50
l 7, 7 × 7 = 49
m 7, 7 × 3 = 21

2 a 11 b 4 c 72 d 12 e 6 f 50 g 56 h 7 i 10

3 a 27 ÷ 3 = 9 $\begin{array}{r}9\\3\overline{)2\ 7}\end{array}$
b 36 ÷ 12 = 3 $\begin{array}{r}3\\12\overline{)3\ 6}\end{array}$
c 10 ÷ 2 = 5 $\begin{array}{r}5\\2\overline{)1\ 0}\end{array}$
d 56 ÷ 8 = 7 $\begin{array}{r}7\\8\overline{)5\ 6}\end{array}$
e 60 ÷ 6 = 10 $\begin{array}{r}1\ 0\\6\overline{)6\ 0}\end{array}$

Different Ways To Write Division

Page 170 – Your Turn

a $\frac{21}{4}$ b $8\overline{)3\ 2}$ c $8\overline{)6\ 4}$ d $\frac{20}{5}$ e $\frac{47}{3}$ f $3\overline{)9\ 2}$

Page 171 – Practice

1 a $\frac{62}{4}$ b $\frac{90}{9}$ c $\frac{84}{4}$ d $\frac{36}{6}$ e $\frac{81}{4}$ f $\frac{21}{3}$ g $\frac{26}{2}$ h $\frac{54}{5}$ i $\frac{29}{5}$ j $\frac{101}{10}$ k $\frac{72}{9}$

2 a $4\overline{)2\ 1}$ b $9\overline{)8\ 3}$ c $4\overline{)7\ 0}$ d $4\overline{)2\ 9}$ e $8\overline{)3\ 3}$ f $11\overline{)121}$ g $10\overline{)9\ 1}$ h $10\overline{)6\ 3}$ i $3\overline{)8\ 7}$ j $6\overline{)3\ 2}$ k $3\overline{)2\ 7}$

3 a 37 ÷ 6 b 43 ÷ 7 c 46 ÷ 5 d 53 ÷ 4 e 84 ÷ 5 f 24 ÷ 4 g 27 ÷ 8 h 75 ÷ 5 i 89 ÷ 6 j 39 ÷ 3 k 42 ÷ 6

4

	Fraction	$\overline{)\ }$	÷
a	$\frac{33}{3}$	$3\overline{)3\ 3}$	33 ÷ 3
b	$\frac{27}{9}$	$9\overline{)2\ 7}$	27 ÷ 9
c	$\frac{63}{10}$	$10\overline{)6\ 3}$	63 ÷ 10
d	$\frac{36}{4}$	$4\overline{)3\ 6}$	36 ÷ 4
e	$\frac{81}{9}$	$9\overline{)8\ 1}$	81 ÷ 9
f	$\frac{72}{8}$	$8\overline{)7\ 2}$	72 ÷ 8
g	$\frac{32}{5}$	$5\overline{)3\ 2}$	32 ÷ 5
h	$\frac{50}{7}$	$7\overline{)5\ 0}$	50 ÷ 7
i	$\frac{56}{7}$	$7\overline{)5\ 6}$	56 ÷ 7
j	$\frac{44}{11}$	$11\overline{)4\ 4}$	44 ÷ 11
k	$\frac{70}{10}$	$10\overline{)7\ 0}$	70 ÷ 10

Division With Remainders

Page 172 – Your Turn

a 8 remainder 1

Page 173 – Practice

1 a 3 r 4 b 8 r 1 c 6 r 3 d 3 r 3 e 10 r 2 f 7 r 1 g 8 r 9
2 a 6 r 4 b 10 r 3 c 4 r 3 d 6 r 6 e 5 r 3 f 8 r 3 g 12 r 2 h 6 r 1 i 7 r 1
3 a 4 r 6 b 6 r 4 c 5 r 3
4 a 23 ÷ 3 = 7 r 2
b 37 ÷ 5 = 7 r 2
c 26 ÷ 7 = 3 r 5
d 29 ÷ 8 = 3 r 5
e 74 ÷ 9 = 8 r 2
f 69 ÷ 10 = 6 r 9
g 87 ÷ 7 = 12 r 3
h 153 ÷ 12 = 12 r 9
i 189 ÷ 13 = 14 r 7

Division Of 2-Digit Numbers

Page 174 – Your Turn

a 24 r 1 b 19 c 14 r 3 d 14 r 3 e 14 f 3 r 2 g 19 r 3 h 11 r 5

Page 175 – Practice

1 a 11 b 13 c 12 d 7 e 7 f 37 g 9
2 a 12 r 3 b 13 r 1 c 15 r 5 d 12 r 3 e 12 f 9 r 5 g 14 h 10 r 3 i 12 r 2 j 15 r 3 k 28 r 1 l 11 r 6 m 16 r 3 n 47 o 13 r 2
3 a 12 × 4 + 3 = 51
b 13 × 7 + 1 = 92
c 15 × 6 + 5 = 95
d 12 × 5 + 3 = 63
e 12 × 7 = 84
f 9 × 9 + 5 = 86
g 14 × 3 = 42
h 10 × 9 + 3 = 93
i 12 × 6 + 2 = 74
j 15 × 4 + 3 = 63
k 28 × 2 + 1 = 57
l 11 × 8 + 6 = 94
m 16 × 4 + 3 = 67
n 47 × 2 = 94
o 13 × 5 + 2 = 67

Division Of 3-Digit Numbers

Page 176 – Your Turn

a 179 r 1 b 141 r 2 c 366 r 1 d 107 r 2 e 70 r 3 f 70 r 5 g 101 r 4 h 123 r 5

Page 177 – Practice

1 a 91 r 6 b 81 r 5 c 69 r 3 d 93 r 1 e 64 r 1 f 73 g 95 h 138 r 2 i 149 r 3 j 371 r 1 k 139 r 3 l 109 r 3 m 284 r 2 n 130 r 3

5 DIVISION CONTINUED

2 a $91 \times 7 + 6 = 643$
b $81 \times 6 + 5 = 491$
c $69 \times 5 + 3 = 348$
d $93 \times 8 + 1 = 745$
e $64 \times 4 + 1 = 257$
f $73 \times 2 = 146$
g $95 \times 3 = 285$
h $138 \times 5 + 2 = 692$
i $149 \times 4 + 3 = 599$
j $371 \times 2 + 1 = 743$
k $139 \times 6 + 3 = 837$
l $109 \times 9 + 3 = 984$
m $284 \times 3 + 2 = 854$
n $130 \times 5 + 3 = 653$

3 a 120, 60, 300, 400, 50, 80, 150
b 20, 30, 40, 80, 100, 70, 50
c 100, 120, 40, 50, 60, 70, 110

Recording Remainders As Fractions And Decimals

Page 178 – Your Turn

1 a 98 r 3, $98\frac{3}{5}$
b 260 r 1, $260\frac{1}{2}$

2 a 128 r 1, 128.25
b 420 r 1, 420.5

Page 179 – Practice

1 a 8, 1, $8\frac{1}{2}$, 8.5
b 6, 3, $6\frac{3}{5}$, 6.6
c 14, 1, $14\frac{1}{4}$, 14.25
d 12, 1, $12\frac{1}{3}$, 12.33

2 a 36 r 3, 36.75
b 89 r 4, 89.8
c 104 r 4, 104.5
d 418 r 1, 418.5
e 245 r 1, 245.25

3 a 185 r 1, $185\frac{1}{4}$
b 84 r 1, $84\frac{1}{4}$
c 99 r 1, $99\frac{1}{5}$
d 177 r 1, $177\frac{1}{3}$
e 95 r 1, $95\frac{1}{2}$

Averages

Page 180 – Your Turn

a $24 \div 4 = 6$
b $25 \div 5 = 5$
c $45 \div 5 = 9$

Page 181 – Practice

1 a $35 \div 5 = 7$
b $32 \div 4 = 8$
c $36 \div 6 = 6$
d $8 \div 8 = 1$
e $60 \div 5 = 12$
f $50 \div 5 = 10$
g $32 \div 4 = 8$
h $50 \div 5 = 10$
i $35 \div 5 = 7$
j $40 \div 5 = 8$
k $30 \div 5 = 6$
l $188 \div 4 = 47$
m $42 \div 6 = 7$
n $119 \div 7 = 17$
o $280 \div 8 = 35$

Dividing By 10, 100 And 1000

Page 182 – Your Turn

	Number	÷ 10	÷ 100	÷ 1000
a	45	4.5	0.45	0.045
b	124	12.4	1.24	0.124
c	865	86.5	8.65	0.865

Page 183 – Practice

1 a 9.6, 0.96, 0.096
b 0.8, 0.08, 0.008
c 15.4, 1.54, 0.154
d 2.7, 0.27, 0.027
e 35.7, 3.57, 0.357
f 421.5, 42.15, 4.215
g 638.7, 63.87, 6.387
h 0.9, 0.09, 0.009
i 1435.9, 143.59, 14.359

2 a 7.2
b 15.9
c 268.7
d 5.3
e 47.5
f 682.4
g 7437.2
h 594.5
i 163.8
j 44
k 8243.7

3 a 0.04
b 2.49
c 19.43
d 0.44
e 742.73
f 815.10
g 0.17
h 725.93
i 5.60
j 694.20
k 8.26

4 a 0.006
b 0.182
c 7.111
d 75.264
e 0.95
f 8.256
g 0.019
h 0.028
i 6.359
j 42.003
k 0.778

Division Review Page 184

1 a 8 b 6 c 9 d 9 e 7 f 10 g 7 h 4

2 $8 \div 2$, $40 \div 10$, $48 \div 12$, $16 \div 4$, $20 \div 5$, $24 \div 6$, $36 \div 9$, $32 \div 8$

3

	Fraction	$\overline{)\ \ }$	÷
a	$\frac{63}{2}$	$2\overline{)63}$	$63 \div 2$
b	$\frac{47}{5}$	$5\overline{)47}$	$47 \div 5$
c	$\frac{23}{8}$	$8\overline{)23}$	$23 \div 8$
d	$\frac{41}{8}$	$8\overline{)41}$	$41 \div 8$
e	$\frac{22}{4}$	$4\overline{)22}$	$22 \div 4$
f	$\frac{63}{7}$	$7\overline{)63}$	$63 \div 7$

4 a 72 b 5 c 11 d 56 e 108

5 a 15, $15 \div 3 = 5$, $3\overline{)15}$ = 5
b 12, $12 \div 2 = 6$, $2\overline{)12}$ = 6
c 56, $56 \div 8 = 7$, $8\overline{)56}$ = 7
d 110, $110 \div 10 = 11$, $10\overline{)110}$ = 11
e 84, $84 \div 12 = 7$, $12\overline{)84}$ = 7
f 18, $18 \div 9 = 2$, $9\overline{)18}$ = 2

6 a 12 r 1 b 6 r 2 c 8 r 2 d 6 r 1 e 3 r 5 f 7 r 4

7 a 9 r 2 b 2 r 6 c 4 r 8 d 8 r 5

8 a 16
b 13
c 26
d 4
e 21 r 3
f 3 r 5
g 12 r 3
h 9 r 3

9 a $16 \times 4 = 64$
b $13 \times 2 = 26$
c $26 \times 3 = 78$
d $4 \times 7 = 28$
e $21 \times 4 + 3 = 87$
f $3 \times 8 + 5 = 29$
g $12 \times 5 + 3 = 63$
h $9 \times 6 + 3 = 57$

10 a 106
b 117
c 126
d 28
e 129 r 1
f 39 r 4
g 156 r 1
h 84 r 1

11 a $106 \times 4 = 424$
b $117 \times 5 = 585$
c $126 \times 6 = 756$
d $28 \times 9 = 252$
e $129 \times 5 + 1 = 646$
f $39 \times 7 + 4 = 277$
g $156 \times 4 + 1 = 625$
h $84 \times 8 + 1 = 673$

12 a 50, 100, 70, 60, 20, 120, 110
b 40, 20, 50, 100, 120, 3, 60

13 a $181\frac{2}{4}$ b $170\frac{1}{3}$ c $382\frac{1}{2}$ d $280\frac{2}{3}$ e $174\frac{3}{5}$ f $142\frac{2}{3}$

14 a 247.33
b 212.25
c 151.2
d 422.5
e 24.4
f 83.75

15 a 3, 5, $3\frac{5}{8}$, 3.625
b 9, 1, $9\frac{1}{4}$, 9.25
c 28, 1, $28\frac{1}{3}$, 28.33
d 39, 1, $39\frac{1}{2}$, 39.5
e 9, 3, $9\frac{3}{5}$, 9.6

16 a 4 b 5 c 36 d 4 e 4 f 21

17 a 0.7
b 4.2
c 73.9
d 149.3
e 7267.3
f 378.1
g 331.3
h 9257.8
i 0.6

18 a 0.05
b 0.63
c 9.73
d 94.41
e 818.1
f 52.43
g 26.24
h 491
i 0.09

19 a 0.001
b 0.034
c 0.387
d 6.64
e 91.113
f 8.756
g 4.6
h 82.403
i 0.003

20 a 0.4, 0.04, 0.004
b 7.8, 0.78, 0.078
c 46.9, 4.69, 0.469
d 37.1, 3.71, 0.371

6 FRACTIONS

Halves – Fractions and Collections

Page 188 – Your Turn

1 a Circle **c**, **d**, **e**

2 a Circle 3. Half of 6 = 3

b Circle 6. Half of 12 = 6

Page 189 – Practice

1 Cross **a** and **c**

2 Sample answers:

a   b  c  d  e 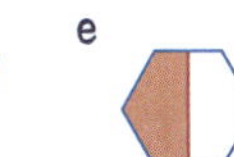

3 a 8, 16
16, 8
$\frac{1}{2}$ of 16 = 8

b 12, 24
24, 12
$\frac{1}{2}$ of 24 = 12

c 14, 28
28, 14
$\frac{1}{2}$ of 28 = 14

4 a 14, 7
14, 7

b 20, 10
20, 10

5 a 1 b 3 c 50 d 24 e 13 f 21 g 15 h 11 i 25 j 28 k 30 l 33 m 32 n 60 o 72 p 62 q 65 r 200 s 48

Quarters and Eighths – Fractions and Collections

Page 190 – Your Turn

1 Circle eighths in red: **b**, **c**, **e**.
Circle quarters in green: **a**, **d**.

2 a Circle 10. $\frac{1}{4}$ of 40 = 10

b Circle 5. $\frac{1}{8}$ of 40 = 5

Page 191 – Practice

1 a  

b 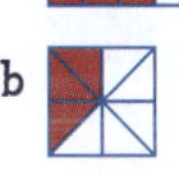

c

d

2 a 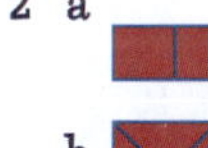

b 

c 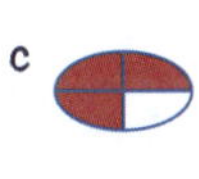

d

3

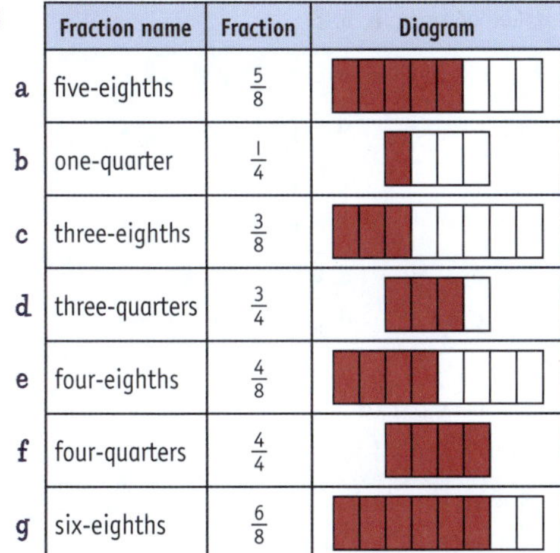

	Fraction name	Fraction	Diagram
a	five-eighths	$\frac{5}{8}$	
b	one-quarter	$\frac{1}{4}$	
c	three-eighths	$\frac{3}{8}$	
d	three-quarters	$\frac{3}{4}$	
e	four-eighths	$\frac{4}{8}$	
f	four-quarters	$\frac{4}{4}$	
g	six-eighths	$\frac{6}{8}$	

4 a 4 b 1 c 7 d 10 e 3 f 5 g 8

5 a 5 b 6 c 11 d 12 e 25 f 8 g 9

6 a 6 b 3 c 4 d 9 e 10 f 7 g 10

Thirds And Fifths – Fractions And Collections

Page 192 – Your Turn

1 Circle fifths in green: **a**, **c**, **e**.
Circle thirds in red: **b**, **d**.

2 a Circle 8. $\frac{1}{3}$ of 24 = 8

b Circle 4. $\frac{1}{5}$ of 20 = 4

Page 193 – Practice

1

	Fraction name	Fraction	Diagram
a	two-fifths	$\frac{2}{5}$	
b	three-fifths	$\frac{3}{5}$	
c	two-thirds	$\frac{2}{3}$	
d	three-thirds	$\frac{3}{3}$	
e	four-fifths	$\frac{4}{5}$	

2 a $\frac{1}{3}$ b $\frac{3}{3}$ c $\frac{3}{5}$ d $\frac{1}{3}$ e $\frac{5}{5}$ f $\frac{2}{5}$ g $\frac{2}{3}$

3 a $\frac{2}{3}$ b $\frac{0}{3}$ c $\frac{2}{5}$ d $\frac{2}{3}$ e $\frac{0}{5}$ f $\frac{3}{5}$ g $\frac{1}{3}$

4 a 10 b 6 c 5 d 12 e 20 f 1 g 2 h 8 i 3

5 a 4 b 8 c 22 d 7 e 10 f 30 g 12 h 11 i 3

6 a 6 b 7 c 9 d 9 e 11 f 23 g 15

Equivalent Fractions

Page 194 – Your Turn

a $\frac{6}{8}$ 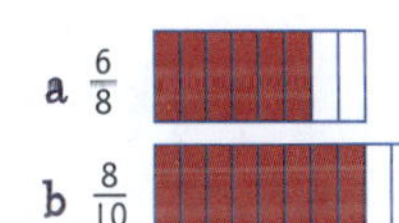

b $\frac{8}{10}$

c $\frac{4}{6}$ 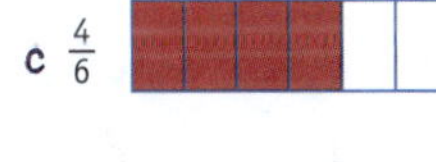

Page 195 – Practice

1 a 4 b 9 c 1 d 4 e 16 f 15 g 70

2 a 40 b 8 c 1 d 6 e 5 f 6 g 60 h 25 i 6

3 a $\frac{3}{4} = \frac{6}{8} = \frac{9}{12} = \frac{15}{20}$

b $\frac{2}{5} = \frac{4}{10} = \frac{6}{15} = \frac{40}{100}$

c $\frac{3}{5} = \frac{6}{10} = \frac{12}{20} = \frac{60}{100}$

d $\frac{1}{4} = \frac{2}{8} = \frac{4}{16} = \frac{25}{100}$

e $\frac{1}{5} = \frac{2}{10} = \frac{4}{20} = \frac{20}{100}$

4 a $\frac{1}{4}, \frac{2}{8}, \frac{4}{16}, \frac{5}{20}$

b $\frac{2}{6}, \frac{1}{3}, \frac{3}{9}$

c $\frac{1}{5}, \frac{2}{10}$

d $\frac{4}{8}, \frac{6}{12}, \frac{1}{2}$

e $\frac{10}{100}, \frac{1}{10}, \frac{20}{200}$

5 a False b True c True d True e True f False g True h True i True

Comparing Fractions

Page 196 – Your Turn

Adult to check.

Page 197 – Practice

1 a 5, 4, 2, 1, 6, 3 b 3, 2, 1, 4, 5, 6 c 2, 6, 1, 5, 4, 3

6 FRACTIONS CONTINUED

2

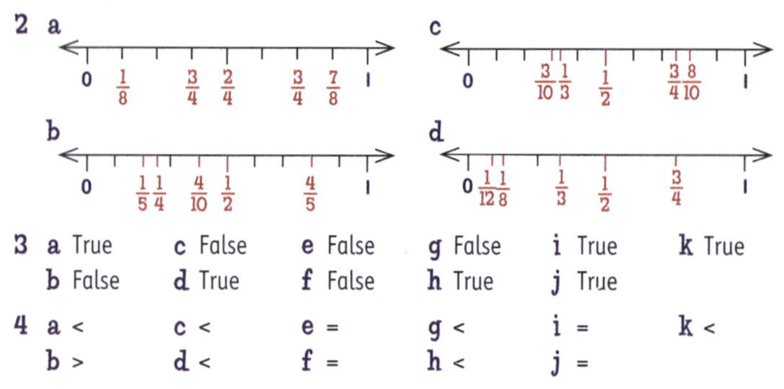

3 a True b False c False d True e False f False g False h True i True j True k True

4 a < b > c < d < e = f = g < h < i = j = k <

Proper and Improper Fractions

Page 198 – Your Turn

Proper fractions:

$\frac{2}{3}$ $\frac{10}{12}$ $\frac{3}{4}$ $\frac{4}{10}$ $\frac{89}{100}$ $\frac{1}{8}$ $\frac{9}{10}$ $\frac{2}{6}$ $\frac{5}{12}$ $\frac{1}{5}$ $\frac{3}{5}$ $\frac{4}{7}$ $\frac{9}{12}$ $\frac{4}{5}$ $\frac{4}{8}$

Improper fractions:

$\frac{5}{4}$ $\frac{6}{1}$ $\frac{8}{5}$ $\frac{10}{9}$ $\frac{2}{1}$ $\frac{100}{97}$ $\frac{5}{2}$ $\frac{7}{2}$ $\frac{5}{3}$ $\frac{16}{8}$ $\frac{3}{2}$ $\frac{12}{7}$ $\frac{3}{1}$ $\frac{15}{12}$ $\frac{4}{3}$ $\frac{8}{6}$ $\frac{25}{20}$

Page 199 – Practice

1 a $\frac{8}{3}$ b $\frac{10}{9}$ c $\frac{12}{10}$ d $\frac{6}{3}$ e $\frac{3}{2}$

2 a $\frac{1}{3}$ b $\frac{27}{30}$ c $\frac{14}{20}$ d $\frac{16}{35}$ e $\frac{9}{13}$

3 Adult to check

4 Adult to check

Mixed Numbers

Page 200 – Your Turn

1 a $\frac{23}{4}$, $5\frac{3}{4}$ b $\frac{55}{8}$, $6\frac{7}{8}$ c $\frac{9}{2}$, $4\frac{1}{2}$

Page 201 – Practice

1 a $2\frac{2}{4}$ b $4\frac{1}{3}$ c $6\frac{1}{2}$ d $1\frac{5}{20}$ e $1\frac{7}{40}$ f $1\frac{9}{50}$ g $4\frac{3}{4}$ h $2\frac{5}{6}$ i $6\frac{1}{7}$ j $4\frac{4}{12}$ k $7\frac{1}{5}$

2 a $\frac{23}{4}$ b $\frac{23}{6}$ c $\frac{13}{5}$ d $\frac{19}{5}$ e $\frac{74}{7}$ f $\frac{128}{10}$ g $\frac{95}{7}$ h $\frac{124}{6}$ i $\frac{75}{8}$ j $\frac{67}{9}$ k $\frac{83}{12}$

3 a False b True c False d True e False f True g False h False

4 a $7\frac{3}{4}$ b $14\frac{1}{2}$ c $4\frac{1}{3}$ d $2\frac{4}{10}$ e $4\frac{3}{4}$

5 a $\frac{39}{8}$ b $\frac{88}{9}$ c $\frac{38}{5}$ d $\frac{70}{8}$ e $\frac{43}{4}$

Add and Subtract Fractions with the Same Denominator

Page 202 – Your Turn

a $\frac{6}{12}$ b $\frac{6}{15}$ c $\frac{15}{20}$ d $\frac{13}{10} = 1\frac{3}{10}$ e $\frac{12}{6} = 2$ f $\frac{12}{8} = 1\frac{4}{8}$ g $\frac{6}{5} = 1\frac{1}{5}$ h $\frac{2}{12}$ i $\frac{3}{15}$

Page 203 – Practice

1 a $\frac{7}{10}$ b $\frac{5}{6}$ c $\frac{9}{10}$ d $\frac{8}{10}$ e $\frac{10}{12}$ f $\frac{7}{8}$ g $\frac{6}{10}$

2 a $\frac{8}{6} = 1\frac{2}{6}$ b $\frac{11}{10} = 1\frac{1}{10}$ c $\frac{18}{12} = 1\frac{6}{12}$ d $\frac{8}{4} = 2$ e $\frac{6}{5} = 1\frac{1}{5}$

3 a $\frac{5}{12}$ b $\frac{1}{2}$ c $\frac{2}{12}$ d $\frac{3}{9}$ e $\frac{5}{20}$ f $\frac{15}{100}$ g $\frac{14}{20}$

4 a $\frac{13}{12} = 1\frac{1}{12}$ b $\frac{5}{5} = 1$ c $\frac{14}{10} = 1\frac{4}{10}$ d $\frac{9}{8} = 1\frac{1}{8}$ e $\frac{8}{6} = 1\frac{2}{6}$ f $\frac{11}{8} = 1\frac{3}{8}$ g $\frac{4}{3} = 1\frac{1}{3}$ h $\frac{14}{12} = 1\frac{2}{12}$

5

	Mixed Numbers		Add	Subtract
a	$1\frac{3}{10}$	$1\frac{2}{10}$	$2\frac{5}{10}$	$\frac{1}{10}$
b	$1\frac{2}{12}$	$2\frac{5}{12}$	$3\frac{7}{12}$	$1\frac{3}{12}$
c	$3\frac{7}{8}$	$1\frac{3}{8}$	$5\frac{2}{8}$	$2\frac{4}{8}$
d	$3\frac{24}{100}$	$5\frac{32}{100}$	$8\frac{56}{100}$	$2\frac{8}{100}$
e	$1\frac{11}{12}$	$\frac{4}{12}$	$4\frac{3}{12}$	$1\frac{7}{12}$
f	$3\frac{5}{6}$	$1\frac{3}{6}$	$5\frac{2}{6}$	$2\frac{2}{6}$
g	$2\frac{3}{8}$	$5\frac{5}{8}$	8	$3\frac{2}{8}$
h	$3\frac{1}{4}$	$5\frac{3}{4}$	9	$2\frac{2}{4}$
i	$7\frac{3}{5}$	$5\frac{2}{5}$	13	$2\frac{1}{5}$

Add and Subtract Fractions with Different Denominators

Page 204 – Your Turn

a $\frac{2}{5} = \frac{4}{10}$
$\frac{7}{10} - \frac{4}{10} = \frac{3}{10}$

b $\frac{3}{5} = \frac{12}{20}$
$\frac{12}{20} - \frac{7}{20} = \frac{5}{20}$

Page 205 – Practice

1 a $\frac{9}{8} = 1\frac{1}{8}$ b $\frac{7}{10}$ c $\frac{7}{8}$ d $\frac{17}{20}$ e $\frac{17}{20}$

2 a $\frac{4}{20}$ b $\frac{2}{8}$ c $\frac{1}{12}$ d $\frac{5}{12}$ e $\frac{3}{8}$

3 a $4\frac{4}{10}$ b $3\frac{7}{10}$

Fractions Review Page 206

1 Sample answers:

a c e g b d f h

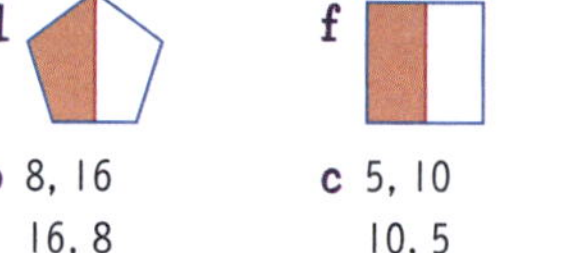

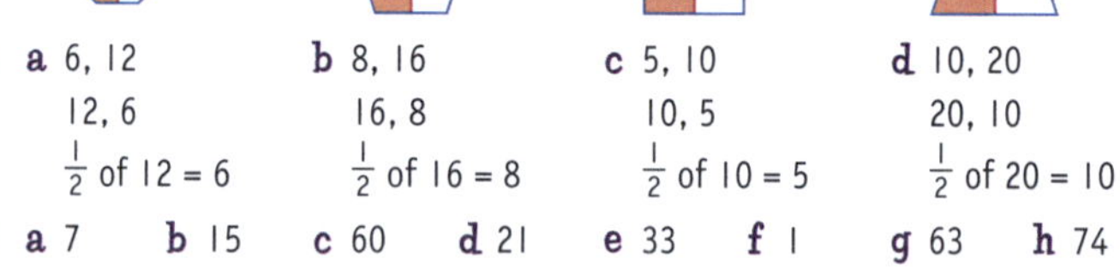

2 a 6, 12
12, 6
$\frac{1}{2}$ of 12 = 6

b 8, 16
16, 8
$\frac{1}{2}$ of 16 = 8

c 5, 10
10, 5
$\frac{1}{2}$ of 10 = 5

d 10, 20
20, 10
$\frac{1}{2}$ of 20 = 10

3 a 7 b 15 c 60 d 21 e 33 f 1 g 63 h 74

4 Colour red: c, d, e, g. Colour green: a, b, f

5 a Circle 1: $\frac{1}{8}$ of 8 = 1
b Circle 4: $\frac{1}{8}$ of 32 = 4
c Circle 3: $\frac{1}{8}$ of 24 = 3
d Circle 2: $\frac{1}{8}$ of 16 = 2

6 a Circle 4: $\frac{1}{4}$ of 16 = 4
b Circle 5: $\frac{1}{4}$ of 20 = 5
c Circle 7: $\frac{1}{4}$ of 28 = 7
d Circle 3: $\frac{1}{4}$ of 12 = 3

7 a 10 b 9 c 7 d 4 e 8

8 a 2 b 6 c 9 d 11 e 13

9 a 7 b 5 c 10 d 4 e 12 f 10 g 6 h 12

10 Colour red: a, e, f. Colour blue: b, c, d

11 a Circle 3: $\frac{1}{3}$ of 9 = 3
b Circle 5: $\frac{1}{3}$ of 15 = 5
c Circle 4: $\frac{1}{3}$ of 12 = 4

CATCH UP MATHS YEAR 6 BOOK A © PASCAL PRESS ISBN: 9781925726183

12 a Circle 3: $\frac{1}{5}$ of 15 = 3
b Circle 5: $\frac{1}{5}$ of 25 = 5
c Circle 7: $\frac{1}{5}$ of 35 = 7

13 a 6 b 1 c 4 d 8 e 12

14 a 10 b 1 c 7 d 12 e 16

15 a 2 b 6 c 10 d 15 e 9 f 9 g 8 h 11 i 10 j 4 k 12 l 20

16 a 6 b 5 c 6 d 6 e 2 f 14 g 10 h 4 i 6 j 4 k 5 l 1 m 4 n 3 o 1 p 4

17 a 1, 4, 2, 3, 5 b 3, 2, 5, 4, 1 c 4, 2, 3, 5, 1 d 5, 1, 3, 4, 2

18 a Number line: 0, $\frac{1}{5}$, $\frac{1}{4}$, $\frac{3}{8}$, $\frac{1}{2}$, $\frac{4}{4}$
b Number line: 0, $\frac{1}{10}$, $\frac{2}{5}$, $\frac{2}{3}$, $\frac{3}{4}$, $\frac{7}{8}$, 1

19 Adult to check

20 a $4\frac{1}{3}$ b $3\frac{3}{7}$ c $10\frac{9}{10}$ d $4\frac{1}{4}$ e $4\frac{2}{5}$ f $11\frac{1}{2}$ g $6\frac{4}{10}$ h $12\frac{2}{3}$ i $45\frac{1}{2}$ j $13\frac{1}{4}$ k $9\frac{2}{4}$ l $8\frac{3}{7}$ m $12\frac{3}{5}$ n $7\frac{1}{4}$ o $3\frac{9}{10}$ p $5\frac{2}{8}$

21 a $\frac{5}{3}$ b $\frac{53}{12}$ c $\frac{77}{9}$ d $\frac{13}{4}$ e $\frac{62}{8}$ f $\frac{70}{12}$ g $\frac{48}{5}$ h $\frac{31}{12}$ i $\frac{59}{5}$ j $\frac{35}{8}$ k $\frac{52}{5}$ l $\frac{68}{10}$ m $\frac{38}{3}$ n $\frac{73}{10}$ o $\frac{39}{4}$ p $\frac{114}{10}$

22 a $\frac{4}{8}$ b $\frac{4}{4}$ c $\frac{8}{8}$ d $\frac{5}{10}$ e $\frac{7}{10}$ f $\frac{7}{12}$ g $\frac{7}{8}$ h $\frac{3}{10}$

23 a $\frac{8}{12}$ b $\frac{6}{10}$ c $\frac{2}{8}$ d $\frac{4}{12}$ e $\frac{7}{10}$ f $\frac{6}{12}$ g $\frac{1}{8}$ h $\frac{2}{5}$

24 a $1\frac{4}{8}$ b $2\frac{5}{12}$ c $4\frac{2}{4}$ d $7\frac{5}{10}$ e $5\frac{3}{12}$ f $5\frac{2}{5}$ g $2\frac{3}{4}$ h $4\frac{4}{5}$ i $7\frac{6}{10}$ j $10\frac{9}{12}$ k $8\frac{3}{5}$ l $7\frac{6}{8}$

25 a $\frac{10}{20}$ b $\frac{2}{10}$ c $\frac{5}{10}$ d $\frac{12}{16}$ e $\frac{3}{10}$ f $\frac{4}{8}$ g $\frac{19}{20}$ h $\frac{8}{8}$ i $\frac{11}{20}$ j $\frac{8}{10}$ k $\frac{4}{10}$ l $\frac{11}{8} = 1\frac{3}{8}$

26 a $2\frac{2}{12}$ b $3\frac{3}{10}$ c $7\frac{4}{8}$ d $3\frac{7}{10}$ e $5\frac{5}{8}$ f $8\frac{8}{10}$

7 DECIMALS

Writing Decimals

Page 212 – Your Turn

1 a 1.46 b 2.37 c 0.43 d 18.62

2 a 3 b 8 c 0

3 a 6 b 2 c 0

4 a 6 b 2 c 0

Page 213 – Practice

1 a twelve point zero seven
b sixty-three point four five
c nineteen point three
d zero point three six
e seventy-eight point five nine

2 a 0.36, 5.62, 12.07, 19.3, 63.45, 78.59

3

	Tens	Ones	.	Tenths	Hundredths
a	6	5	.	0	9
b	3	0	.	7	2
c	0	8	.	0	6
d	7	4	.	1	2
e	5	0	.	0	9
f	1	3	.	5	8
g	4	2	.	4	9

4 74.12, 65.09, 50.09, 42.49, 13.58, 9.43, 8.06, 3.72

5 a 2 ones + 1 tenth + 7 hundredths
b 4 ones + 5 hundredths
c 1 ten + 3 ones + 8 tenths + 9 hundredths
d 1 ten + 7 ones + 5 tenths
e 8 tenths + 5 hundredths

Thousandths

Page 214 – Your Turn

1 a 206.1573 b 35.9824 c 5962.1374

2 a 6493.1275 b 147.2495 c 8753.6219

3 a 2468.1357 b 201.5946 c 67.3258

4 a 145.6032 b 34.9876 c 9082.4137

5 a 950.134 b 5490.376 c 4156.372

Page 215 – Practice

1 a ones b tens c tenths d hundreds e thousands f hundredths g tens

2

	Decimal	Thous.	Hund.	Tens	Ones	.	Tenths	Hund'ths	Thous'ths
a	247.376	0	2	4	7	.	3	7	6
b	5891.423	5	8	9	1	.	4	2	3
c	149.380	0	1	4	9	.	3	8	0
d	62.413	0	0	6	2	.	4	1	3
e	8.625	0	0	0	8	.	6	2	5
f	4963.820	4	9	6	3	.	8	2	0
g	853.621	0	8	5	3	.	6	2	1
h	1954.191	1	9	5	4	.	1	9	1
i	9.009	0	0	0	9	.	0	0	9
j	0.392	0	0	0	0	.	3	9	2

3 a 702.603 b 5743.026 c 300.004

4 a 14.926 b 5.493 c 73.498 d 0.826 e 0.492 f 0.109 g 0.660

Partitioning Decimals

Page 216 – Your Turn

a $5 + \frac{75}{100}$
b $2 + \frac{5}{10} + \frac{6}{100}$
c $10 + \frac{2}{10} + \frac{5}{100} + \frac{4}{1000}$
d $9 + \frac{3}{10} + \frac{4}{100}$
e $4 + \frac{19}{100}$
f $8 + \frac{3}{10} + \frac{5}{100} + \frac{4}{1000}$

Page 217 – Practice

1 a 493 + 6 tenths, 2 hundredths, 1 thousandth, $493 + \frac{6}{10} + \frac{2}{100} + \frac{1}{1000}$
b 5243 + 781 thousandths, $5243 + \frac{781}{1000}$
c 6 + 73 hundredths + 8 thousandths, $6 + \frac{73}{100} + \frac{8}{1000}$

7 DECIMALS CONTINUED

2 a 8.043, $8\frac{43}{1000}$
b 16.475, 16, 4, 7, 5
c 9.742, $9\frac{742}{1000}$
d $5\frac{37}{1000}$, 5, 0, 3, 7
e 6.201, $6\frac{201}{1000}$
f 11.230, 11, 2, 3, 0
g $23\frac{573}{1000}$, 23, 5, 7, 3

3 a 9.342, $9\frac{342}{1000}$
b 15.642, 15, $\frac{642}{1000}$
c 10.999, $10\frac{999}{1000}$
d 37.109, $37\frac{109}{1000}$
e $5\frac{337}{1000}$, 5, $\frac{337}{1000}$
f 16.673, $16\frac{673}{1000}$
g 1.541, 1, $\frac{541}{1000}$

4 a 24.357 b 4937.569 c 53.027 d 702.064 e 836.43
f 5.35 g 17.008 h 19.47 i 27.009 j 86.301
k 93.04 l 184.438 m 109.257 n 246.228 o 6290.478
p 18.008 q 124.05 r 72.077 s 150.809

Adding and Subtracting Decimals

Page 219 – Your Turn

a 19.148 b 67.162 c 1.525 d 66.574 e 84.052

Page 220 – Practice

1 a 1915.78 b 37 689.06 c 4036.05 d 537.766 e 2426.785 f 8693.198
2 a 157.722 b 216.849 c 871.567 d 307.819
3 a 7391.923 b 647.941 c 241.016 d 809.243 e 318.179 f 5921.335 g 173.229 h 1558.58
4 a 1.021 b 86.72 c 4709.349 d 834.872 e 4481.957 f 7306.763

Multiplying Decimals

Page 222 – Your Turn

1 a 59.2 b 7.180 c 99.4
2 a 36.525 b 164.626

Page 223 – Practice

1

	x	3	2	9	12
a	1.3	3.9	2.6	11.7	15.6
b	2.25	6.75	4.5	20.25	27
c	4.32	12.96	8.64	38.88	51.84
d	5.535	16.605	11.07	49.815	66.42
e	1.116	3.348	2.232	10.044	13.392

	x	6	8	7	13
a	1.3	7.8	10.4	9.1	16.9
b	2.25	13.5	18	15.75	29.25
c	4.32	25.92	34.56	30.24	56.16
d	5.535	33.21	44.28	38.745	71.955
e	1.116	6.696	8.928	7.812	14.508

2 a $48.96 b $130.76 c $143.79 d $114.05

3 a 14.92 m b 91.92 m c 230.4 m d 1158.78 m e 1325.52 m f 5743.92 m

Dividing Decimals

Page 224 – Your Turn

a 26.3 b 2.154 c 52.14 d 11.5 e 15.47 f 2.587 g 6.528 h 6.523 i 0.874 j 1.258 k 125.8

Page 225 – Practice

1 a 7.5 b 17.35 c 502.15 d 575.1 e 11.111 f 569.2 g 36.2 h 177.56 i 36.52 j 235.6 k 9.65

2 a 7.5 × 5 = 37.5
b 17.35 × 9 = 156.25
c 502.15 × 5 = 2510.75
d 575.1 × 7 = 4025.7
e 11.111 × 8 = 88.888
f 569.2 × 8 = 4553.6
g 36.2 × 7 = 253.4
h 177.56 × 6 = 1065.36
i 36.52 × 6 = 219.12
j 235.6 × 9 = 2120.4
k 9.65 × 7 = 67.55

Dividing Decimals by Powers of 10

Page 226 – Your Turn

a 27.9 b 0.862 c 0.001 932 d 6300 e 849.3 f 0.0009 g 0.001 825

Page 227 – Practice

1 a 314, 3140, 31 400
b 1227, 12 270, 122 700
c 99, 990, 9900
d 5.4, 54, 540
e 12.46, 124.6, 1246
f 387.20, 3872.0, 38 720
g 3.29, 32.9, 329.0
h 1798.55, 17 985.5, 179 855

2 a 2.13, 0.213, 0.0213
b 80.24, 8.024, 0.8024
c 0.09, 0.009, 0.0009
d 0.037, 0.0037, 0.000 37
e 0.0994, 0.009 94, 0.000 994
f 1.4816, 0.148 16, 0.014 816
g 0.5049, 0.050 49, 0.005 049
h 73.6863, 7.368 63, 0.736 863

3 a 902 b 300 c 0.173 d 0.830 205 e 0.675 f 62.03 g 360.3 h 0.703 24 i 6.099 09

Percentages

Page 228 – Your Turn

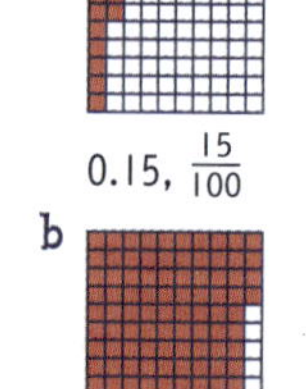

a 0.15, $\frac{15}{100}$

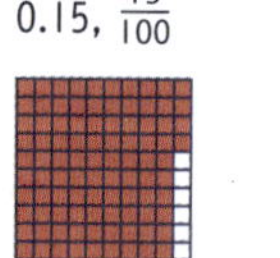

b 0.94, $\frac{94}{100}$

c 0.57, $\frac{57}{100}$

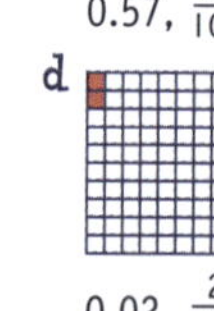

d 0.02, $\frac{2}{100}$

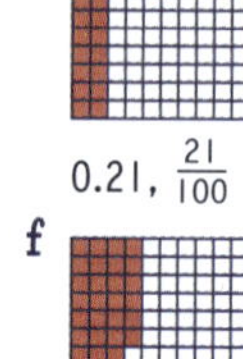

e 0.21, $\frac{21}{100}$

f 0.36, $\frac{36}{100}$

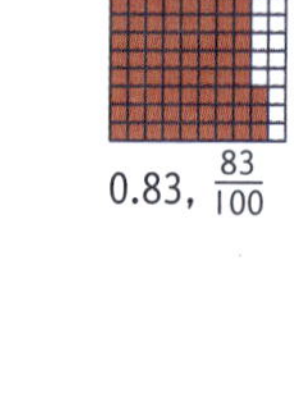

g 0.83, $\frac{83}{100}$

Page 229 – Practice

1 a 62% b 90% c 14% d 27% e 44% f 53% g 79%
2 a 64% b 99% c 17% d 22% e 41% f 56% g 77%
3 a 0.20, 20% b 0.3, 30% c 0.4, 40% d 0.5, 50% e 0.6, 60% f 0.7, 70% g 0.8, 80% h 0.9, 90% i 1.0, 100% j 0.25, 25% k 0.50, 50% l 0.75, 75% m 0.20, 20% n 0.33, $33\frac{1}{3}$% o 0.125, 12.5%
4 a 20c b 25c c $1.00 d $1.50 e $1.80 f 50c g $4.00 h $12.00 i $16.00 j $120.00 k $33.33 l $1.33 m $1.25

CATCH UP MATHS YEAR 6 BOOK A © PASCAL PRESS ISBN: 9781925726183

ANSWERS

Equivalent Fractions, Decimals and Percentages

Page 230 – Your Turn

1 a 0.24 b 0.015

2 a $\frac{127}{100}$ b $\frac{61}{100}$

3 a 103% b 801%

Page 231 – Practice

1 a $\frac{83}{100}$, 83% b $\frac{76}{100}$, 0.76 c 0.03, 3% d $\frac{49}{100}$, 0.49 e $\frac{50}{100}$, 50% f 0.10, 10% g $\frac{13}{100}$, 0.13 h 1.49, 149% i $\frac{262}{100}$, 262% j $\frac{501}{100}$, 5.01 k 8.30, 830% l $\frac{487}{100}$, 487% m $\frac{990}{100}$, 9.90 n $\frac{109}{100}$, 109% o 7.47, 747%

2 a 295% b 436% c 80% d 700% e 310% f 895% g 606%

3 a 7.17 b 1.24 c 0.33 d 0.01 e 6.06 f 0.07 g 7.36

4 a $\frac{229}{100}$ b $\frac{43}{100}$ c $\frac{6}{100}$ d $\frac{440}{100}$ e $\frac{386}{100}$ f $\frac{591}{100}$ g $\frac{1680}{100}$ h $\frac{72}{100}$ i $\frac{780}{100}$ j $\frac{600}{100}$ k $\frac{2395}{100}$ l $\frac{188}{100}$ m $\frac{99}{100}$ n $\frac{63}{100}$ o $\frac{489}{100}$

Discounts

Page 232 – Your Turn

a 170 × 25% = $42.50
0.25 × 170 = $42.50
$170 – $42.50 = $127.50

b 15 × 20% = $3
0.2 × 15 = $3
$15 – $3 = $12

Page 233 – Practice

1 a $12.50, $112.50 b $11.25, $33.75 c $457.20, $1066.80 d $199.50, $199.50 e $62, $93 f $9.80, $39.20

2 Dress Me $91

3 Dom

Item	Original price	Disc.	Disc. price
T-shirt	$45.00	10%	$40.50
Jeans	$50.00	20%	$40.00
Shoes	$150.00	30%	$105.00
Total:	$245.00	Total:	$185.50

Livvy

Item	Original price	Disc.	Disc. price
T-shirt	$50.00	20%	$40.00
Jeans	$60.00	30%	$42.00
Shoes	$170.00	20%	$136.00
Total:	$280.00	Total:	$218.00

a Livvy b $59.50 c $62

Decimals Review Page 234

1 a zero point five eight
b two point six three
c eight point four nine five

2 a 8.07 b 5034.236 c 4.568 d 0.334 e 5000.05

3 a 2 ones + 7 tenths + 2 hundredths
b 3 tenths + 9 hundredths
c 9 ones + 1 tenth
d 5 tens + 7 ones + 4 tenths + 8 hundredths + 5 thousandths

4 a tenths b tens c hundredths d thousands e ones f tens g hundreds h thousandths i thousands j tens

5 a 146.437 b 33.03 c 302.657 d 16.008 e 857.069 f 9.205 g 8.709 h 41.362

6 a 559.46 b 997.13 c 459.68 d 2095.11 e 222.091 f 553.225 g 5258.333

7 a 51.499 b 575.855

8 a 2412.7 b 6504.221 c 351.755

9 a 81.331 b 5711.774 c 696.675

10 a 7.492 b 26.262 c 16.584 d 64.078 e 28.29 f 24.941

11 a 639.53 b 652.3 c 158.56 d 147.25 e 369.258 f 125.256 g 521.40 h 147.258 i 123.456

12 a 5755.77 b 2609.2 c 1109.92 d 883.50 e 1846.290 f 876.792 g 1564.20 h 1178.064 i 1111.104

13 a 6, 60, 600
b 154, 1540, 15 400
c 1845.9, 18 459, 184 590
d 44 325.73, 443 257.3, 4 432 573
e 8.95, 89.5, 895
f 0.08, 0.008, 0.0008
g 3.647, 0.3647, 0.036 47
h 16.486, 1.6486, 0.164 86
i 598.1246, 59.812 46, 5.981 246
j 0.0458, 0.004 58, 0.000 458

14 a 931 b 7311.6 c 8097 d 0.245 e 5.842 51 f 0.000 953 g 1756.9 h 1.495 231 i 0.0702 j 200 k 5.924 l 0.64

15 a 749% b 520% c 895% d 60% e 800% f 410% g 637% h 704%

16 a 5.43 b 1.04 c 5.6 d 8.56 e 9.1 f 7.9 g 0.08 h 0.17

17 a $\frac{495}{100}$ b $\frac{519}{100}$ c $\frac{820}{100}$ d $\frac{15}{100}$ e $\frac{711}{100}$ f $\frac{2}{100}$ g $\frac{437}{100}$ h $\frac{48}{100}$

18 a $\frac{21}{100}$, 21% b 0.34, 34% c $\frac{57}{100}$, 0.57 d $\frac{40}{100}$, 40% e 0.79, 79% f $\frac{69}{100}$, 0.69 g $\frac{157}{100}$, 1.57 h $\frac{506}{100}$, 506%

19 a 31, 100 b 72, 100 c 544, 100 d 801, 100

20 a 0.3, 30% b 0.25, 25% c 0.75, 75% d 0.4, 40% e 0.67, 66.67% f 0.125, 12.5% g 1.0, 100% h 0.2, 20%

21 a 20c b $2.50 c $7.50 d $60.00 e $7.00 f $8.50 g $20.00 h $88 i $3.33 j $72

22 a $19, $171 b $30, $30 c $36, $84 d $84, $56 e $15, $60 f $63, $27 g $18, $27

23 Max

Item	Original price	Disc.	Disc. price
Dress	$120.00	40%	$72.00
Shoes	$140.00	30%	$98.00
Hat	$55.00	20%	$44.00
Jacket	$190.00	30%	$133.00
Total:	$505.00	Total:	$347.00

Zoe

Item	Original price	Disc.	Disc. price
Dress	$150.00	50%	$75.00
Shoes	$160.00	20%	$128.00
Hat	$40.00	10%	$36.00
Jacket	$180.00	20%	$144.00
Total:	$530.00	Total:	$383.00

a Zoe b $505 c $530 d $158 e $147

8 PATTERNS AND ALGEBRA

Number Patterns

Page 240 – Your Turn

1 a 56, 49, 35 b 25, 21, 9 c 5, 125, 3125

2 a – 9 b + 4 c ÷ 2 d × 3

Page 241 – Practice

1 a 9, 17, 25 b 99, 95, 93 c 1356, 1372, 1380 d 12 503, 12 471, 12 423

2 a 53, 44, 35. Rule: – 9
b 625, 3125, 15 625. Rule: × 5
c 183, 179, 175. Rule: – 4
d 1024, 512, 256. Rule: ÷ 2
e 6, 24, 192. Rule: × 2

3 a H = G × I b C = A ÷ B c L = J – K

4 a 7, 7, 6, 8 b 8, 12, 10, 15 c 12, 13, 15, 16 d 16, 11, 20, 13 e 10, 20, 5

8 PATTERNS AND ALGEBRA CONTINUED

Pattern Grids

Page 242 – Your Turn

a 20, 30, 40, 50, 60, 70
b 8, 14, 20, 26, 32, 38
c 0, 4, 8, 12, 16, 20

Page 243 – Practice

1 a 21, 26, 31, 36, 41, 46
b 38, 40, 42, 44, 46, 48
c 54, 51, 48, 45, 42, 39

2 a 4, 2 b 3, 4 c 3, 2

3 a

Rule	1	3	5	7	9	11
– 5	–4	–2	0	2	4	6
× 2	2	6	10	14	18	22

b

Rule	5	6	7	8	9	10
× 2	10	12	14	16	18	20
+ 4	9	10	11	12	13	14
– 6	–1	0	1	2	3	4
× 11	55	66	77	88	99	110

c

Rule	5	6	7	8	9	10
× 1	5	6	7	8	9	10
+ 8	13	14	15	16	17	18
– 10	–5	–4	–3	–2	–1	0
÷ 2	2.5	3	3.5	4	4.5	5

d

Rule	6	8	10	12	14	16
× 3	18	24	30	36	42	48
+ 15	21	23	25	27	29	31
÷ 2	3	4	5	6	7	8
× 10	60	80	100	120	140	160
– 6	0	2	4	6	8	10

Equivalent Number Sentences

Page 244 – Your Turn

Tick a, c, d, e

Page 245 – Practice

1 a 2 b 7 c 8 d 17 e 45 f 2 g 2 h 48 i 8 j 3 k 3

2 Adult to check

3 a 8 b 9 c 3 d 9 e 5 f 10 g 8 h 12 i 38 j 4 k 4

Patterns with Fractions and Decimals

Page 246 – Your Turn

a $\frac{1}{32}, \frac{3}{4}, \frac{1}{64}, \frac{2}{4}$
b 1.6, 2.0, 1.9, 1.8
c 11.1, 11.3, 11.5, 11.7

Page 247 – Practice

1 a $\frac{5}{4}, \frac{6}{4}, \frac{7}{4}, \frac{8}{4}$
b $\frac{5}{8}, \frac{6}{8}, \frac{7}{8}, \frac{8}{8}$
c $\frac{5}{6}, \frac{6}{6}, \frac{7}{6}, \frac{8}{6}$
d $\frac{9}{12}, \frac{11}{12}, \frac{13}{12}, \frac{15}{12}$
e $2\frac{2}{3}, 3, 3\frac{1}{3}, 3\frac{2}{3}$

2 a $\frac{3}{5}, \frac{5}{12}, \frac{2}{5}, \frac{4}{12}$ b $\frac{6}{8}, \frac{1}{4}, \frac{5}{8}, 0$

3 a $\frac{4}{3}, \frac{5}{3}, \frac{6}{3}, \frac{7}{3}$ b $\frac{5}{7}, \frac{4}{7}, \frac{3}{7}, \frac{2}{7}$ c $\frac{1}{16}, \frac{1}{32}, \frac{1}{64}, \frac{1}{128}$

4 a 31.7, 31.5, 31.3, 31.1
b 10.1, 10.5, 10.9, 11.3
c 6.8, 6.2, 5.6, 5.0
d 65.6, 131.2, 262.4, 524.8
e 54.9, 45.9, 36.9, 27.9

5 a 6.3, 4.9, 3.5, 2.1
b 86.4, 259.2, 777.6, 2332.8

Terms in Sequences

Page 248 – Your Turn

1 a 10, 30 b 6, 18 c 9, 27 d 20, 30 e 1, 3 f 7, 21 g 11, 33

2 a 20 b 12 c 18 d 25 e 2 f 14 g 22

Page 249 – Practice

1 a 5th b 2nd c 1st d 3rd e 4th

2 a 32, 36
b 35, 40, 45
c 90, 72, 54
d 96, 108, 132
e 63, 70, 77

3 a Increase, 4
b Increase, 5
c Decrease, 9
d Increase, 12
e Increase, 7

4 a 23, 19, 15, 11
b 3.9, 3.5, 3.1, 2.7
c 2.6, 3.2, 3.8, 4.4
d 10, 13, 16, 19
e 4.5, 3.8, 3.1, 2.4

5 a 32, 24, 16 b 45, 36, 27 c 47, 57, 67

6 a 12, 15, 18, 21, 24, 27
b 30
c 45
d 60

The Cartesian Number Plane

Page 250 – Your Turn

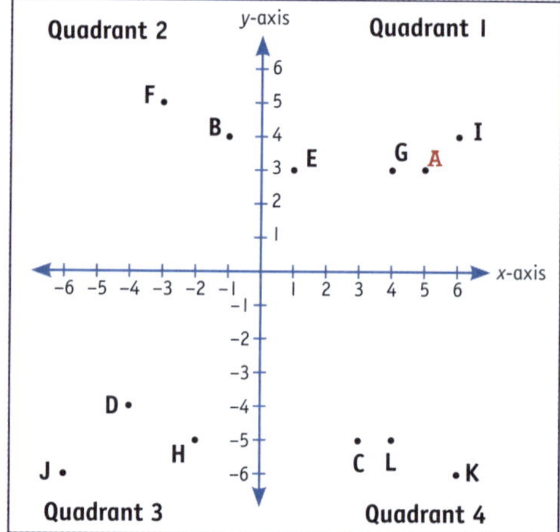

Page 251 – Practice

1

	Ordered pair	Quadrant
A	(1,6)	1
B	(1,–3)	4
C	(–1,1)	2
D	(6,1)	1
E	(–2,3)	2
F	(–2,–4)	3
G	(–4,–2)	3
H	(–2,–1)	3
I	(4,–2)	4
J	(6,–6)	4

	Ordered pair	Quadrant
K	(3,–4)	4
L	(2,1)	1
M	(4,3)	1
N	(6,5)	1
O	(–4,2)	2
P	(–6,6)	2
Q	(–5,4)	2
R	(2,–2)	4
S	(–2,–3)	3
T	(–6,–5)	3

CATCH UP MATHS YEAR 6 BOOK A © PASCAL PRESS ISBN: 9781925726183

2

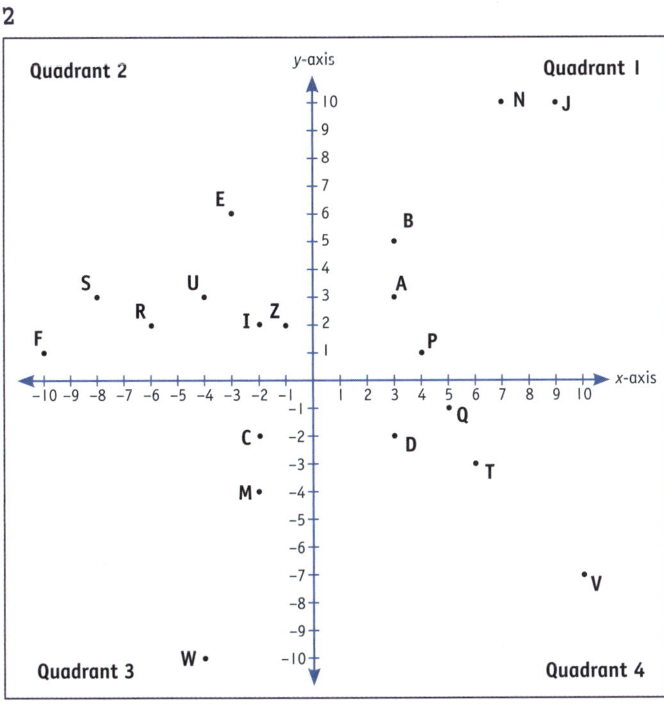

Patterns & Algebra Review Page 252

1 a 35, 175, 875, 4375, 21 875
b 10, 1, 0.1, 0.01, 0.001
c 20, 26, 32, 38, 44
d 42, 35, 28, 21, 14
e 6, 12, 24, 48, 96
f 12, 48, 192, 768, 3072
g 12, 72, 432, 2592, 15 552
h 121, 226, 331, 436, 541

2 a + 7
b − 3
c × 2
d − 5
e − 0.5
f + 0.25
g − 6
h + 7
i − 0.15
j + 0.1

3 a 10, 4, 11, 3, 12, 2
b 27, 7, 29, 9, 31, 11
c 12, 16, 11, 19, 10, 22
d 17, 18, 22, 22, 27, 26
e $2\frac{1}{4}, \frac{7}{8}, 2\frac{1}{2}, \frac{9}{8}, 2\frac{3}{4}, \frac{11}{8}$
f 19, 36, 24, 45, 29, 54
g 32, 16, 36, 12, 40, 8

4 a ÷ b × c − d +

5 a 7, 13, 19, 25, 31, 37, 43, 49, 55, 61, 67
b 27, 33, 39, 45, 51, 57, 63, 69, 75, 81, 87
c 85, 77, 69, 61, 53, 45, 37, 29, 21, 13, 5
d 3, 18, 33, 48, 63, 78, 93, 108, 123, 138, 153

6 a ÷ 2, + 4 b × 3, + 4 c × 3, + 5 d ÷ 2, − 3

7 a

Rule	5	6	7	8	9	10
× 2	10	12	14	16	18	20
+ 3	8	9	10	11	12	13
− 5	0	1	2	3	4	5
× 8	40	48	56	64	72	80
− 3	2	3	4	5	6	7

b

Rule	10	9	8	7	6	5
× 3	30	27	24	21	18	15
+ 4	14	13	12	11	10	9
− 6	4	3	2	1	0	−1
× 9	90	81	72	63	54	45
+ 5	15	14	13	12	11	10

8 a 6
b 7
c 4
d 24
e 11
f 9
g 11
h 1
i 9
j 27

9 Adult to check

10 a $\frac{8}{8}, \frac{10}{8}, \frac{12}{8}, \frac{14}{8}$
b $\frac{4}{4}, \frac{2}{4}, 0, -\frac{2}{4}$
c $\frac{4}{3}, \frac{5}{3}, \frac{6}{3}, \frac{7}{3}$
d 1.2, 1.5, 1.8
e 7.8, 8.0, 8.2
f 7.2, 6.7, 6.2

11 a 5.8, 4.6, 5.6, 4.9, 5.4, 5.2
b 9.6, 13.2, 10, 13.5, 10.4, 13.8

12 a 2.9, 3.2, 3.5, 3.8
b $1\frac{3}{8}, 1\frac{4}{8}, 1\frac{5}{8}, 1\frac{6}{8}$
c 8.4, 7, 5.6, 4.2
d $3\frac{2}{3}, 3\frac{1}{3}, 3, 2\frac{2}{3}$
e 2.2, 2.7, 3.2, 3.7
f 5.4, 4.6, 3.8, 3.0

13 a 1, 3, 5, 7, 9
b 2, 4, 6, 8, 10
c 25, 20, 15, 10, 5
d 12, 9, 6, 3, 0
e 7, 17, 27, 37, 47
f 24, 26, 28, 30, 32

14 a 4th
b 2nd
c 1st
d 4th
e 2nd
f 5th
g 3rd
h 5th
i 2nd
j 4th

15 a Increase, 2
b Decrease, 10
c Increase, 4
d Increase, 7
e Decrease, 4
f Increase, 7
g Increase, 6
h Decrease, 1.2
i Increase, 0.4
j Increase, 15

16 a 20, 25, 30 b 24, 30, 36 c 40, 32, 24 d 80, 70, 60

17 a 28, 35, 42, 49, 56, 63
b 70
c 140
d 700

18 a 89, 83, 77, 71, 65, 59
b 116, 123, 130, 137, 144, 151
c 45, 33, 21, 9, −3, −15
d 11.75, 11.5, 11.25, 11, 10.75, 10.5
e 4.5, 3.9, 3.3, 2.7, 2.1, 1.5
f 95, 101, 107, 113, 119, 125
g 49, 45, 41, 37, 33, 29
h 18, 11, 4, −3, −10, −17
i $9\frac{3}{4}, 9\frac{1}{2}, 9\frac{1}{4}, 9, 8\frac{3}{4}, 8\frac{1}{2}$
j 71.1, 70.7, 70.3, 69.9, 69.5, 69.1
k $5\frac{2}{4}, 5\frac{1}{4}, 5, 4\frac{3}{4}, 4\frac{2}{4}, 4\frac{1}{4}$
l 1.60, 1.55, 1.50, 1.45, 1.40, 1.35
m 94.5, 97, 99.5, 102, 104.5, 107
n 16.84, 17.04, 17.24, 17.44, 17.64, 17.84
o 91, 90.5, 90, 89.5, 89, 88.5

19

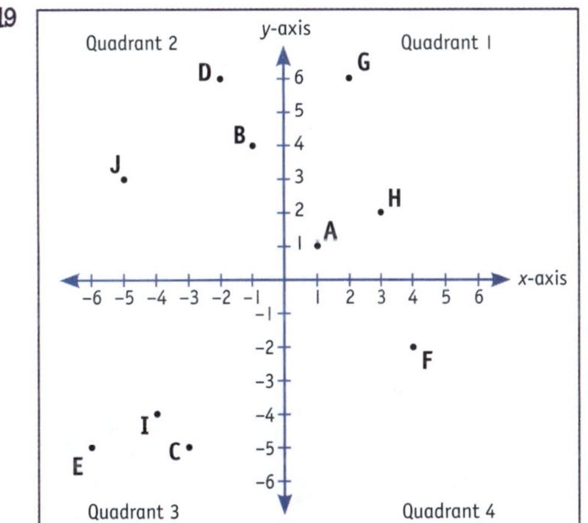

20

	Ordered pair	Quadrant
Z	(1,2)	1
X	(−4,−4)	3
A	(1,−8)	4
S	(−4,3)	2
C	(6,−2)	4
K	(8,5)	1
J	(−1,7)	2
Y	(−2,−1)	3
D	(2,−4)	4
B	(−7,−8)	3
L	(5,−5)	4
M	(6,1)	1
T	(−7,1)	2
R	(3,7)	1

9 CHANCE

Probability

Page 258 – Your Turn

$\frac{0}{100}$	$\frac{10}{100}$	$\frac{20}{100}$	$\frac{30}{100}$	$\frac{40}{100}$	$\frac{50}{100}$	$\frac{60}{100}$	$\frac{70}{100}$	$\frac{80}{100}$	$\frac{90}{100}$	$\frac{100}{100}$
0	0.1	0.2	0.3	0.4	0.5	0.6	0.7	0.8	0.9	1
0%	10%	20%	30%	40%	50%	60%	70%	80%	90%	100%

Page 259 – Practice

1 a 0.33 b 0.7 c 0.5 d 1.0 e 0 f 0.5 g 0.8

2 a 0.5 b 0.1 c 0.5 d 0.3 e 0.3 f 0.7

3 Adult to check

Outcomes

Page 260 – Your Turn

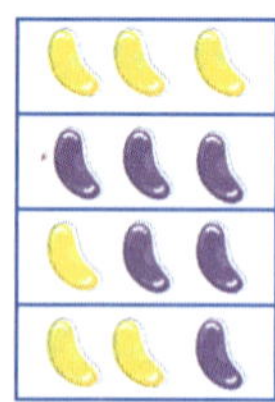

Page 261 – Practice

1 a Adult to check b Adult to check c 24

2 RPBY, RYBP, PBRY, BYRP, BPYR, YPBR,
RBYP, RPYB, PBYR, BYPR, BPRY, YPRB,
RYPB, PRBY, PYRB, BRPY, YRPB, YBRP,
RBPY, PRYB, PYBR, BRYP, YRBP, YBPR

Equally Likely Outcomes

Page 262 – Your Turn

Adult to check.
There should be the same number of each colour in each bag.

Page 263 – Practice

1 Sample answers:

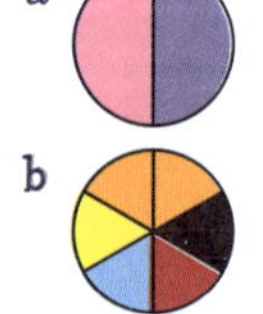

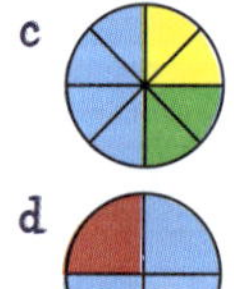

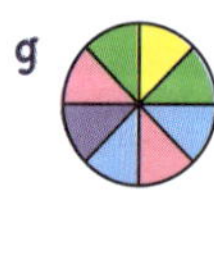

2 Adult to check.
There should be the same number of each colour in each jar.

3 a ✓ b ✗ c ✓ d ✓ e ✗

Unequal Chances

Page 264 – Your Turn

a unequal b equal c equal d unequal e unequal

Page 265 – Practice

1 a $\frac{4}{10}$ b $\frac{3}{10}$ c $\frac{1}{10}$ d $\frac{1}{10}$ e 0

2 a 40 b 30 c 10 d 10 e 0

3 a 5, 20, $\frac{5}{20}$, unequal c 2, 20, $\frac{2}{20}$, unequal
b 3, 20, $\frac{3}{20}$, unequal d 4, 20, $\frac{4}{20}$, equal

4 a orange b yellow

Probability As Decimals, Fractions And Percentages

Page 266 – Your Turn

	Marble Colour	Probability of picking out a colour		
		Decimals	Fractions	Percentages
	Orange	0.15	$\frac{2}{20}$	15%
a	Purple	0.30	$\frac{6}{20}$	30%
b	Blue	0.25	$\frac{5}{20}$	25%
c	Green	0.20	$\frac{4}{20}$	20%
d	Pink	0.10	$\frac{2}{20}$	10%

Page 267 – Practice

1 a 0.3, $\frac{30}{100}$, 30% e 0.8, $\frac{80}{100}$, 80% i 0.75, $\frac{75}{100}$, 75%
b 0.5, $\frac{50}{100}$, 50% f 1.0, $\frac{100}{100}$, 100% j 0, $\frac{0}{100}$, 0%
c 0.4, $\frac{40}{100}$, 40% g 0.2, $\frac{20}{100}$, 20%
d 0.6, $\frac{60}{100}$, 60% h 0.25, $\frac{25}{100}$, 25%

2 a 12 d 9
b 3 e 12
c 18 f 0.2, 0.05, 0.3, 0.15, 0.2

Chance Review Page 268

1 a 0.5, $\frac{1}{2}$, 50% d 0.33, $\frac{1}{3}$, 33% g 0.1, $\frac{1}{10}$, 10% j 0.2, $\frac{2}{10}$, 20%
b 0.5, $\frac{1}{2}$, 50% e 0.4, $\frac{4}{10}$, 40% h 0.25, $\frac{1}{4}$, 25%
c 0.33, $\frac{1}{3}$, 33% f 0, 0, 0% i 0.75, $\frac{3}{4}$, 75%

2 RBGY, RYBG, BGYR, GYRB, GBYR, YBRG, RBYG, RYGB, BGRY, GYBR, GBRY, YBGR, RGBY, BRGY, BYRG, GRBY, YRBG, YGRB, RGYB, BRYG, BYGR, GRYB, YRGB, YGBR

3

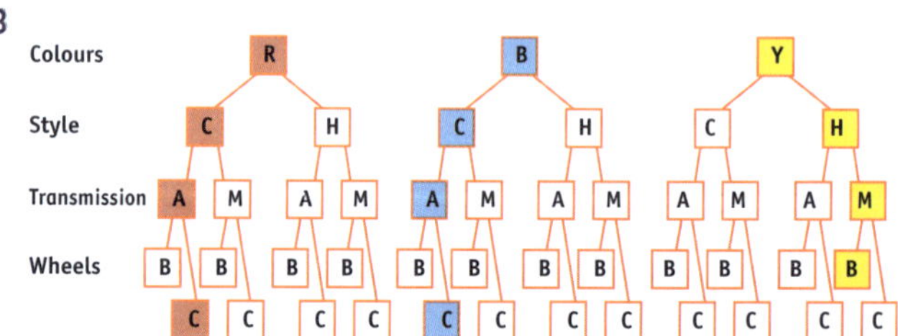

4 Circle **b, c, e**

5 Sample answers:

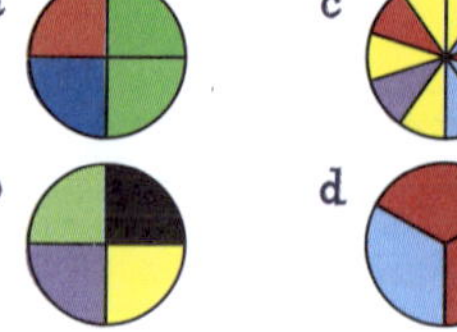

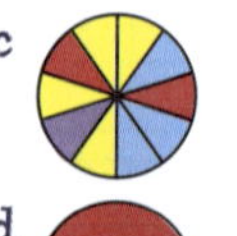

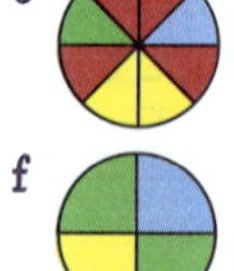

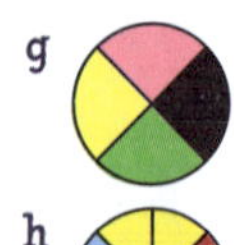

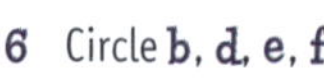

6 Circle **b, d, e, f**

7 a 5, 20, $\frac{5}{20}$ b 7, 20, $\frac{7}{20}$ c 1, 20, $\frac{1}{20}$ d Green e Orange

8 a 4 b 6 c 8 d 12 e 10

CATCH UP MATHS YEAR 6 BOOK A © PASCAL PRESS ISBN: 9781925726183

9

	Fraction	Decimal	Percentage
a	$\frac{30}{100}$	0.3	30%
b	$\frac{100}{100}$	1.0	100%
c	$\frac{70}{100}$	0.70	70%
d	$\frac{60}{100}$	0.60	60%
e	$\frac{90}{100}$	0.90	90%
f	$\frac{25}{100}$	0.25	25%
g	$\frac{40}{100}$	0.40	40%
h	$\frac{90}{100}$	0.90	90%

	Fraction	Decimal	Percentage
i	$\frac{10}{100}$	0.10	10%
j	$\frac{33}{100}$	0.33	33%
k	$\frac{50}{100}$	0.50	50%
l	$\frac{20}{100}$	0.2	20%
m	$\frac{80}{100}$	0.80	80%
n	$\frac{75}{100}$	0.75	75%
o	$\frac{70}{100}$	0.70	70%
p	$\frac{67}{100}$	0.67	67%

10

Jellybeans	Fraction	Decimal	Percentage
Red	$\frac{5}{20} = \frac{1}{4}$	0.25	20%
Blue	$\frac{5}{20} = \frac{1}{4}$	0.25	25%
Orange	$\frac{2}{20} = \frac{1}{10}$	0.1	10%
Green	$\frac{8}{20} = \frac{4}{10}$	0.4	40%

10 DATA

Tables

Page 272 – Your Turn

6T's Favourite Food

Food	Total
Burgers	4
Pasta	5
Pizza	10
Sushi	9
Tacos	8
	36

a 9
b burgers
c pizza
d 6

Page 274 – Practice

1 a 28
b Dance
c Classical
d 6
e 7
f 1

2

Gummy Bears

Colour	Tally	Total
Black	||||	4
Blue	𝍸 𝍸	10
Yellow	𝍸 |||	8
Green	𝍸 |	6
Red	𝍸	5
Orange	𝍸 ||	7
Purple	𝍸 |	6
Pink	||||	4
		50

3 a 8
b 6
c 6
d black, pink
e 50

Two-Way Tables

Page 275 – Your Turn

a

Takeaway choice	People		Total
	Boys	Girls	
Pizza	3	2	5
Burger	5	3	8
Hot chips	2	6	8
Kebab	4	3	7
Sushi	6	4	10
Total	20	18	38

b 20
c 38
d sushi
e hot chips
f hot chips

Page 276 – Practice

1 a email
b phonecall
c women
d men
e women
f men
g 200

2 a 52
b 116
c less
d 3
e 12

3 a males
b ice hockey
c basketball
d golf
e golf
f 21
g 20
h golf

4 a afternoon
b grouper
c cardinalfish
d 34
e 167
f parrotfish

Picture Graphs

Page 278 – Your Turn

a June
b 60
c March
d April
e 180
f 190

Page 280 – Practice

1 a 6500
b 2021
c 2018
d 4000
e 30 000

2 a Cookies and cream
b 2
c 3
d Cookies and cream
e Cookies and cream
f 4
g 20
h Adult to check

Column Graphs

Page 281 – Your Turn

a Thursday
b 37
c 4

Page 283 – Practice

1 a Seals
b 1
c 34
d 15

2 a Students at Annie's Dance School

Dance style	Group		Total
	Boys	Girls	
Jazz	6	2	8
Hip Hop	12	12	24
Tap	11	7	18
Ballet	9	6	15
Country	5	10	15
Total	43	37	80

b Boys
c Hip Hop
d Jazz
e 5
f 3
g 43
h 37

Divided Bar Graphs

Page 284 – Your Turn

a art
b less
c more
d less

Page 285 – Practice

1

Clothes	Shoes	Cap

a $100
b $50
c $50

10 DATA CONTINUED

2 a For example:
Pets Owned by 6C
b Cats and dogs
c Less
d More
e 5
f 90, 30, 60

3 From left to right: 16, 16, 8, 8, 8, 4, 4

Dot Plots

Page 287 – Your Turn

a 53 b 8 c 28 d 6

Page 288 – Practice

1 a 75 b 66 c 120 d 160 e 204

2

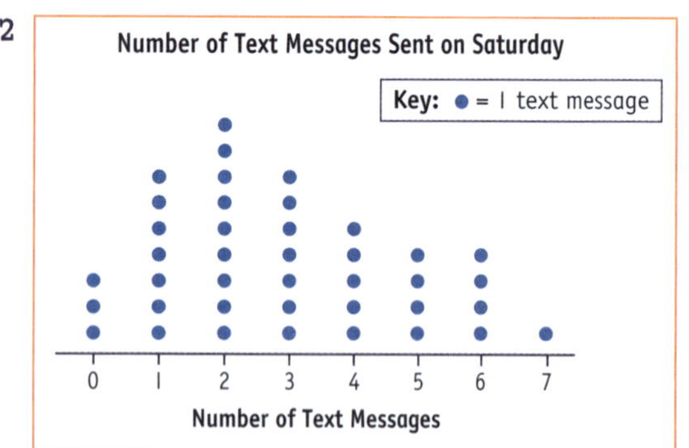

a 7
b 3
c 4
d 2

Line Graphs

Page 289 – Your Turn

a

Temperature on Saturday 12 June

Time	10 am	11 am	12 pm	1 pm	2 pm	3 pm	4 pm	5 pm	6 pm	7 pm	8 pm	9 pm	10 pm	11 pm	12 am
Temp. (°C)	10	14	16	20	18	16	15	13	10	9	9	6	4	2	1

b May c 350 d 1150 e 700 f 250 g 3350

Page 291 – Practice

1

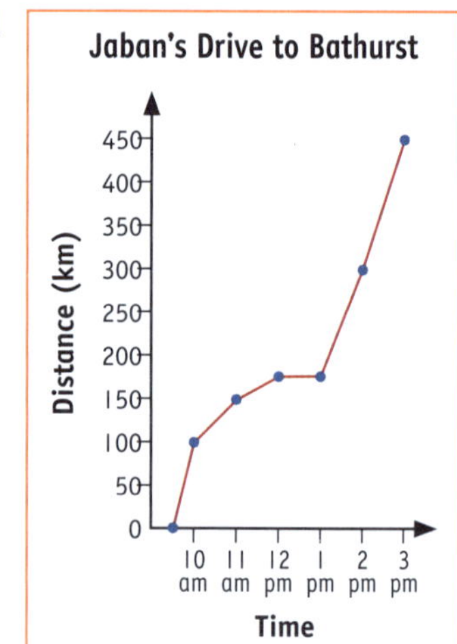

a 12 pm–1 pm
b 2 pm–3 pm
c 450 km

2 a

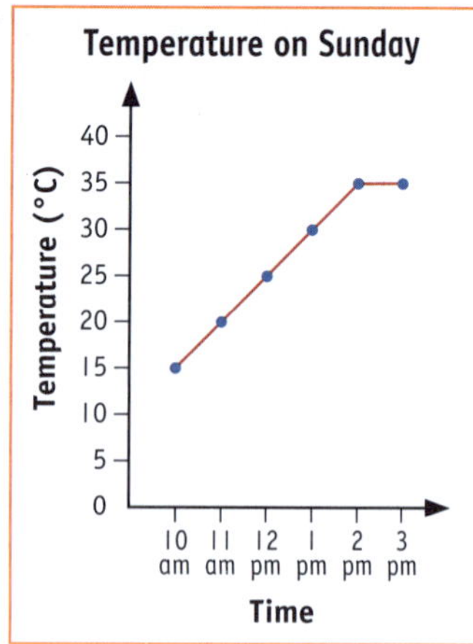

b 35 °C
c 15 °C
d 30 °C
e 35 °C
f 35 °C
g 20 °C

Pie Graphs

Page 292 – Your Turn

Note: The colours for Chocolate cake and Cheese cake can be reversed because they show the same percentage.

- Jelly 50%
- Ice cream 25%
- Chocolate cake 10%
- Cheese cake 10%
- Caramel slice 5%

Page 293 – Practice

1 a 2
b 2
c Antonia
d Annie
e Antonia
f Dani
g Annie
h Christian
i Annie

2 a 3
b 5
c c
d d
e h
f 80
g 160
h h

Spreadsheets

Page 294 – Your Turn

	C
1	SUBTOTAL
2	$1000
3	$2000
4	$3200
5	$3700
6	$4700

a $4700
b 5
c $3700

Page 295 – Practice

1 a D4 = $3250
D5 = $4800
D6 = $5450
D7 = $7100
b 10 June
c $1650
d $1550
e $7100

2

Lilly Pilly Football Club Expenses

	A	B	C	D
1	DATE	ITEM	COST	BALANCE
6	19 JULY	Trophies bought	$1550	$4700
7	23 JULY	Socks bought	$290	$4410
8	27 JULY	Fundraiser food bought	$575	$3835
9	29 JULY	Goals bought	$1200	$2635

CATCH UP MATHS YEAR 6 BOOK A © PASCAL PRESS ISBN: 9781925726183

Data Review Page 296

1 **6K's Favourite Food**

Food	Tally	Total
Fish	////	4
Pasta	~~////~~ /	6
Sausages	//	2
Chicken	~~////~~ ////	9
Pizza	~~////~~ //	7
Steak	///	3
		31

a 31
b 6
c 3
d Chicken
e Sausages
f 1

2

Hair colour	Hair type: Straight	Hair type: Curly	Total
Brown	3	5	8
Blond	7	9	16
Black	9	2	11
Red	4	4	8
Total	23	20	43

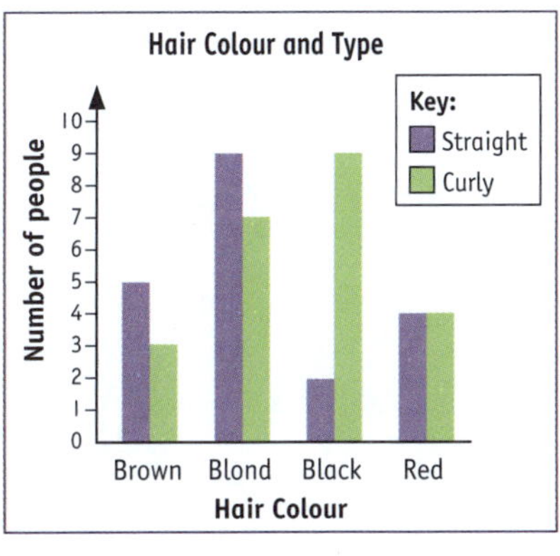

a 23
b 20
c black
d straight
e brown and red
f blond
g curly
h 11
i more

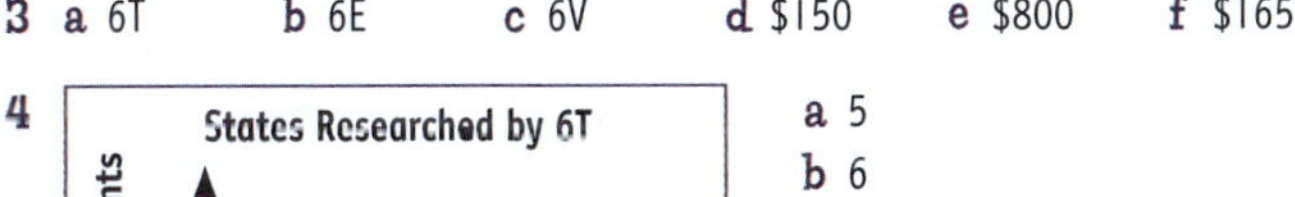
3 a 6T b 6E c 6V d $150 e $800 f $1650

4

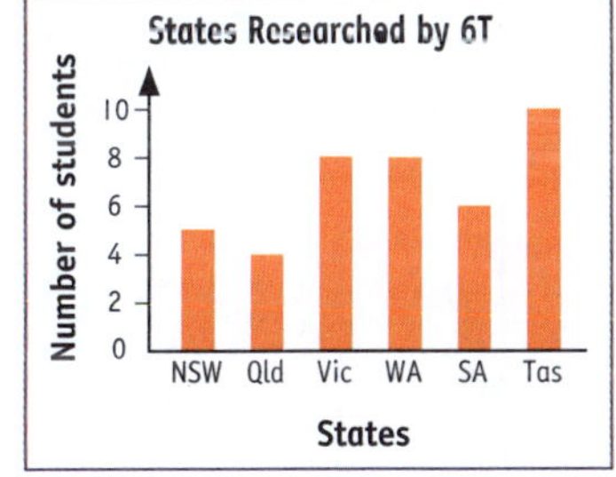

a 5
b 6
c 10
d 12
e 2
f Victoria, Western Australia
g 41

5

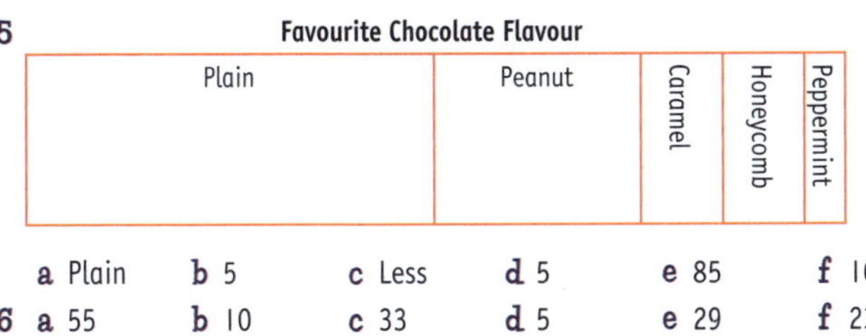

a Plain b 5 c Less d 5 e 85 f 100

6 a 55 b 10 c 33 d 5 e 29 f 23

7

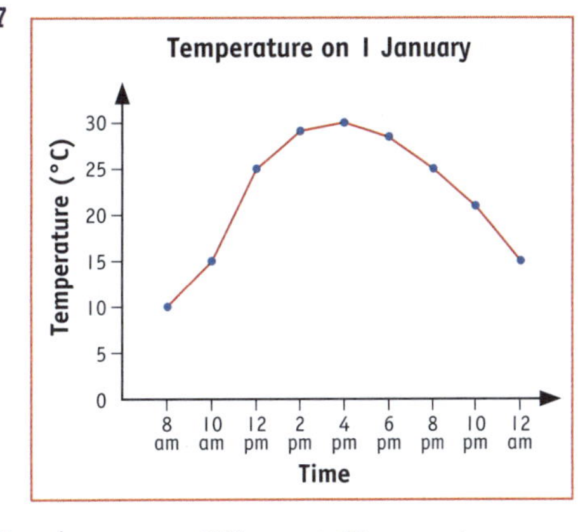

a 15 °C
b 30 °C, 4 pm
c 10 °C, 8 am
d 10 pm
e 2 pm
f 4 pm

8 a b
b 75%
c 25%
d a
e 25
f 300
g c
h h

9 C2 = $1250
C3 = $1925
C4 = $2174
C5 = $3394
C6 = $4857
C7 = $7282
C8 = $8027

a $675
b $3394
c 7
d $8027

10 D3 = $13 985.80
D4 = $12 531.30
D5 = $9677.50
D6 = $9114.30
D7 = $8511.80
D8 = $8361.80
D9 = $8009.05
D10 = $6485.42

a Insurance
b 20 August
c $1454.50
d $8361.80
e $12 531.30
f Tennis shoes
g 8

  ISBN: 9781925726183

Catch-Up Maths Book 6A

ISBN: 9781925726183

Published by Pascal Press
PO Box 250
Glebe NSW 2037
www.pascalpress.com.au
contact@pascalpress.com.au

Design: Janice Bowles
Author: Deborah Frendo-Toman
Publisher: Lynn Dickinson
Typesetter: Ruth Schultz
Editor: Rosemary Peers, Vaishali Batra

Printed by Wai Man Book Binding (China) Ltd.